T0297044

LONDON MATHEMATICAL SOCIETY LECTURE NOTE SERIES

Managing Editor: Professor N.J. Hitchin, Mathematical Institute,
University of Oxford, 24–29 St Giles, Oxford OX1 3LB, United Kingdom

The titles below are available from booksellers, or from Cambridge University Press at www.cambridge.org/mathematics.

189 Locally presentable and accessible categories, J. ADAMEK & J. ROSICKY
190 Polynomial invariants of finite groups, D. J. BENSON
191 Finite geometry and combinatorics, F. DE CLERCK et al.
192 Symplectic geometry, D. SALAMON (ed.)
194 Independent random variables and rearrangement invariant spaces, M. BRAVERMAN
195 Arithmetic of blowup algebras, W. VASCONCELOS
196 Microlocal analysis for differential operators, A. GRIGIS & J. SJÖSTRAND
197 Two-dimensional homotopy and combinatorial group theory, C. HOG-ANGELONI et al.
198 The algebraic characterization of geometric 4-manifolds, J. A. HILLMAN
199 Invariant potential theory in the unit ball of C^n, M. STOLL
200 The Grothendieck theory of dessins d'enfant, L. SCHNEPS (ed.)
201 Singularities, J.-P. BRASSELET (ed.)
202 The technique of pseudodifferential operators, H. O. CORDES
203 Hochschild cohomology of von Neumann algebras, A. SINCLAIR & R. SMITH
204 Combinatorial and geometric group theory, A. J. DUNCAN, N. D. GILBERT & J. HOWIE (eds)
205 Ergodic theory and its connections with harmonic analysis, K. PETERSEN & I. SALAMA (eds)
206 Groups of Lie type and their geometries, W. M. KANTOR & L. DI MARTINO (eds)
208 Vector bundles in algebraic geometry, N. J. HITCHIN, P. NEWSTEAD & W. M. OXBURY (eds)
209 Arithmetic of diagonal hypersurfaces over infite fields, F. Q. GOUVÉA & N. YUI
210 Hilbert C*-modules, E. C. LANCE
211 Groups 93 Galway / St Andrews I, C. M. CAMPBELL et al. (eds)
212 Groups 93 Galway / St Andrews II, C. M. CAMPBELL et al. (eds)
214 Generalised Euler–Jacobi inversion formula and asymptotics beyond all orders, V. KOWALENKO et al.
215 Number theory 1992–93, S. DAVID (ed.)
216 Stochastic partial differential equations, A. ETHERIDGE (ed.)
217 Quadratic forms with applications to algebraic geometry and topology, A. PFISTER
218 Surveys in combinatorics, 1995, P. ROWLINSON (ed.)
220 Algebraic set theory, A. JOYAL & I. MOERDIJK
221 Harmonic approximation, S. J. GARDINER
222 Advances in linear logic, J.-Y. GIRARD, Y. LAFONT & L. REGNIER (eds)
223 Analytic semigroups and semilinear initial boundary value problems, KAZUAKI TAIRA
224 Computability, enumerability, unsolvability, S. B. COOPER, T. A. SLAMAN & S. S. WAINER (eds)
225 A mathematical introduction to string theory, S. ALBEVERIO et al.
226 Novikov conjectures, index theorems and rigidity I, S. FERRY, A. RANICKI & J. ROSENBERG (eds)
227 Novikov conjectures, index theorems and rigidity II, S. FERRY, A. RANICKI & J. ROSENBERG (eds)
228 Ergodic theory of Z^d actions, M. POLLICOTT & K. SCHMIDT (eds)
229 Ergodicity for infinite dimensional systems, G. DA PRATO & J. ZABCZYK
230 Prolegomena to a middlebrow arithmetic of curves of genus 2, J. W. S. CASSELS & E. V. FLYNN
231 Semigroup theory and its applications, K. H. HOFMANN & M. W. MISLOVE (eds)
232 The descriptive set theory of Polish group actions, H. BECKER & A. S. KECHRIS
233 Finite fields and applications, S. COHEN & H. NIEDERREITER (eds)
234 Introduction to subfactors, V. JONES & V. S. SUNDER
235 Number theory 1993–94, S. DAVID (ed.)
236 The James forest, H. FETTER & B. G. DE BUEN
237 Sieve methods, exponential sums, and their applications in number theory, G. R. H. GREAVES et al.
238 Representation theory and algebraic geometry, A. MARTSINKOVSKY & G. TODOROV (eds)
240 Stable groups, F. O. WAGNER
241 Surveys in combinatorics, 1997, R. A. BAILEY (ed.)
242 Geometric Galois actions I, L. SCHNEPS & P. LOCHAK (eds)
243 Geometric Galois actions II, L. SCHNEPS & P. LOCHAK (eds)
244 Model theory of groups and automorphism groups, D. EVANS (ed.)
245 Geometry, combinatorial designs and related structures, J. W. P. HIRSCHFELD et al.
246 p-Automorphisms of finite p-groups, E. I. KHUKHRO
247 Analytic number theory, Y. MOTOHASHI (ed.)
248 Tame topology and o-minimal structures, L. VAN DEN DRIES
249 The atlas of finite groups: ten years on, R. CURTIS & R. WILSON (eds)
250 Characters and blocks of finite groups, G. NAVARRO
251 Gröbner bases and applications, B. BUCHBERGER & F. WINKLER (eds)
252 Geometry and cohomology in group theory, P. KROPHOLLER, G. NIBLO, R. STÖHR (eds)
253 The q-Schur algebra, S. DONKIN
254 Galois representations in arithmetic algebraic geometry, A. J. SCHOLL & R. L. TAYLOR (eds)
255 Symmetries and integrability of difference equations, P. A. CLARKSON & F. W. NIJHOFF (eds)
256 Aspects of Galois theory, H. VÖLKLEIN et al.
257 An introduction to noncommutative differential geometry and its physical applications 2ed, J. MADORE
258 Sets and proofs, S. B. COOPER & J. TRUSS (eds)
259 Models and computability, S. B. COOPER & J. TRUSS (eds)
260 Groups St Andrews 1997 in Bath, I, C. M. CAMPBELL et al.
261 Groups St Andrews 1997 in Bath, II, C. M. CAMPBELL et al.
262 Analysis and logic, C. W. HENSON, J. IOVINO, A. S. KECHRIS & E. ODELL
263 Singularity theory, B. BRUCE & D. MOND (eds)
264 New trends in algebraic geometry, K. HULEK, F. CATANESE, C. PETERS & M. REID (eds)
265 Elliptic curves in cryptography, I. BLAKE, G. SEROUSSI & N. SMART
267 Surveys in combinatorics, 1999, J. D. LAMB & D. A. PREECE (eds)
268 Spectral asymptotics in the semi-classical limit, M. DIMASSI & J. SJÖSTRAND
269 Ergodic theory and topological dynamics, M. B. BEKKA & M. MAYER
270 Analysis on Lie Groups, N. T. VAROPOULOS & S. MUSTAPHA

271 Singular perturbations of differential operators, S. ALBEVERIO & P. KURASOV
272 Character theory for the odd order theorem, T. PETERFALVI
273 Spectral theory and geometry, E. B. DAVIES & Y. SAFAROV (eds)
274 The Mandelbrot set, theme and variations, TAN LEI (ed.)
275 Descriptive set theory and dynamical systems, M. FOREMAN et al.
276 Singularities of plane curves, E. CASAS-ALVERO
277 Computational and geometric aspects of modern algebra, M. D. ATKINSON et al.
278 Global attractors in abstract parabolic problems, J. W. CHOLEWA & T. DLOTKO
279 Topics in symbolic dynamics and applications, F. BLANCHARD, A. MAASS & A. NOGUEIRA (eds)
280 Characters and automorphism groups of compact Riemann surfaces, T. BREUER
281 Explicit birational geometry of 3-folds, A. CORTI & M. REID (eds)
282 Auslander–Buchweitz approximations of equivariant modules, M. HASHIMOTO
283 Nonlinear elasticity, Y. FU & R. OGDEN (eds)
284 Foundations of computational mathematics, R. DEVORE, A. ISERLES & E. SÜLI (eds)
285 Rational points on curves over finite fields, H. NIEDERREITER & C. XING
286 Clifford algebras and spinors 2ed, P. LOUNESTO
287 Topics on Riemann surfaces and Fuchsian groups, E. BUJALANCE et al.
288 Surveys in combinatorics, 2001, J. HIRSCHFELD (ed.)
289 Aspects of Sobolev-type inequalities, L. SALOFF-COSTE
290 Quantum groups and Lie theory, A. PRESSLEY (ed.)
291 Tits buildings and the model theory of groups, K. TENT (ed.)
292 A quantum groups primer, S. MAJID
293 Second order partial differential equations in Hilbert spaces, G. DA PRATO & J. ZABCZYK
294 Introduction to operator space theory, G. PISIER
295 Geometry and Integrability, L. MASON & YAVUZ NUTKU (eds)
296 Lectures on invariant theory, I. DOLGACHEV
297 The homotopy category of simply connected 4-manifolds, H.-J. BAUES
298 Higher operands, higher categories, T. LEINSTER
299 Kleinian Groups and Hyperbolic 3-Manifolds, Y. KOMORI, V. MARKOVIC & C. SERIES (eds)
300 Introduction to Möbius Differential Geometry, U. HERTRICH-JEROMIN
301 Stable Modules and the D(2)-Problem, F. E. A. JOHNSON
302 Discrete and Continuous Nonlinear Schrödinger Systems, M. J. ABLOWITZ, B. PRINARI & A. D. TRUBATCH
303 Number Theory and Algebraic Geometry, M. REID & A. SKOROBOOATOV (eds)
304 Groups St Andrews 2001 in Oxford Vol. 1, C. M. CAMPBELL, E. F. ROBERTSON & G. C. SMITH (eds)
305 Groups St Andrews 2001 in Oxford Vol. 2, C. M. CAMPBELL, E. F. ROBERTSON & G. C. SMITH (eds)
306 Peyresq lectures on geometric mechanics and symmetry, J. MONTALDI & T. RATIU (eds)
307 Surveys in Combinatorics 2003, C. D. WENSLEY (ed.)
308 Topology, geometry and quantum field theory, U. L. TILLMANN (ed.)
309 Corings and Comdules, T. BRZEZINSKI & R. WISBAUER
310 Topics in Dynamics and Ergodic Theory, S. BEZUGLYI & S. KOLYADA (eds)
311 Groups: topological, combinatorial and arithmetic aspects, T. W. MÜLLER (ed.)
312 Foundations of Computational Mathematics, Minneapolis 2002, FELIPE CUCKER et al. (eds)
313 Transcendantal aspects of algebraic cycles, S. MÜLLER-STACH & C. PETERS (eds)
314 Spectral generalizations of line graphs, D. CVETKOVIC, P. ROWLINSON & S. SIMIC
315 Structured ring spectra, A. BAKER & B. RICHTER (eds)
316 Linear Logic in Computer Science, T. EHRHARD et al. (eds)
317 Advances in elliptic curve cryptography, I. F. BLAKE, G. SEROUSSI & N. SMART
318 Perturbation of the boundary in boundary-value problems of Partial Differential Equations, DAN HENRY
319 Double Affine Hecke Algebras, I. CHEREDNIK
320 L-Functions and Galois Representations, D. BURNS, K. BUZZARD & J. NEKOVÁŘ (eds)
321 Surveys in Modern Mathematics, V. PRASOLOV & Y. ILYASHENKO (eds)
322 Recent perspectives in random matrix theory and number theory, F. MEZZADRI, N. C. SNAITH (eds)
323 Poisson geometry, deformation quantisation and group representations, S. GUTT et al. (eds)
324 Singularities and Computer Algebra, C. LOSSEN & G. PFISTER (eds)
325 Lectures on the Ricci Flow, P. TOPPING
326 Modular Representations of Finite Groups of Lie Type, J. E. HUMPHREYS
328 Fundamentals of Hyperbolic Manifolds, R. D. CANARY, A. MARDEN & D. B. A. EPSTEIN (eds)
329 Spaces of Kleinian Groups, Y. MINSKY, M. SAKUMA & C. SERIES (eds)
330 Noncommutative Localization in Algebra and Topology, A. RANICKI (ed.)
331 Foundations of Computational Mathematics, Santander 2005, L. PARDO, A. PINKUS, E. SULI & M. TODD (eds)
332 Handbook of Tilting Theory, L. ANGELERI HÜGEL, D. HAPPEL & H. KRAUSE (eds)
333 Synthetic Differential Geometry 2ed, A. KOCK
334 The Navier–Stokes Equations, P. G. DRAZIN & N. RILEY
335 Lectures on the Combinatorics of Free Probability, A. NICA & R. SPEICHER
336 Integral Closure of Ideals, Rings, and Modules, I. SWANSON & C. HUNEKE
337 Methods in Banach Space Theory, J. M. F. CASTILLO & W. B. JOHNSON (eds)
338 Surveys in Geometry and Number Theory, N. YOUNG (ed.)
339 Groups St Andrews 2005 Vol. 1, C. M. CAMPBELL, M. R. QUICK, E. F. ROBERTSON & G. C. SMITH (eds)
340 Groups St Andrews 2005 Vol. 2, C. M. CAMPBELL, M. R. QUICK, E. F. ROBERTSON & G. C. SMITH (eds)
341 Ranks of Elliptic Curves and Random Matrix Theory, J. B. CONREY, D. W. FARMER, F. MEZZADRI & N. C. SNAITH (eds)
342 Elliptic Cohomology, H. R. MILLER & D. C. RAVENEL (eds)
343 Algebraic Cycles and Motives Vol. 1, J. NAGEL & C. PETERS (eds)
344 Algebraic Cycles and Motives Vol. 2, J. NAGEL & C. PETERS (eds)
345 Algebraic and Analytic Geometry, A. NEEMAN
346 Surveys in Combinatorics, 2007, A. HILTON & J. TALBOT (eds)
347 Surveys in Contemporary Mathematics, N. YOUNG & Y. CHOI (eds)
348 Transcendental Dynamics and Complex Analysis, P. RIPPON & G. STALLARD (eds)
349 Model Theory with Applications to Algebra and Analysis Vol 1, Z. CHATZIDAKIS, D. MACPHERSON, A. PILLAY & A. WILKIE (eds)
350 Model Theory with Applications to Algebra and Analysis Vol 2, Z. CHATZIDAKIS, D. MACPHERSON, A. PILLAY & A. WILKIE (eds)
351 Finite von Neumann Algebras and Masas, A. SINCLAIR & R. SMITH
352 Number Theory and Polynomials, J. MCKEE & C. SMYTH (eds)

Groups and Analysis

The legacy of Hermann Weyl

Edited by

KATRIN TENT

Universität Bielefeld

CAMBRIDGE
UNIVERSITY PRESS

University Printing House, Cambridge CB2 8BS, United Kingdom

One Liberty Plaza, 20th Floor, New York, NY 10006, USA

477 Williamstown Road, Port Melbourne, VIC 3207, Australia

314-321, 3rd Floor, Plot 3, Splendor Forum, Jasola District Centre, New Delhi - 110025, India

103 Penang Road, #05-06/07, Visioncrest Commercial, Singapore 238467

Cambridge University Press is part of the University of Cambridge.

It furthers the University's mission by disseminating knowledge in the pursuit of
education, learning and research at the highest international levels of excellence.

www.cambridge.org
Information on this title: www.cambridge.org/9780521717885

© Cambridge University Press 2008

First published 2008

A catalogue record for this publication is available from the British Library

ISBN 978-0-521-71788-5 Hardback

Contents

Preface   *page* vii

1   *Roe Goodman*
    Harmonic analysis on compact symmetric spaces   1

2   *Erik van den Ban*
    Weyl, eigenfunction expansions, symmetric spaces   24

3   *W.N. Everitt and H. Kalf*
    Weyl's work on singular Sturm–Liouville operators   63

4   *Markus J. Pflaum*
    From Weyl quantization to modern algebraic index theory   84

5   *A.M. Hansson and A. Laptev*
    Sharp spectral inequalities for the Heisenberg Laplacian   100

6   *Ursula Hamenstädt*
    Equidistribution for quadratic differentials   116

7   *Werner Müller*
    Weyl's law in the theory of automorphic forms   133

8   *Daniel W. Stroock*
    Weyl's Lemma, one of many   164

9   *Christopher Deninger*
    Analysis on foliated spaces and arithmetic geometry   174

10 *R.E. Howe, E.-C. Tan and J.F. Willenbring*
    Reciprocity algebras and branching   191

11 *Jens Carsten Jantzen*
    Character formulae from Hermann Weyl to the present   232

12 *Richard M. Weiss*
    The Classification of affine buildings   271

13 *Peter Roquette*
    Emmy Noether and Hermann Weyl   285

Preface

This volume grew out of the conference in honour of Hermann Weyl that took place in Bielefeld in September 2006.

Weyl was born in 1885 in Elmshorn, a small town near Hamburg. He studied mathematics in Göttingen and Munich, and obtained his doctorate in Göttingen under the supervision of Hilbert. After taking a teaching post for a few years, he left Göttingen for Zürich to accept a Chair of Mathematics at the ETH Zürich, where he was a colleague of Einstein just at the time when Einstein was working out the details of the theory of general relativity. Weyl left Zürich in 1930 to become Hilbert's successor at Göttingen, moving to the new Institute for Advanced Study in Princeton, New Jersey after the Nazis took power in 1933. He remained there until his retirement in 1951. Together with his wife, he spent the rest of his life in Princeton and Zürich, where he died in 1955.

The Collaborative Resarch Centre (SFB 701) *Spectral Structures and Topological Methods in Mathematics* has manifold connections with the areas of mathematics that were founded or influenced by Weyl's work. These areas include geometric foundations of manifolds and physics, topological groups, Lie groups and representation theory, harmonic analysis and analytic number theory as well as foundations of mathematics.

In 1913, Weyl published *Die Idee der Riemannschen Fläche* ('The Concept of a Riemann Surface'), giving a unified treatment of Riemann surfaces.

He described the development of relativity theory in his *Raum, Zeit, Materie* ('Space, Time, Matter') from 1918, which reached a fourth edition in 1922. In 1918, he introduced the concept of gauge and gave the first example of what is now known as a gauge theory.

From 1923 to 1938, Weyl developed the theory of compact groups in terms of matrix representations and proved a fundamental character formula for compact Lie groups. His book *Classical Groups* opened new directions in invariant theory. It covered symmetric groups, general linear groups, orthogonal groups, and symplectic groups, and results on their invariants and representations.

In *The Continuum*, Weyl developed the logic of classical analysis along the lines of Brouwer's intuitionism. However, he later decided that this radical constructivism puts too much of a restriction on his mathematics and reconciled himself with the more formalistic ideas of Hilbert.

Weyl also showed how to use exponential sums in diophantine approximation, with his criterion for uniform distribution modulo one, which was a fundamental contribution to analytic number theory.

During the conference, his lasting influence on current mathematics became evident through a series of impressive talks often connecting theorems of Weyl with the most current results in dynamical systems, invariant theory, or partial differential equations. We are happy that so many speakers agreed to contribute to this volume.

The conference was funded by the Collaborative Research Center (SFB 701) 'Spectral structures and topological methods in mathematics'. We gratefully acknowledge support by the German Research Foundation (DFG). Thanks are also due to Philip Herrmann for editing this volume, and to Markus Rost and Ulf Rehmann.

Bielefeld, December 2007

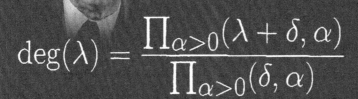

$$\deg(\lambda) = \frac{\prod_{\alpha>0}(\lambda + \delta, \alpha)}{\prod_{\alpha>0}(\delta, \alpha)}$$

Hermann Weyl Conference

Bielefeld   September 10–13, 2006

Speakers include:

E. van den Ban / *Utrecht*   G. Huisken / *Potsdam*   P. Roquette / *Heidelberg*
M. Broué / *Paris*   J. Jantzen / *Aarhus*   D. Salamon / *ETHZ*
C. Deninger / *Münster*   T. Lyons / *Oxford*   W. Schmid / *Harvard*
R. Goodman / *Rutgers*   A. Macintyre / *Royal Society* D. Stroock / *MIT*
U. Hamenstädt / *Bonn*   W. Müller / *Bonn*   R. Weiss / *Tufts*

SFB 701 · Mathematik · Universität Bielefeld   www.math.uni-bielefeld.de/weyl

List of speakers and talks

M. Broue (Paris)
Complex Reflection Groups as Weyl Groups

R. Goodman (Rutgers)
Harmonic Analysis on Compact Symmetric Spaces - the Legacy of H. Weyl and E. Cartan

G. Huisken (Potsdam)
The concept of Mass in General Relativity

H. Kalf (München)
Weyl's work on singular Sturm-Liouville operators

T. Lyons (Oxford)
Inverting the signature of a path-extensions of a theorem of Chen

W. Müller (Bonn)
Weyl's law and the theory of automorphic forms

M. Pflaum (Frankfurt)
From Weyl quantization and Weyl asymptotics to modern index theory.

M. Rost (Bielefeld)
On Galois cohomology, norm functions and cycles

T.A. Springer (Utrecht)
A short history of the theory of Weyl groups

E. van den Ban (Utrecht)
Weyl, eigenfunction expansions and analysis on non-compact symmetric spaces

C. Deninger (Münster)
Determinants in von Neumann algebras and the entropy of noncommutativ group actions

U. Hamenstädt (Bonn)
Mixing properties of the Teichmüller flow

J.C. Jantzen (Aarhus)
Character formulae from Weyl to the present

A. Laptev (Stockholm)
Lieb–Thirring Inequalities. Recent results

A. Macintyre (London)
Some model theory of Lie groups

N. Nadirashvili (Marseille)
Rearrangements and eigenvalues

P. Roquette (Heidelberg)
Hermann Weyl and Emmy Noether (Some observations from the correspondence Hasse–Noether and other documents)

D. Salamon (Zürich)
Pseudoholomorphic curves in symplectic topology

D.W. Stroock (MIT)
Weyl's Lemma, the original elliptic regularity result

R.M. Weiss (Tufts)
Affine Buildings

1

Harmonic Analysis on Compact Symmetric Spaces: the Legacy of Elie Cartan and Hermann Weyl

Roe Goodman

Department of Mathematics
Rutgers, The State University of New Jersey

1 Introduction

In his lecture *Relativity theory as a stimulus in mathematical research* [Wey4], Hermann Weyl says that "Frobenius and Issai Schur's spadework on finite and compact groups and Cartan's early work on semi-simple Lie groups and their representations had nothing to do with it [relativity theory]. But for myself I can say that the wish to understand what really is the mathematical substance behind the formal apparatus of relativity theory led me to the study of representations and invariants of groups, and my experience in this regard is probably not unique."

Weyl's first encounter with Lie groups and representation theory as a tool to understand relativity theory occurred in connection with the Helmholtz-Lie space problem and the problem of decomposing the tensor product $\otimes^k \mathbb{C}^n$ under the mutually commuting actions of the general linear group $\mathrm{GL}(n, \mathbb{C})$ (on each copy of $\mathbb{C}^n$) and the symmetric group $\mathfrak{S}_k$ (in permuting the k copies of $\mathbb{C}^n$).[1] He later described the tensor decomposition problem in general terms [Wey3] as "an epistemological principle basic for all theoretical science, that of projecting the actual upon the background of the possible." Mathematically, the issue was to find subspaces of tensor space that are invariant and irreducible under all transformations that commute with $\mathfrak{S}_k$. This had already been done by Frobenius and Schur around 1900, but apparently Weyl first became aware of these results in the early 1920's. The subspaces in question, which are the ranges of minimal projections in the group algebra of $\mathfrak{S}_k$, are exactly the irreducible (polynomial) representations of $\mathrm{GL}(n, \mathbb{C})$, and all irreducible representations arise this way for varying k by including multiplication by integral powers of $\det(g)$ in the action. It seems clear

---

1 see [Haw, §11.2-3]

from his correspondence with Schur at this time that these results were Weyl's starting point for his later work in representation theory and invariant theory.

Near the end of his monumental paper on representations of semisimple Lie groups [Wey1, Kap. IV, §4], Weyl considers the problem of constructing all the irreducible representations of a simply-connected simple Lie group G such as $SL(n, \mathbb{C})$. This had been done on a case-by-case basis by Cartan [Car1], starting with the defining representations for the classical groups (or the adjoint representation for the exceptional groups) and building up a general irreducible representation by forming tensor products. By contrast, Weyl, following the example of Frobenius for finite groups, says that "the correct starting point for building representations does not lie in the adjoint group, but rather in the *regular representation*, which through its reduction yields *in one blow* all irreducible representations." He introduces the infinite-dimensional space $C(U)$ of all continuous functions on the compact real form U of G ($U = SU(n)$ when $G = SL(n, \mathbb{C})$) and the right translation representation of U on $C(U)$. He then obtains the irreducible representations of U and their characters by using the eigenspaces of compact integral operators given by left convolution with positive-definite functions in $C(U)$, in analogy with the decomposition of tensor spaces for $GL(n, \mathbb{C})$ using elements of the group algebra of $\mathfrak{S}_k$. The details are spelled out in the famous Peter–Weyl paper [Pe-We], which proves that the normalized matrix entries of the irreducible unitary representations of U furnish an orthonormal basis for $L^2(U)$, and that every continuous function on U is a uniform limit of linear combinations of these matrix entries.

In the introduction to [Car2], É. Cartan says that his paper was inspired by the paper of Peter and Weyl, but he points out that for a compact Lie group their use of integral equations "gives a transcendental solution to a problem of an algebraic nature" (namely, the completeness of the set of finite-dimensional irreducible representations of the group). Cartan's goal is "to give an algebraic solution to a problem of a transcendental nature, more general than that treated by Weyl." Namely, to find an explicit decomposition of the space of all L^2 functions on a homogeneous space into an orthogonal direct sum of group-invariant irreducible subspaces.

Cartan's paper [Car2] then stimulated Weyl [Wey2] to treat the same problem again and write "the systematic exposition by which I should like to replace the two papers Peter–Weyl [Pe-We] and Cartan [Car2]." In his characteristic style of finding the core of a problem through generalization, Weyl takes the finite-dimensional irreducible subspaces of functions (which he calls the *harmonic sets* by analogy with the case of spherical harmonics) on the compact homogeneous space X as his starting point.[2] Using the invariant measure on the homogenous space, he constructs integral operators that intertwine the representation of the compact group U on $C(X)$ with the left regular representation on $C(U)$.

In this paper we approach the Weyl–Cartan results by way of algebraic groups. The *finite* functions on a homogeneous space for a compact connected Lie group (that is, the functions whose translates span a finite-dimensional subspace) can be viewed as *regular* functions on the complexified group (a complex reductive algebraic group). Irreducible subspaces of functions under the action of the compact group correspond to irreducible subspaces of regular functions on the complex reductive group—this is Weyl's *unitarian trick*. We describe the algebraic group version of the Peter–Weyl decomposition and geometric criterion for simple spectrum of a homogeneous space (due to E. Vinberg and B. Kimelfeld). We present R. Richardson's algebraic group version of the Cartan embedding of a symmetric space, and the celebrated results of Cartan and S. Helgason concerning finite-dimensional spherical representations.

We then turn to more recent results of J.-L. Clerc [Cle] concerning the complexified Iwasawa decomposition and zonal spherical functions on a compact symmetric space, and S. Gindikin's construction ([Gin1], [Gin2], [Gin3]) of the *horospherical Cauchy–Radon transform*, which shows that compact symmetric spaces have canonical dual objects that are complex manifolds.

We make frequent citations to the extraordinary books of A. Borel [Bor] and T. Hawkins [Haw], which contain penetrating historical accounts of the contributions of Weyl and Cartan. Borel's book also describes the development of algebraic groups by C. Chevalley that is basic to our approach. For a survey of other developments in harmonic analysis on symmetric spaces from Cartan's paper to the mid 1980's see Helgason [Hel3]. Thanks go to the referee for pointing out some notational inconsistencies and making suggestions for improving the organization of this paper.

---

2 Weyl's emphasis on function spaces, rather than the underlying homogeneous space, is in the spirit of the recent development of *quantum groups*; his immediate purpose was to make his theory sufficiently general to include also J. von Neumann's theory of almost-periodic functions on groups, in which the functions determine a compactification of the underlying group.

2 Algebraic Group Version of Peter–Weyl Theorem

2.1 Isotypic Decomposition of $\mathcal{O}[X]$

The paper [Pe-We] of Peter and Weyl considers compact Lie groups U; because the group is compact left convolution with a continuous function is a compact operator. Hence such an operator, if self-adjoint, has finite-dimensional eigenspaces that are invariant under right translation by elements of U. The finiteness of the invariant measure on U also guarantees that every finite-dimensional representation of U carries a U-invariant positive-definite inner product, and hence is *completely reducible* (decomposes as the direct sum of irreducible representations).[3]

Turning from Weyl's transcendental methods to the more algebraic and geometric viewpoint preferred by Cartan, we recall that a subgroup $G \subset \mathrm{GL}(n, \mathbb{C})$ is an *algebraic group* if it is the zero set of a collection of polynomials in the matrix entries. The *regular functions* $\mathcal{O}[G]$ are the restrictions to G of polynomials in matrix entries and $\det^{-1}$. In particular, G is a complex Lie group and the regular functions on G are holomorphic. A finite-dimensional complex representation (π, V) of G is *rational* if the matrix entries of the representation are regular functions on G. The group G is *reductive* if every rational representation is completely reducible.

Let $\mathfrak{g}$ be a complex semisimple Lie algebra. From the work of Cartan, Weyl, and Chevalley, one knows the following:

(1) There is a simply-connected complex linear algebraic group G with Lie algebra $\mathfrak{g}$.

(2) The finite-dimensional representations of $\mathfrak{g}$ correspond to rational representations of G.

(3) There is a real form $\mathfrak{u}$ of $\mathfrak{g}$ and a simply-connected compact Lie group $U \subset G$ with Lie algebra $\mathfrak{u}$.

(4) The finite-dimensional unitary representations of U extend uniquely to rational representations of G, and U-invariant subspaces correspond to G-invariant subspaces.[4]

(5) The irreducible rational representations of G are parameterized by the positive cone in a lattice of rank l (Cartan's theorem of the *highest weight*).[5]

---

3 This is the Hurwitz "trick" (*Kunstgriff*) that Weyl learned from I. Schur; see Hawkins [Haw, §12.2].

4 This is Weyl's *unitary trick*.

5 The first algebraic proofs of this that did not use case-by-case considerations were found by Chevalley and Harish-Chandra in 1948; see [Bor, Ch. VII, §3.6-7].

The highest weight construction is carried out as follows: Fix a Borel subgroup $B = HN^+$ of G (a maximal connected solvable subgroup). Here $H \cong (\mathbb{C}^\times)^l$, with $l = \operatorname{rank}(G)$, is a maximal algebraic torus in G, and N^+ is the unipotent radical of B associated with a set of positive roots of H on $\mathfrak{g}$. Let $\bar{B} = HN^-$ be the opposite Borel subgroup. We can always arrange the embedding $G \subset \operatorname{GL}(n, \mathbb{C})$ so that H consists of the diagonal matrices in G, N^+ consists of the upper-triangular unipotent matrices in G, and N^- consists of the lower-triangular unipotent matrices in G. Let $\mathfrak{h}$ be the Lie algebra of H and $\Phi \subset \mathfrak{h}^*$ the roots of $\mathfrak{h}$ on $\mathfrak{g}$. Write $P(\Phi) \subset \mathfrak{h}^*$ for the *weight lattice* of H and $P_{++} \subset P(\Phi)$ for the *dominant weights*, relative to the system of positive roots determined by N^+. For $\lambda \in P(\Phi)$ we denote by $h \mapsto h^\lambda$ the corresponding character of H. It extends to a character of B by $(hn)^\lambda = h^\lambda$ for $h \in H$ and $n \in N^+$.

An irreducible rational representation (π, E) of G is then determined (up to equivalence) by its *highest weight*. The subspace E^{N^+} of N^+-fixed vectors in E is one-dimensional, and H acts on it by a character $h \mapsto h^\lambda$ where $\lambda \in P_{++}$. The subspace E^{N^-} of N^--fixed vectors in E is also one-dimensional, and H acts on it by the character $h \mapsto h^{-\lambda^*}$ where $\lambda^* = -w_0 \cdot \lambda$. Here w_0 is the element of the Weyl group of $(\mathfrak{g}, \mathfrak{h})$ that interchanges positive and negative roots.

For each $\lambda \in P_{++}$ we fix a model (π_λ, E_λ) for the irreducible rational representation with highest weight λ. Then $(\pi_{\lambda^*}, E_\lambda^*)$ is the contragredient representation. Fix a highest weight vector $e_\lambda \in E_\lambda$ and a lowest weight vector $f_{\lambda^*} \in E_\lambda^*$, normalized so that

$$\langle e_\lambda, f_{\lambda^*} \rangle = 1.$$

Here we are using $\langle v, v^* \rangle$ to denote the tautological duality pairing between a vector space and its dual (in particular, this pairing is complex linear in both arguments). For dealing with matrix entries as regular functions on the complex algebraic group G this is more convenient than using a U-invariant inner product on E_λ and identifying E_λ^* with E_λ via a conjugate-linear map.

Let X be an irreducible affine algebraic G space. Denote the regular functions on X by $\mathcal{O}[X]$. There is a representation ρ of G on $\mathcal{O}[X]$:

$$\rho(g)f(x) = f(g^{-1}x) \quad \text{for } f \in \mathcal{O}[X] \text{ and } g \in G.$$

Because the G-action is algebraic, $\operatorname{Span}\{\rho(G)f\}$ is a finite-dimensional rational G-module for $f \in \mathcal{O}[X]$. There is a tautological G-intertwining map

$$E_\lambda \otimes \operatorname{Hom}_G(E_\lambda, \mathcal{O}[X]) \to \mathcal{O}[X],$$

given by $v \otimes T \mapsto Tv$. For $\lambda \in P_{++}$ let

$$\mathcal{O}[X]^{N^+}(\lambda) = \{f \in \mathcal{O}[X] \ : \ \rho(hn)f = h^\lambda f \quad \text{for } h \in H \text{ and } n \in N^+\}. \tag{2.1}$$

The key point is that the choice of a highest weight vector e_λ gives an isomorphism

$$\mathrm{Hom}_G\,(E_\lambda, \mathcal{O}[X]) \cong \mathcal{O}[X]^{N^+}(\lambda). \tag{2.2}$$

Here a G-intertwining map T applied to the highest weight vector gives the function $\varphi = Te_\lambda \in \mathcal{O}[X]^{N^+}(\lambda)$, and conversely every such function φ defines a unique intertwining map T by this formula.[6]   From (2.2) we see that the highest weights of the G-irreducible subspaces of $\mathcal{O}[X]$ comprise the set

$$\mathrm{Spec}(X) = \{\lambda \in P_{++} \ : \ \mathcal{O}[X]^{N^+}(\lambda) \neq 0\} \quad \text{(the } G \text{ spectrum of } X)$$

Using the isomorphism (2.2) and the reductivity of G, we obtain the decomposition of $\mathcal{O}[X]$ under the action of G, as follows:

Theorem 2.1 *The isotypic subspace of type* (π_λ, E_λ) *in* $\mathcal{O}[X]$ *is the linear span of the G-translates of* $\mathcal{O}[X]^{N^+}(\lambda)$. *Furthermore,*

$$\mathcal{O}[X] \cong \bigoplus_{\lambda \in \mathrm{Spec}(X)} E_\lambda \otimes \mathcal{O}[X]^{N^+}(\lambda) \quad \text{(algebraic direct sum)} \tag{2.3}$$

as a G-module, with action $\pi_\lambda(g) \otimes 1$ *on the λ summand.*

The action of G on $\mathcal{O}[X]$ is not only linear; it also preserves the algebra structure. Since $\mathcal{O}[X]^{N^+}(\lambda) \cdot \mathcal{O}[X]^{N^+}(\mu) \subset \mathcal{O}[X]^{N^+}(\lambda+\mu)$ under pointwise multiplication and $\mathcal{O}[X]$ has no zero divisors (X is irreducible), it follows from (2.3) that

$$\mathrm{Spec}(X) \text{ is an additive subsemigroup of } P_{++}.$$

The multiplicity of π_λ in $\mathcal{O}[X]$ is $\dim \mathcal{O}[X]^{N^+}(\lambda)$ (which may be infinite). All of this was certainly known (perhaps in less precise form) by Cartan and Weyl at the time [Pe-We] appeared. We now consider Cartan's goal in [Car2] to determine the decomposition (2.3) when G acts transitively on X; especially, when X is a symmetric space. This requires determining the *spectrum* and the *multiplicities* in this decomposition.

---

6 Weyl uses a similar construction in [Wey2], defining intertwining maps by integration over a compact homogeneous space.

2.2 Multiplicity Free Spaces

We say that an irreducible affine G-space X is *multiplicity free* if all the irreducible representations of G that occur in $\mathcal{O}[X]$ have multiplicity one. Thanks to the theorem of the highest weight, this property can be translated into a geometric statement (see [Vi-Ki]). For a subgroup $K \subset G$ and $x \in X$ write $K_x = \{k \in L : k \cdot x = x\}$ for the isotropy group at x.

Theorem 2.2 (Vinberg–Kimelfeld) *Suppose there is a point $x_0 \in X$ such that $B \cdot x_0$ is open in X. Then X is multiplicity free. In this case, if $\lambda \in \mathrm{Spec}(X)$ then $h^\lambda = 1$ for all $h \in H_{x_0}$.*

Proof If $B \cdot x_0$ is open in X, then it is Zariski dense in X (since X is irreducible). Hence $f \in \mathcal{O}[X]^{N^+}(\lambda)$ is determined by $f(x_0)$, since on the dense set $B \cdot x_0$ it satisfies $f(b \cdot x_0) = b^{-\lambda} f(x_0)$. In particular, if $f \neq 0$ then $f(x_0) \neq 0$, and hence $h^\lambda = 1$ for all $h \in H_{x_0}$. Thus

$$\dim \mathcal{O}[X]^{N^+}(\lambda) \leq 1 \quad \text{for all } \lambda \in P_{++}.$$

Now apply Theorem 2.1. □

Remark. The converse to Theorem 2.2 is true; this depends on some results of Rosenlicht [Ros] and is the starting point for the classification of multiplicity free spaces (see [Be-Ra]).

Example: Algebraic Peter–Weyl Decomposition

Theorem 2.2 implies the algebraic version of the Peter-Weyl decomposition of the regular representation of G. Consider the reductive group $G \times G$ acting on $X = G$ by left and right translations. Denote this representation by ρ:

$$\rho(y, z)f(x) = f(y^{-1}xz), \quad \text{for} \quad f \in \mathcal{O}[G] \text{ and } x, y, z \in G.$$

Take $H \times H$ as the Cartan subgroup and $\bar{B} \times B$ as the Borel subgroup of $G \times G$. Let $x_0 = I$ (the identity in G). The orbit of x_0 under the Borel subgroup is

$$(\bar{B} \times B) \cdot x_0 = N^- H N^+ \quad \text{(Gauss decomposition)} \qquad (2.4)$$

This orbit is open in G since $\mathfrak{g} = \mathfrak{n}^- + \mathfrak{h} + \mathfrak{n}^+$. Hence G is multiplicity free as a $G \times G$ space. The $G \times G$ highest weights (relative to this choice of Borel subgroup) are pairs $(w_0\mu, \lambda)$, with $\lambda, \mu \in P_{++}$. The diagonal subgroup $\widetilde{H} = \{(h, h) : h \in H\}$ fixes x_0, so if $(w_0\mu, \lambda)$ occurs as a highest weight in $\mathcal{O}[X]$, then

$$h^{w_0\mu+\lambda} = 1 \quad \text{for all } h \in H.$$

This means that $\mu = -w_0\lambda = \lambda^*$; hence $E_\mu = E_{\lambda^*}$ is the contragredient representation of G.

Now set $\psi_\lambda(g) = \langle \pi_\lambda(g)e_\lambda, f_{\lambda^*} \rangle$. This function satisfies $\psi_\lambda(x_0) = 1$ and

$$\psi_\lambda(\bar{b}^{-1}gb) = \langle \pi_\lambda(g)\pi_\lambda(b)e_\lambda, \pi_{\lambda^*}(\bar{b})f_{\lambda^*} \rangle = b^\lambda \bar{b}^{w_0\lambda^*} \psi_\lambda(g)$$

for $b \in B$ and $\bar{b} \in \bar{B}$. Hence ψ_λ is a $B \times \bar{B}$ highest weight vector for $G \times G$ of weight $(w_0\lambda^*, \lambda)$. This proves that $\text{Spec}(X) = \{(w_0\lambda^*, \lambda) : \lambda \in P_{++}\}$.

Theorem 2.3 *For $\lambda \in P_{++}$ let $V_\lambda = \text{Span}\{\rho(G \times G)\psi_\lambda\}$. Then $V_\lambda \cong E_{\lambda^*} \otimes E_\lambda$ as a $G \times G$ module. Furthermore,*

$$\mathcal{O}[G] = \bigoplus_{\lambda \in P_{++}} V_\lambda. \tag{2.5}$$

In particular, $\mathcal{O}[G]$ is multiplicity free as a $G \times G$ module, while under the action of $G \times 1$ it decomposes into the sum of $\dim E_\lambda$ copies of E_λ for all $\lambda \in P_{++}$.

The function ψ_λ in Theorem 2.3 is called the *generating function* [Žel] for the representation π_λ. Since $\psi_\lambda(n^-hn^+) = h^\lambda$ and N^-HN^+ is dense in G, it is clear that

$$\psi_\lambda(g)\psi_\mu(g) = \psi_{\lambda+\mu}(g). \tag{2.6}$$

The semigroup P_{++} of dominant integral weights is free with generators $\lambda_1, \ldots, \lambda_l$, called the *fundamental weights*.

Proposition 2.4 (Product Formula)   *Set $\psi_i(g) = \psi_{\lambda_i}(g)$. Let $\lambda \in P_{++}$ and write $\lambda = m_1\lambda_1 + \cdots + m_l\lambda_l$ with $m_i \in \mathbb{N}$. Then*

$$\psi_\lambda(g) = \psi_1(g)^{m_1} \cdots \psi_l(g)^{m_l} \quad \text{for } g \in G. \tag{2.7}$$

Remark. From the product formula it is evident that the existence of a rational representation with highest weight λ is equivalent to the property that the functions $n^-hn^+ \mapsto h^{\lambda_i}$ on N^-HN^+ extend to regular functions on G for $i = 1, \ldots, l$.

Example. Suppose $G = \mathrm{SL}(n, \mathbb{C})$. Take B as the group of upper-triangular matrices. We may identify P with $\mathbb{Z}^n$, where $\lambda = [\lambda_1, \ldots, \lambda_n]$ gives the character

$$h^\lambda = x_1^{\lambda_1} \cdots x_n^{\lambda_n}, \quad h = \mathrm{diag}[x_1, \ldots, x_n].$$

Then P_{++} consists of the monotone decreasing n-tuples and is generated by

$$\lambda_i = [\underbrace{1, \ldots, 1}_{i}, 0, \ldots, 0] \quad \text{for } i = 1, \ldots, n-1.$$

The fundamental representations are the exterior powers $E_{\lambda_i} = \bigwedge^i \mathbb{C}^n$ of the defining representation, for $i = 1, \ldots, n-1$. The generating function $\psi_i(g)$ is the ith principal minor of g. The Gauss decomposition (2.4) is the familiar LDU matrix factorization from linear algebra, and

$$N^- H N^+ = \{g \in \mathrm{SL}(n, \mathbb{C}) : \psi_i(g) \neq 0 \quad \text{for } i = 1, \ldots, n-1 \}.$$

Let $K \subset G$ be a subgroup and let $\mathcal{O}[G]^{R(K)}$ be the *right* K-invariant regular functions on G (those functions f such that $f(gk) = f(g)$ for all $k \in K$). This subspace of $\mathcal{O}[G]$ is invariant under left translations by G.

Corollary 2.1 *Let E_λ^K be the subspace of K-fixed vectors in E_λ. Then*

$$\mathcal{O}[G]^{R(K)} \cong \bigoplus_{\lambda \in P_{++}} E_\lambda \otimes E_{\lambda^*}^K \tag{2.8}$$

as a G module under left translations, with G acting by $\pi_\lambda \otimes 1$ on the λ-isotypic summand. Thus the multiplicity of π_λ in $\mathcal{O}[G]^{R(K)}$ is $\dim E_{\lambda^*}^K$.

For any closed subgroup K of G whose Lie algebra is a complex subspace of $\mathfrak{g}$, the coset space G/K is a complex manifold on which G acts holomorphically, and the elements of $\mathcal{O}[G]^{R(K)}$ are holomorphic functions on G/K. When K is a *reductive* algebraic subgroup, then the manifold G/K also has the structure of an affine algebraic G-space such that the regular functions are exactly the elements of $\mathcal{O}[G]^{R(K)}$ (a result of Matsushima [Mat]; see also Borel and Harish-Chandra [Bo-Ha]). Also, when K is reductive then $\dim E_{\lambda^*}^K = \dim E_\lambda^K$, since the identity representation is self-dual.

The pair (G, K) is called *spherical* if

$$\dim E_\lambda^K \leq 1 \quad \text{for all } \lambda \in P_{++}.$$

In this case, we refer to K as a *spherical subgroup* of G. When K is reductive, this property is equivalent to G/K being a multiplicity-free G-space, by Corollary 2.1.

3 Complexifications of Compact Symmetric Spaces

3.1 Algebraic Version of Cartan Embedding

Cartan's paper [Car2] studies the decomposition of $C(U/K_0)$, where U is a compact real form of the simply-connected complex semisimple group G and $K_0 = U^\theta$ is the fixed-point set of an involutive automorphism θ of U. The compact symmetric space $X = U/K_0$ is simply-connected and hence the group K_0 is connected.[7] The involution extends uniquely to an algebraic group automorphism of G that we continue to denote as θ. The algebraic subgroup group $K = G^\theta$ is connected and is the complexification of K_0 in G, hence reductive. By Matsushima's theorem G/K is an affine algebraic variety. It can be embedded into G as an affine algebraic subset as follows (see [Ric1], [Ric2]):

Define

$$g \star y = gy\theta(g)^{-1}, \quad \text{for } g, y \in G.$$

We have $(g \star (h \star y)) = (gh) \star y$ for $g, h, y \in G$, so this gives an action of G on itself which we will call the *θ-twisted conjugation* action. Let

$$Q = \{y \in G : \theta(y) = y^{-1}\}.$$

Then Q is an algebraic subset of G. Since $\theta(g \star y) = \theta(g)y^{-1}g^{-1} = (g \star y)^{-1}$, we have $G \star Q = Q$.

Theorem 3.1 (Richardson) *The θ-twisted action of G is transitive on each irreducible component of Q. Hence Q is a finite union of Zariski-closed θ-twisted G-orbits.*

The proof consists of showing that the tangent space to a twisted G-orbit coincides with the tangent space to Q.

Corollary 3.1 *Let $P = G \star 1 = \{g\theta(g)^{-1} : g \in G\}$ be the orbit of the identity element under the θ-twisted conjugation action. Then P is a Zariski-closed irreducible subset of G isomorphic to G/K as an affine G-space (relative to the θ-twisted conjugation action of G).*

---

7 This theorem of Cartan extends Weyl's results for compact semisimple groups–see Borel [Bor, Chap. IV, §2].

There is a θ-stable noncompact real form G_0 of G so that K_0 is a maximal compact subgroup of G_0. The symmetric space G_0/K_0 is the *noncompact dual* to U/K_0. The *Cartan embedding* is the map $G_0/K_0 \rightarrow P_0 \subset G_0$, where $P_0 = G_0 \star 1 = \exp \mathfrak{p}_0$ and $\mathfrak{p}_0$ is the -1 eigenspace of θ in $\mathfrak{g}_0$ (P_0 is Cartan's space $\mathcal{E}$–see Borel [Bor, Ch. IV, §2.4]).

3.2 Classical Examples

Let $G \subset \mathrm{GL}(n,\mathbb{C})$ be a connected classical group whose Lie algebra is simple. The involutions and associated symmetric spaces G/K for G can be described in terms of the following three kinds of geometric structures on $\mathbb{C}^n$ (in the second and third type, G is the isometry group of the form and K is the subgroup preserving the indicated decomposition of $\mathbb{C}^n$):

(1) **nondegenerate bilinear forms** $G = \mathrm{SL}(n,\mathbb{C})$ and $K = \mathrm{SO}(n,\mathbb{C})$ or $\mathrm{Sp}(n,\mathbb{C})$
(2) **polarizations** $\mathbb{C}^n = V_+ \oplus V_-$ with $V_\pm$ totally isotropic subspaces for a bilinear form (zero or nondegenerate)
(3) **orthogonal decompositions** $\mathbb{C}^n = V_+ \oplus V_-$ with $V_\pm$ nondegenerate subspaces for a nondegenerate bilinear form

The proof that these structures give all the possible involutive automorphisms of the classical groups (up to inner automorphisms) can be obtained from following characterization of automorphisms of the classical groups:

Proposition 3.2 *Let σ be a regular automorphism of the classical group* G.
(1) *If $G = \mathrm{SL}(n,\mathbb{C})$ then there exists $s \in G$ so that σ is either $\sigma(g) = sgs^{-1}$ or $\sigma(g) = s(g^t)^{-1}s^{-1}$.*
(2) *If G is $\mathrm{Sp}(n,\mathbb{C})$ then there exists $s \in G$ so that $\sigma(g) = sgs^{-1}$.*
(3) *If G is $\mathrm{SO}(n,\mathbb{C})$ with $n \neq 2,4$, then there exists $s \in \mathrm{O}(n,\mathbb{C})$ so that $\sigma(g) = sgs^{-1}$.*

Proof The Weyl dimension formula implies that the defining representation (and its dual, in the case $G = \mathrm{SL}(n,\mathbb{C})$) is the unique representation of smallest dimension. So this representation is sent to an equivalent representation (or its dual) by σ. The existence of the element s follows from this equivalence (see [Go-Wa, §11.2.4] for details).[8] □

---

8 This type of result was one motivation for Weyl to learn Cartan's theory of representations of semisimple Lie groups–see Borel[Bor, Chap. III, §1] for more details.

Example. Let $G = \mathrm{SL}(n, \mathbb{C})$ and $\theta(g) = (g^t)^{-1}$. Then

$$K = \mathrm{SO}(n, \mathbb{C}), \quad U = \mathrm{SU}(n), \quad K_0 = \mathrm{SO}(n), \quad G_0 = \mathrm{SL}(n, \mathbb{R}).$$

Also $g \star y = g y g^t$ and $Q = \{y \in G : y^t = y\} = P$, so there is one orbit. Hence the map $gK \mapsto gg^t$ gives the algebraic embedding

$$\mathrm{SL}(n, \mathbb{C})/\mathrm{SO}(n, \mathbb{C}) \cong \{y \in M_n(\mathbb{C}) : y = y^t, \det y = 1\}.$$

For the other classical examples, see Goodman–Wallach [Go-Wa, §11.2.5].

3.3 Complexified Iwasawa Decomposition

The real semisimple Lie algebra $\mathfrak{g}_0$ has a Cartan decomposition $\mathfrak{g}_0 = \mathfrak{k}_0 + \mathfrak{p}_0$ into $+1$ and -1 eigenspaces of the Cartan involution θ. The non-compact real group G_0 has an Iwasawa decomposition [9] $G_0 = K_0 A_0 N_0$. Here $A_0 = \exp \mathfrak{a}_0$ is a vector group with $\mathfrak{a}_0$ a maximal abelian subspace of $\mathfrak{p}_0$, and N_0 is a nilpotent subgroup normalized by A_0. Let A and N be the complexifications of A_0 and N_0 in G, respectively. Then A is a complex algebraic torus of rank l (the *rank* of G/K) and N is a unipotent subgroup. There is a θ-stable Cartan subgroup H of G such that $A \subset H$ and the following holds (see Vust [Vus] for the general case and Goodman-Wallach [Go-Wa, §12.3.1] for the classical groups):

(1) KAN is a Zariski-dense subset of G.
(2) The subgroup $M = \mathrm{Cent}_K(A)$ is reductive and normalizes N.
(3) Let $T = H \cap K$. Then $H = AT$ and $A \cap T$ is finite.
(4) There exists a Borel subgroup B with $HN \subset B \subset MAN$.

Thus MAN is a parabolic subgroup of G with reductive Levi component MA and unipotent radical N. We will give a more precise description of the set KAN in the next section.

4 Representations on Symmetric Spaces

4.1 Spherical Representations

We continue with the same setting and notation as in Section 3.3; in particular, P_{++} is the set of B-dominant weights. If $\lambda \in P_{++}$ and $E_\lambda^K \neq 0$ then λ will be called a K *spherical* highest weight and E_λ a K *spherical* representation.

[9] When $G_0 = \mathrm{SL}(n, \mathbb{R})$ this decomposition is the so-called QR factorization of a matrix obtained by the Gram-Schmidt orthogonalization algorithm.

Proposition 4.1

(i)   *K is a spherical subgroup of G.*

(ii) *Let $T = H \cap K$. If $\lambda \in P_{++}$ is a K spherical highest weight, then*

$$t^\lambda = 1 \quad \text{for all } t \in T. \tag{4.1}$$

Proof Since B contains AN, the Iwasawa decomposition shows that BK is dense in G, so B has an open orbit on G/K. Hence K is a spherical subgroup by Theorem 2.2. Since T is the stabilizer in H of the point $K \in G/K$, condition (4.1) likewise holds. $\qquad\square$

We say that λ is *θ-admissible* if it satisfies (4.1).

Example.   Let $G = \mathrm{SL}(n, \mathbf{C})$ and $\theta(g) = (g^t)^{-1}$. Here $A = H$ (diagonal matrices in G), $N =$ all upper-triangular unipotent matrices, and $M = T \cong (\mathbf{Z}/2\mathbf{Z})^{n-1}$ consists of all matrices

$$t = \mathrm{diag}[\delta_1, \dots, \delta_n], \quad \delta_i = \pm 1, \det(t) = 1.$$

Hence the θ-admissible highest weights $\lambda = [\lambda_1, \dots, \lambda_{n-1}, 0]$ are those with λ_i *even* for all i. $\qquad\square$

Remark.   In general, the subgroup $F = T \cap A$ is finite and consists of elements of order 2, since $h = \theta(h) = h^{-1}$ for $h \in F$. Thus a θ-admissible highest weight λ is trivial on T and its restriction to A is *even*, in the sense that $h^\lambda = 1$ for $h \in F$.

Cartan [Car2] proved the implication (i) $\implies$ (iii) in the following theorem and gave some indications for the proof of the converse (see Borel [Bor, Chap. IV §4.4-5]). Thus the following result is sometimes called the *Cartan–Helgason* theorem, although part (ii) and the first complete proof of the theorem is due to Helgason [Hel1].

Theorem 4.2 *Let (π_λ, E_λ) be an irreducible rational representation of G with highest weight λ (relative to B). The following are equivalent:*

(i) $E_\lambda^K \neq 0$.

(ii) *MN fixes the B-highest weight vector in E_λ.*

(iii) *λ is θ-admissible.*

Proof The equivalence of (ii) and (iii) follows by a Lie algebra argument using $\mathfrak{sl}_2$ representation theory (see [Hel1] or [Go-Wa, §12.3.3]), and the implication (i) $\implies$ (iii) comes from Proposition 4.1. We give Helgason's analytic proof that (iii) $\implies$ (i).[10] Let λ be θ-admissible. Define

$$v_0 = \int_{K_0} \pi_\lambda(k) e_\lambda \, dk. \tag{4.2}$$

Then $v_0 \in (E_\lambda)^K$ by the unitarian trick, since K is connected. To show $v_0 \neq 0$, let ψ_λ be the generating function for π_λ. Then

$$\langle v_0, f_{\lambda^*} \rangle = \int_{K_0} \psi_\lambda(k) \, dk. \tag{4.3}$$

We use the following properties:

(1) Let σ be the complex conjugation of G whose fixed-point set is G_0. Then $\chi(\sigma(a)) = \overline{\chi(a)}$ for any regular character χ of A.

(2) If $h \in H \cap G_0 = (T \cap G_0)A_0$, then $h^\lambda > 0$ by (1), since $h = t \exp(x)$ with $t \in T$ and $x \in \mathfrak{a}_0$.

(3) $\psi_\lambda(g) \geq 0$ for $g \in G_0$ by (2) and the Gauss decomposition.

Since $\psi_\lambda(1) = 1$ and $K_0 \subset G_0$, property (3) shows that the integral (4.3) is nonzero. $\square$

Example.   Let $G = \mathrm{SL}(n, \mathbb{C})$ and $\theta(g) = (g^t)^{-1}$. Here A_0 consists of real diagonal matrices, $G_0 = \mathrm{SL}(n, \mathbb{R})$, and

$$\psi_\lambda(g) = \det_1(g)^{m_1} \cdots \det_{n-1}(g)^{m_{n-1}},$$

where $\det_i$ is the ith principal minor and $m_i = \lambda_i - \lambda_{i+1}$. Since λ is θ-admissible iff all λ_i are even, condition (3) in the proof of Theorem 4.2 obviously holds. For example, the highest weight $\lambda = [2, 0, \ldots, 0]$ is admissible, and the corresponding spherical representation $E_\lambda = S^2(\mathbb{C}^n)$. The K-fixed vector in E_λ is $\sum_i e_i \otimes e_i$, where $\{e_i\}$ is the standard basis for $\mathbb{C}^n$. $\square$

The l *fundamental* K-spherical highest weights $\mu_1, \ldots, \mu_r$ (with $l = \dim A$ the rank of G/K) are linearly independent, and the general spherical highest weight is $\mu = m_1 \mu_1 + \cdots + m_l \mu_l$ with $m_i \in \mathbb{N}$ (see [Hel2, Ch. V, §4]). Let $\Lambda \subset P_{++}$ be the subsemigroup of spherical highest weights. Since K is reductive and the identity representation is self-dual, $E_\lambda^K \neq 0$ if and only if $E_{\lambda^*}^K \neq 0$. Hence Λ is invariant under the map $\lambda \mapsto \lambda^*$ on P_{++}.

Corollary 4.1 *As a G-module,* $\mathcal{O}[G/K] \cong \bigoplus_{\mu \in \Lambda} E_\mu$.

---

10 An algebraic-geometric proof was given later by Vust [Vus].

4.2 *Zonal Spherical and Horospherical Functions*

For each $\mu \in \Lambda$ choose a K-fixed *spherical vector* $e_\mu^K \in E_\mu$ and a MN-fixed *conical vector* $e_\mu \in E_\mu$, normalized so that

$$\langle e_\mu, e_{\mu^*}^K \rangle = 1, \qquad \langle e_\mu^K, e_{\mu^*}^K \rangle = 1. \tag{4.4}$$

The *zonal spherical function* $\varphi_\mu \in \mathcal{O}[G]$ is the representative function determined by pairing the K-fixed vectors in E_μ and E_{μ^*}:

$$\varphi_\mu(g) = \langle \pi_\mu(g)e_\mu^K, e_{\mu^*}^K \rangle.$$

From the definition it is clear that

$$\varphi_\mu(kgk') = \varphi_\mu(g) \quad \text{and} \quad \varphi_\mu(1) = 1$$

for $k, k' \in K$ and $g \in G$. Thus φ_μ is a regular function on G/K that is constant on the K-orbits.

The *zonal horospherical function* $\Delta^\mu \in \mathcal{O}[G]$ is the representative function determined by pairing the MN-fixed vector in E_μ with the K-fixed vector in E_{μ^*}:

$$\Delta^\mu(g) = \langle \pi_\mu(g)e_\mu, e_{\mu^*}^K \rangle.$$

From the definition it is clear that

$$\Delta^\mu(kgman) = a^\mu \Delta^\mu(g) \quad \text{and} \quad \Delta(1) = 1 \tag{4.5}$$

for $k \in K$, $g \in G$, and $man \in MAN$. Properties (4.5) with $g = 1$ determine Δ^μ uniquely, since KAN is dense in G. We can view Δ^μ as a holomorphic function on the affine symmetric space $K\backslash G$ that transforms by the character $man \mapsto a^\mu$ along the MAN orbits. The existence of a regular function on G with these transformation properties is equivalent to the existence of the K-spherical representation π_μ (just as for the generating functions ψ_λ in Section 2.2, which are the zonal horospherical functions associated with the diagonal embedding of G as a spherical subgroup of $G \times G$). Let μ and ν be K-spherical highest weights. From (4.5) and the density of the set $KMAN$ it follows that

$$\Delta^\mu(g)\Delta^\nu(g) = \Delta^{\mu+\nu}(g) \quad \text{for } g \in G. \tag{4.6}$$

Let $\mu_1, \ldots, \mu_r$ be the fundamental K-spherical highest weights, and define[11]

$$\Delta_j(g) = \Delta^{\mu_j}(g).$$

---

11 Gindikin [Gin3] calls $\{\Delta_j\}$ the *Sylvester functions*; Theorem 4.3 shows they play the same role for the KAN decomposition as the generating functions $\{\psi_j\}$ for the N^-HN^+ decomposition.

For a general K-spherical highest weight $\mu = m_1\mu_1 + \cdots + m_r\mu_r$ formula (4.6) implies the *product formula*

$$\Delta^\mu(g) = \Delta_1(g)^{m_1} \cdots \Delta_r(g)^{m_r}. \qquad (4.7)$$

Set

$$\Omega = \{g \in G : \Delta_j(g) \neq 0 \quad \text{for } j = 1, \ldots, r\}.$$

The weight μ is *regular* if $m_i \neq 0$ for $i = 1, \ldots, r$. If μ is regular, then we see from (4.7) that $\Omega = \{g \in G : \Delta^\mu(g) \neq 0\}$. Using techniques originating with Harish-Chandra [H-C], Clerc [Cle] obtained the following precise description of the complexified Iwasawa decomposition:

Theorem 4.3 *One has* $\Omega = KAN$. *Let* $g = k(g)a(g)n(g)$ *be the Iwasawa factorization in* G_0.

(i) *The function* $g \mapsto n(g)$ *extends holomorphically to a map from* Ω *to* N.

(ii) *The functions* $g \mapsto k(g)$ *and* $g \mapsto a(g)$ *extend to multivalent holomorphic functions on* Ω, *with values in* K *and* A, *respectively. The branches are related by elements of the finite subgroup* $F = T \cap A$.

(iii) *Let* $g \mapsto \mathcal{H}(g)$ *be the multivalent* $\mathfrak{a}$-*valued function on* Ω *such that* $a(g) = \exp \mathcal{H}(g)$. *Then*

$$\Delta^\mu(g) = e^{\langle \mathcal{H}(g), \mu \rangle} \quad \text{for } g \in \Omega \text{ and } \mu \in \Lambda.$$

Theorem 4.3 and (4.2) yield a formula analogous to Harish-Chandra's integral formula [H-C] for zonal spherical functions on the noncompact symmetric space G_0/K_0:

Corollary 4.2 *For* $g \in G$ *let* $K_g = \{k \in K_0 : gk \in \Omega\}$. *Then* K_g *is an open set in* K_0 *whose complement has measure zero. For* $\mu \in \Lambda$ *one has*

$$\varphi_\mu(g) = \int_{K_g} e^{\langle \mathcal{H}(gk), \mu \rangle} \, dk.$$

Clerc, elaborating on methods introduced by E. P. Van den Ban [VdBan], uses this integral representation and the method of complex stationary phase to determine the asymptotic behavior of $\varphi_\mu(u)$ as $\mu \to \infty$ in a suitable cone when u is a regular element of U; see [Cle, Théorème 3.4] for details.

4.3 Horospherical Cauchy–Radon Transform

By Theorem 4.2 the G-modules $\mathcal{O}[G]^{R(K)}$ and $\mathcal{O}[G]^{R(MN)}$ are multiplicity free and have the same spectrum (the set Λ of K-spherical highest weights). Using the normalized K-fixed vectors and MN-fixed highest weight vectors, we can thus define bijective G-intertwining maps

$$T : \bigoplus_{\mu \in \Lambda} E_\mu \xrightarrow{\cong} \mathcal{O}[G]^{R(K)}, \qquad \sum_{\mu \in \Lambda} v_\mu \mapsto \sum_{\mu \in \Lambda} d(\mu) \langle v_\mu, \pi_{\mu^*}(g) e_{\mu^*}^K \rangle$$

and

$$S : \bigoplus_{\mu \in \Lambda} E_\mu \xrightarrow{\cong} \mathcal{O}[G]^{R(MN)}, \qquad \sum_{\mu \in \Lambda} v_\mu \mapsto \sum_{\mu \in \Lambda} \langle v_\mu, \pi_{\mu^*}(g) e_{\mu^*} \rangle.$$

In both cases we assume that the components $v_\mu = 0$ for all but finitely many $\mu \in \Lambda$.

Let $f \in \mathcal{O}[G]^{R(K)}$. The (algebraic) Peter–Weyl expansion of f is

$$f(g) = \sum_{\mu \in \Lambda} d(\mu) \langle v_\mu, \pi_{\mu^*}(g) e_{\mu^*}^K \rangle \tag{4.8}$$

where $v_\mu \in E_\mu$ and $v_\mu = 0$ for all but finitely many μ. Here $d(\mu) = \dim E_\mu$. Following Gindikin [Gin2], we define the *horospherical Cauchy–Radon transform* $f \mapsto \hat{f}$ by

$$\hat{f}(g) = \sum_{\mu \in \Lambda} \langle v_\mu, \pi_{\mu^*}(g) e_{\mu^*} \rangle$$

Note that the dimension factor is removed, and the spherical vector is replaced by the conical vector in E_{μ^*}. It is easy to check that this definition does not depend on the choice of spherical and conical vectors, subject to the normalizations (4.4). We can express this transform in terms of the maps S and T just introduced as follows: If $v \in \bigoplus_\mu E_\mu$ and $f = Tv$, then $\hat{f} = Sv$. Since S and T are G-module isomorphisms, it follows that the map $f \mapsto \hat{f}$ gives a G-module isomorphism between the function spaces $\mathcal{O}[G]^{R(K)}$ and $\mathcal{O}[G]^{R(MN)}$. We now express this isomorphism in a more analytic form.

Theorem 4.4 *The horospherical Cauchy–Radon transform is given by the integral formula*

$$\hat{f}(g) = \sum_{\mu \in \Lambda} \int_U f(u) \Delta^\mu(u^{-1}g) \, du \qquad \textit{for } g \in G \tag{4.9}$$

(the integrals are zero for all but finitely many μ).

Remark. The integrands in (4.9) are invariant under $u \mapsto uk$ with $k \in K_0$, so the integrals can be viewed as taken over the compact symmetric space $X = U/K_0$.

Proof Let f be given by (4.8) and let $\mu \in \Lambda$. Since f is right K-invariant and $E_\mu^K = \mathbb{C}e_\mu^K$, we have

$$\int_U f(u)\pi_\mu(u^{-1}g)e_\mu \, du = c_\mu(g)e_\mu^K \qquad (4.10)$$

for some function $c_\mu(g)$ on G. From the Schur orthogonality relations and (4.4) we find that

$$c_\mu(g) = \langle v_\mu, \, \pi_{\mu^*}(g)e_{\mu^*}\rangle.$$

Evaluating both sides of (4.10) on the vector $e_{\mu^*}^K$, and summing on μ, we obtain (4.9). $\square$

Remarks. 1. The horospherical Cauchy–Radon transform is the representation-theoretic expression of the double fibration

$$
\begin{array}{ccc}
 & G & \\
 \swarrow & & \searrow \\
Z = G/K & & G/MN = \Xi
\end{array}
$$

This sets up a *correspondence* between Z and Ξ: a point gK of Z maps to the *pseudosphere* $gKMN \cong K/M$ in Ξ, and a point gMN in Ξ maps to the *horosphere* $gMNK \cong N$ in Z (see Gindikin [Gin1] for some examples).

2. Let $\bar{N} = \theta(N)$. Then $\bar{N}MAN$ is Zariski-dense in G (the generalized Gauss decomposition) and $A\bar{N}K$ is also Zariski-dense in G (the Iwasawa decomposition). Thus the solvable group $A\bar{N}$ has an open orbit in G/K and in G/MN, but the two homogeneous spaces are not isomorphic as complex manifolds, even though they have the same G spectrum and multiplicities.

An invariant (holomorphic) differential operator $P(D)$ on A has a polynomial symbol $P(\mu)$ such that

$$P(D)a^\mu = P(\mu)a^\mu \quad \text{for } \mu \in \Lambda.$$

If μ is a K-spherical highest weight, then the Weyl dimension formula asserts that

$$d(\mu) = \prod_{\alpha>0} \frac{(\mu+\delta,\alpha)}{(\delta,\alpha)} \qquad \text{where } \delta = \tfrac{1}{2}\sum_{\alpha>0}\alpha.$$

Since $\mu = 0$ on $\mathfrak{t}$, we can view $\mu \mapsto d(\mu)$ as a polynomial function $W(\mu)$ on $\mathfrak{a}^*$. Following Gindikin [Gin3], we define the *Weyl operator* $W(D)$ to be the differential operator on A with symbol $W(\mu)$.

Since A normalizes MN, the space $\mathcal{O}[G]^{R(MN)}$ is stable under $R(A)$. The complex horospherical manifold Ξ is a fiber bundle over the compact flag manifold $F = G/MAN$ (a projective variety), with fiber A. The operator $W(D)$ acts by differentiation along the fibers.

Using the Weyl operator, Gindikin [Gin2] obtains the following inversion formula for the horospherical Cauchy–Radon transform:

Theorem 4.5 *Let* $f \in \mathcal{O}[G]^{R(K)}$. *Then*

$$f(g) = \int_{K_0} (W(D)\hat{f})(gk)\,dk \qquad \text{for } g \in G \tag{4.11}$$

Remark. The integrand in (4.11) is invariant under right translations by M_0, so the integral is taken over the compact flag manifold $K_0/M_0 = G_0/M_0 A_0 N_0$ associated with the dual noncompact symmetric space.

Proof It suffices to prove (4.11) when $f(g) = d(\mu)\langle v_\mu, \pi_{\mu^*}(g)e_{\mu^*}^K\rangle$ with $v_\mu \in E_\mu$. In this case,

$$\hat{f}(ga) = \langle v_\mu, \pi_{\mu^*}(ga)e_{\mu^*}\rangle = a^{\mu^*}\langle v_\mu, \pi_{\mu^*}(g)e_{\mu^*}\rangle.$$

Hence $W(D)\hat{f}(g) = d(\mu)\hat{f}(g)$ since $d(\mu) = d(\mu^*)$. Thus

$$\int_{K_0} (W(D)\hat{f})(gk)\,dk = d(\mu)\int_{K_0} \langle v_\mu, \pi_{\mu^*}(g)\pi_{\mu^*}(k)e_{\mu^*}\rangle\,dk = f(g),$$

since the integration of $\pi_{\mu^*}(k)e_{\mu^*}$ over K_0 yields $e_{\mu^*}^K$. $\qquad\square$

4.4 Cauchy–Radon Transform as a Singular Integral

Denote by $Z = G/K$ the complex symmetric space with origin $x_0 = K$. Let $\zeta_0 = MN$ denote the origin in Ξ. For $z = g{\cdot}x_0 \in Z$ and $\zeta = y{\cdot}\zeta_0 \in \Xi$ we set $\Delta_j(z\mid\zeta) = \Delta_j(g^{-1}y)$. This is well-defined by the transformation properties (4.5), and we have

$$\Delta_j(z\mid\zeta a) = a^{\mu_j}\Delta_j(z\mid\zeta) \quad \text{for } a \in A.$$

Following Gindikin ([Gin2], [Gin3]), we define the *Cauchy–Radon kernel* on $Z \times \Xi$ by

$$K(z \mid \zeta) = \prod_{1 \leq j \leq l} \frac{1}{1 - \Delta_j(z \mid \zeta)}.$$

This function is meromorphic and invariant under the diagonal action of G, since $\Delta_j(g \cdot z \mid g \cdot \zeta) = \Delta_j(z \mid \zeta)$ for $g \in G$. The singular set of $K(z \mid \zeta)$ is the union of the manifolds $\{\Delta_j(z \mid \zeta) = 1\}$ in $Z \times \Xi$ for $j = 1, \ldots, l$.

Recall that $X = U/K_0$ is the compact symmetric space corresponding to θ. Define

$$\Xi(0) = \{\zeta \in \Xi : |\Delta_j(x \mid \zeta)| < 1 \quad \text{for all } x \in X\}.$$

By definition, $U \cdot \Xi(0) = \Xi(0)$. Furthermore, the product formula (4.7) implies that

$$K(x \mid \zeta) = \sum_{\mu \in \Lambda} \Delta^\mu(u^{-1}g) \quad \text{(absolutely convergent series)} \qquad (4.12)$$

for $x = u \cdot x_0 \in X$ and $z = g \cdot \zeta_0 \in \Xi(0)$. Since A normalizes the subgroup MN, the right multiplication action of A on G gives a right action of A on Ξ, denoted by $\zeta, a \mapsto \zeta \cdot a$. This action commutes with the left action of G on Ξ.

Lemma 4.3

(i) *The map* $(U/M_0) \times A \to \Xi$ *given by* $(u, a) \mapsto u \cdot \zeta_0 \cdot a$ *is regular and surjective.*

(ii) *Let* $A_+ = \{a \in A : |a^{\mu_j}| < 1 \text{ for } j = 1, \ldots, l\}$. *Then* $U \cdot \zeta_0 \cdot A_+ \subset \Xi(0)$. *Hence* $\Xi(0)$ *is a nonempty open subset of* Ξ.

Proof Since U is a maximal compact subgroup of G, the Iwasawa decomposition of G shows that $G = UMAN$. This implies (i).

Clerc [Cle, Lemme 2.3], using a representation-theoretic argument originating with Harish-Chandra [H-C], shows that $|\Delta^\mu(u)| \leq 1$ for $\mu \in \Lambda$ and $u \in U$. Let $a \in A$. Then Clerc's estimate implies that

$$|\Delta^\mu(ua)| = |\Delta^\mu(u)| \, |a^\mu| \leq |a^\mu|. \qquad (4.13)$$

If $a \in A_+$ then $|a^\mu| < 1$. Hence for $u, u' \in U$ we have

$$|\Delta^\mu(u \cdot x_0 \mid u'a \cdot \zeta_0)| = |\Delta^\mu(u^{-1}u'a)| < 1$$

by (4.13). This implies (ii). $\qquad \square$

Using Lemma 4.3, we can obtain Gindikin's singular integral formula for the horospherical Cauchy-Radon transform. The noncompact real symmetric space G/U is the space of compact real forms of G, and by the Cartan decomposition of G it is a contractible manifold. For $\nu = gU \in G/U$ we define a compact totally-real cycle $X(\nu) = g \cdot X \subset Z$ and an open set $\Xi(\nu)g \cdot \Xi(0) \subset \Xi$. This furnishes an open covering

$$\Xi = \bigcup_{\nu \in G/U} \Xi(\nu)$$

with a contractible parameter space.

Theorem 4.6 (Gindikin) *For $f \in \mathcal{O}[Z]$ the horospherical Cauchy–Radon transform is given on each set of the covering $\{\Xi(\nu)\}$ by the Cauchy-type singular integral*

$$\hat{f}(\zeta) = \int_{X(\nu)} f(x)K(x \mid \zeta)\,dx \quad for \quad \zeta \in \Xi(\nu) \qquad (4.14)$$

(the integrand is continuous on $X(\nu)$).

Proof Use formula (4.12) for $K(x \mid \zeta)$ when $\zeta \in \Xi(0)$, and then translate by $g \in G$ to get the formula in general. $\square$

5 Concluding Remarks

In this paper we described the harmonic analysis of finitely-transforming functions on a compact symmetric space using algebraic group and Lie group methods, extending the fundamental results of Cartan and Weyl. Our presentation of the horospherical Cauchy-Radon transform has emphasized groups and homogeneous spaces as in [Gin2]; in fact, the integral formulas hold for all holomorphic functions (not just the G-finite functions) on X and Ξ, and also for hyperfunctions. Gindikin's point of view is that a compact symmetric space has a canonical dual object that is a complex manifold, and he develops this transform emphasizing complex analysis and integral geometry (see [Gin3]).

An analytic problem that we have not discussed is the holomorphic extension of real analytic functions on a compact symmetric space. These functions extend holomorphically to complex neighborhoods of the space. The geometric and analytic properties of these neighborhoods were studied by B. Beers and A. Dragt [Be-Dr], L. Frota-Mattos [Fr-Ma] and M. Lasalle [Las].

Bibliography

[Be-Ra] C. Benson and G. Ratcliff, On Multiplicity Free Actions, in *Representations of Real and p-Adic Groups* (Lecture Notes **2**, IMS, National University of Singapore), World Scientific, 2004.

[Be-Dr] B. L. Beers and A. J. Dragt, *New theorems about spherical harmonic expansions on* SU(2), J. Mathematical Phys. **11** (1970), 2313-2328.

[Bor] A. Borel, *Essays in the History of Lie Groups and Algebraic Groups* (History of Mathematics **21**), American Mathematical Society, Providence, 2001.

[Bo-Ha] A. Borel and Harish-Chandra, *Arithmetic subgroups of algebraic groups*, Annals of Mathematics **75** (1962), 485-535.

[Car1] E. Cartan, *Les groupes projectifs qui ne laissent invariante aucune multiplicité plane*, Bull. Soc. Math. de France **41** (1913), 53–96; reprinted in *Oeuvres Complètes* **1**, Part 1, 355–398, Gauthier-Villars, Paris, 1952.

[Car2] E. Cartan, *Sur la détermination d'un système orthogonal complet dans un espace de Riemann symétrique clos*, Rend. Circ. Mat. Palermo **53** (1929), 217-252; reprinted in *Oeuvres Complètes* **1**, Part 2, 1045-1080, Gauthier-Villars, Paris, 1952.

[Cle] J.-L. Clerc, *Fonctions sphériques des espaces symétriques compacts*, Trans. Amer. Math. Soc. **306** (1988), 421-431.

[Fr-Ma] L. A. Frota-Mattos, The complex-analytic extension of the Fourier series on Lie groups, in *Proceedings of Symposia in Pure Mathematics*, Volume 30, Part 2 (1977), 279-282.

[Gin1] S. Gindikin, *Holomorphic horospherical duality "sphere-cone"*, Indag. Mathem, N.S., **16** (2005), 487-497.

[Gin2] S. Gindikin, *Horospherical Cauchy-Radon transform on compact symmetric spaces*, Mosc. Math. J. **6** (2006), no. 2, 299-305, 406.

[Gin3] S. Gindikin, *Harmonic analysis on symmetric Stein manifolds from the point of view of complex analysis*, Jpn. J. Math. **1** (2006), 87-105.

[Go-Wa] R. Goodman and N. R. Wallach, *Representations and Invariants of the Classical Groups* (Encyclopedia of Mathematics and Its Applications, Vol. 68), Cambridge University Press, 1998 (3rd corrected printing 2003).

[H-C] Harish-Chandra, *Spherical functions on a semi-simple Lie group. I*, Amer. J. Math. **80** (1958), 241-310.

[Haw] T. Hawkins, *Emergence of the Theory of Lie Groups: an Essay in the History of Mathematics 1869-1926*, Springer-Verlag, New York, 2000.

[Hel1] S. Helgason, *A duality for symmetric spaces with applications to group representations*, Advances in Math. **5** (1970), 1-154.

[Hel2] S. Helgason, *Groups and Geometric Analysis* (Pure and Applied Mathematics **113**), Academic Press, Orlando, 1984.

[Hel3] S. Helgason, The Fourier transform on symmetric spaces, in *Élie Cartan et les Mathématiques d'Aujourd'hui*, Astérisque No. hors série (1985), Société Mathématique de France, pp. 151-164.

[Las] M. Lasalle, *Series de Laurent des fonctions holomorphes dans la complexification d'un espace symétrique compact*, Ann. Sci. École Norm. Sup. (4) **11** (1978), 167-210.

[Mat] Y. Matsushima, *Espaces homogènes de Stein des groupes de Lie complexes*, Nagoya Math. J. **16** (1960), 205-218.

[Pe-We] F. Peter and H. Weyl, *Die Vollständigkeit der primitiven Darstellungen einer geschlossenen kontinuierlichen Gruppe*, Math. Annalen **97** (1927), 737-755.

[Ric1] R. W. Richardson, *On Orbits of Algebraic Groups and Lie Groups*, Bull. Austral. Math. Soc. **25** (1982), 1-28.

[Ric2] R. W. Richardson, *Orbits, Invariants, and Representations Associated to Involutions of Reductive Groups*, Invent. math. **66** (1982), 287-313 .

[Ros] M. Rosenlicht, *On Quotient Varieties and the Affine Embedding of Certain Homogeneous Spaces*, Trans. Amer. Math. Soc. **101** (1961), 211-223.

[VdBan] E. P. Van den Ban, *Asymptotic expansions and integral formulas for eigenfunctions on a semisimple Lie group*, Thesis, Utrecht, 1983.

[Vi-Ki] E. B. Vinberg and B. N. Kimelfeld, *Homogeneous Domains on Flag Manifolds and Spherical Subgroups*, Func. Anal. Appl. **12** (1978), 168-174.

[Vus] Th. Vust, *Opération de groupes réductifs dans un type de cônes presque homogènes*, Bull. Soc. math. France **102** (1974), 317-333.

[Wey1] H. Weyl, *Theorie der Darstellung kontinuierlicher halfeinfacher Gruppen durch lineare Transformationen, I, II, III, und Nachtrag*, Math. Zeitschrift **23**, 271–309; **24** (1926), 328–376, 377–395, 789–791; reprinted in *Selecta Hermann Weyl*, 262–366, Birkhäuser Verlag, Basel, 1956.

[Wey2] H. Weyl, *Harmonics on homogeneous manifolds*, Annals of Mathematics **35** (1934), 486-499; reprinted in *Hermann Weyl Gesammelte Abhandlungen, Band III*, 386-399, Springer-Verlag, Berlin - Heidelberg, 1968.

[Wey3] H. Weyl, *Elementary algebraic treatment of the quantum mechanical symmetry problem*, Canadian J. Math. **1** (1949), 57-68; reprinted in *Hermann Weyl Gesammelte Abhandlungen, Band IV*, 346-359, Springer-Verlag, Berlin - Heidelberg, 1968.

[Wey4] H. Weyl, *Relativity theory as a stimulus in mathematical research*, Proc. Amer. Phil. Soc. **93** (1949), 535-541; reprinted in *Hermann Weyl Gesammelte Abhandlungen, Band IV*, 394-400, Springer-Verlag, Berlin - Heidelberg, 1968.

[Žel] D. P. Želobenko, *Compact Lie groups and their representations* (Translations of mathematical monographs Vol. 40), American Mathematical Society, Providence, RI, 1973.

2
Weyl, eigenfunction expansions and harmonic analysis on non-compact symmetric spaces

Erik van den Ban
University of Utrecht
ban@math.uu.nl

1 Introduction

This text grew out of an attempt to understand a remark by Harish-Chandra in the introduction of [12]. In that paper and its sequel he determined the Plancherel decomposition for Riemannian symmetric spaces of the non-compact type. The associated Plancherel measure turned out to be related to the asymptotic behavior of the so-called zonal spherical functions, which are solutions to a system of invariant differential eigenequations. Harish-Chandra observed: 'this is reminiscent of a result of Weyl on ordinary differential equations', with reference to Hermann Weyl's 1910 paper, [29], on singular Sturm–Liouville operators and the associated expansions in eigenfunctions.

For Riemannian symmetric spaces of rank one the mentioned system of equations reduces to a single equation of the singular Sturm–Liouville type. Weyl's result indeed relates asymptotic behavior of eigenfunctions to the continuous spectral measure but his result is formulated in a setting that does not directly apply.

In [23], Kodaira combined Weyl's theory with the abstract Hilbert space theory that had been developed in the 1930's. This resulted in an efficient derivation of a formula for the spectral measure, previously obtained by Titchmarsh. In the same paper Kodaira discussed a class of examples that turns out to be general enough to cover all Riemannian symmetric spaces of rank 1.

It is the purpose of this text to explain the above, and to describe later developments in harmonic analysis on groups and symmetric spaces where Weyl's principle has played an important role.

24

2 Sturm–Liouville operators

A Sturm–Liouville operator is a second order ordinary differential operator of the form

$$L = -\frac{d}{dt} p \frac{d}{dt} + q, \tag{2.1}$$

defined on an open interval $]a, b[$, where $-\infty \leq a < b \leq +\infty$. Here p is assumed to be a C^1-function on $]a, b[$ with strictly positive real values; q is assumed to be a real valued continuous function on $]a, b[$.

The operator L is said to be regular at the boundary point a if a is finite, p extends to a C^1-function $[a, b[\to]0, \infty[$ and q extends to a continuous function on $[a, b[$. Regularity at the second boundary point b is defined similarly. The operator L is said to be regular if it is regular at both boundary points. In the singular case, no conditions are imposed on the behavior of the functions p and q towards the boundary points apart from those already mentioned.

The operator L is formally symmetric in the sense that

$$\langle Lf, g \rangle_{[a,b]} = \langle f, Lg \rangle_{[a,b]}$$

for all compactly supported C^2-functions f and g on $]a, b[$. Here we have denoted the standard L^2-inner product on $[a, b]$ by

$$\langle f, g \rangle_{[a,b]} = \int_a^b f(t) \, \overline{g(t)} \, dt.$$

For arbitrary C^2-functions f and g on $]a, b[$ it follows by partial integration that

$$\langle Lf, g \rangle_{[x,y]} - \langle f, Lg \rangle_{[x,y]} = [f, g]_y - [f, g]_x, \tag{2.2}$$

for all $a < x \leq y < b$. Here the sesquilinear form $[\cdot, \cdot]_t$ on $C^1(]a, b[)$ (for $a < t < b$) is defined by

$$[f, g]_t := p(t) [f(t) \, \overline{g'(t)} - f'(t) \, \overline{g(t)}]. \tag{2.3}$$

To better understand the nature of this form, let $\langle \cdot, \cdot \rangle$ denote the standard Hermitian inner product on $\mathbb{C}^2$, and define the (anti-symmetric) sesquilinear form $[\cdot, \cdot]$ on $\mathbb{C}^2$ by

$$[v, w] := \langle Jv, w \rangle, \qquad J = \begin{pmatrix} 0 & -1 \\ 1 & 0 \end{pmatrix}. \tag{2.4}$$

Define the evaluation map $\varepsilon_t : C^1(]a, b[) \to \mathbb{C}^2$ by

$$\varepsilon_t(f) := (f(t), p(t)f'(t)). \tag{2.5}$$

Then the form (2.3) is given by $[f,g]_t = [\varepsilon_t(f), \varepsilon_t(g)]$. We now observe that for ξ a non-zero vector in $\mathbb{R}^2$,

$$\langle \varepsilon_t(f), \xi \rangle = 0 \iff \varepsilon_t(f) \in \mathbb{C} \cdot J\xi. \qquad (2.6)$$

Hence, if f, g are functions in $C^1(\,]a,b[\,)$, then by anti-symmetry of the form $[\,\cdot\,,\,\cdot\,]$ we see that

$$\langle \varepsilon_t(f), \xi \rangle = \langle \varepsilon_t(g), \xi \rangle = 0 \implies [f,g]_t = 0. \qquad (2.7)$$

For a complex number $\lambda \in \mathbb{C}$ we denote by $\mathcal{E}_\lambda$ the space of complex valued C^2-functions f on $\,]a,b[\,$, satisfying the eigenequation $Lf = \lambda f$. This eigenequation is equivalent to a system of two linear first order equations for the function $\varepsilon(f) : t \mapsto \varepsilon_t(f)$. It follows that for every $a < c < b$ and every $v \in \mathbb{C}^2$ there is a unique function $s(\lambda, \cdot\,)v = s_c(\lambda, \cdot\,)v \in C^2(\,]a,b[\,)$ such that

$$s(\lambda, \cdot\,)v \in \mathcal{E}_\lambda, \qquad \text{and} \qquad \varepsilon_c(s(\lambda, \cdot\,)v) = v. \qquad (2.8)$$

By uniqueness, $s(\lambda)v$ depends linearly on v, and by holomorphic parameter dependence of the system, the map $\lambda \mapsto s(\lambda)v$ is entire holomorphic from $\mathbb{C}$ to $C^2(\,]a,b[\,)$.

3 The case of a regular operator

After these preliminaries, we recall the theory of eigenfunction expansions for a regular Sturm–Liouville operator L on $[a,b]$. Let ξ_a, ξ_b be two non-zero vectors in $\mathbb{R}^2$. We consider the linear space $C^2_\xi([a,b])$ of C^2-functions $f : [a,b] \to \mathbb{C}$ satisfying the homogeneous boundary conditions

$$\langle \varepsilon_a(f), \xi_a \rangle = 0, \qquad \langle \varepsilon_b(f), \xi_b \rangle = 0. \qquad (3.1)$$

For all functions f and g in this space, and for $t = a, b$, we now have the conclusion of (2.7). In view of (2.2), this implies that L is symmetric on the domain $C^2_\xi([a,b])$, i.e.,

$$\langle Lf, g \rangle_{[a,b]} = \langle f, Lg \rangle_{[a,b]},$$

for all $f, g \in C^2_\xi([a,b])$. In this setting we have the following result on eigenfunction expansions. Let $\sigma(L, \xi)$ be the set of $\lambda \in \mathbb{C}$ for which the intersection $\mathcal{E}_{\lambda,\xi} := \mathcal{E}_\lambda \cap C^2_\xi([a,b])$ is non-trivial.

Theorem 3.1 *The set $\sigma(L, \xi)$ is a discrete subset of $\mathbb{R}$ without accumulation points. For each $\lambda \in \sigma(L, \xi)$ the space $\mathcal{E}_{\lambda, \xi}$ is one dimensional. Finally,*

$$L^2([a, b]) = \widehat{\bigoplus}_{\lambda \in \sigma(L, \xi)} \mathcal{E}_{\lambda, \xi} \qquad \text{(orthogonal direct sum)}. \qquad (3.2)$$

We will sketch the proof of this result; this allows us to describe what was known about the spectral decomposition associated with a Sturm–Liouville operator when Weyl entered the scene.

For $\lambda \in \mathbb{C}$, let φ_λ be the function in $\mathcal{E}_\lambda$ determined by $\varepsilon_a(\varphi_\lambda) = J\xi_a$. Then $\langle \varepsilon_a(\varphi_\lambda), \xi_a \rangle = 0$, hence $[\varphi_\lambda, \varphi_\lambda]_a = 0$. The function $\lambda \mapsto \varphi_\lambda$ is entire holomorphic with values in $C^2([a, b])$. We observe that $\mathcal{E}_{\lambda, \xi} \neq 0$ if and only if φ_λ belongs to $\mathcal{E}_{\lambda, \xi}$, in which case $\mathcal{E}_{\xi, \lambda} = \mathbb{C}\varphi_\lambda$. We thus see that the condition $\lambda \in \sigma(L, \xi)$ is equivalent to the condition $\langle \varepsilon_b(\varphi_\lambda), \xi_b \rangle = 0$.

The function $\chi : \lambda \mapsto \langle \varepsilon_b(\varphi_\lambda), \xi_b \rangle$ is holomorphic with values in $\mathbb{C}$, and from (2.2) we deduce that

$$(\lambda - \overline{\lambda}) \langle \varphi_\lambda, \varphi_\lambda \rangle_{[a, b]} = [\varphi_\lambda, \varphi_\lambda]_b.$$

In view of (2.7) we now see that the function χ does not vanish for $\mathrm{Im}\, \lambda \neq 0$. Its set of zeros, which equals $\sigma(L, \xi)$, is therefore a discrete subset of $\mathbb{R}$ without accumulation points. Replacing L by a translate $L + \mu$ with $-\mu \in \mathbb{R} \setminus \sigma(L, \xi)$ if necessary, we see that without loss of generality we may assume that $0 \notin \sigma(L, \xi)$. This implies that L is injective on $C_\xi^2([a, b])$. Let $g \in C([a, b])$ and consider the equation $Lf = g$. Writing this equation as a system of first order equations in terms of $\varepsilon(f)$, using a fundamental system for the associated homogeneous equation, and applying variation of the constant one finds a unique solution $f \in C_\xi^2([a, b])$ to the equation. It is expressed in terms of g by an integral transform $\mathcal{G}$ of the form

$$\mathcal{G}g(t) = \int_a^b G(t, \tau)\, g(\tau)\, d\tau,$$

with integral kernel $G \in C([a, b] \times [a, b])$, called Green's function. The operator $\mathcal{G}$ turns out to be a two-sided inverse to the operator L : $C_\xi^2(\,]a, b[\,) \to C([a, b])$.

It follows from D. Hilbert's work on integral equations, [21], that the map $(f, g) \mapsto \langle f, \mathcal{G}g \rangle_{[a, b]}$ may be viewed as a non-degenerate Hermitian form in infinite dimensions, which allows a diagonalization over an orthonormal basis φ_k of $L^2([a, b])$, with associated non-zero diagonal elements λ_k, for $k \in \mathbb{N}$. In today's terminology we would say that the operator $\mathcal{G}$ is symmetric and completely continuous, or compact, and

Hilbert's result has evolved into the spectral theorem for such operators. From this the result follows with $\sigma(L, \xi) = \{\lambda_k^{-1} \mid k \in \mathbb{N}\}$.

4 The singular Sturm–Liouville operator

We now turn to the more general case of a (possibly) singular operator L on $]a, b[$. Weyl had written a thesis with Hilbert, leading to the paper [28], generalizing the theory of integral equations to 'singular kernels.' It was a natural idea to apply this work to singular Sturm–Liouville operators. At the time Weyl started his research it was understood that the regular cases involved discrete spectrum. On the other hand, from his work on singular integral equations it had become clear that continuous spectrum had to be expected.

Also, if one considers the example with $a = 0$, $b = \infty$, and $p = 1, q = 0$, then $L = -d^2/dt^2$ is regular at 0 and singular at ∞. Fix the boundary datum $\xi_0 = (0, 1)$. Then one obtains the eigenfunctions $\cos \sqrt{\lambda} t$ of L, with eigenvalue $\lambda \geq 0$. In this case a function $f \in C_c^2([0, \infty[)$, satisfying the boundary condition $\langle \varepsilon_0(f), \xi_0 \rangle = 0$ admits the decomposition

$$f(t) = \int_0^\infty a(\sqrt{\lambda}) \cos(\sqrt{\lambda} t) \, \frac{d\lambda}{\pi \sqrt{\lambda}}$$

involving the continuous spectral measure $\frac{d\lambda}{\pi \sqrt{\lambda}}$. Here of course, the function a is given by the cosine transform

$$a(\sqrt{\lambda}) = \int_0^\infty f(t) \cos \sqrt{\lambda} t \, dt.$$

Thus, no boundary condition needs to be imposed at infinity. At the time, Weyl faced the task to unify these phenomena, where both discrete and continuous spectrum (in his terminology 'Punktspektrum' and 'Streckenspektrum') could occur, and to clarify the role of the boundary conditions. Finally, the question arose what could be said of the spectral measure.

In [29], Weyl had the important idea to construct a Green operator for the eigenvalue problem $Lf = \lambda f$ with λ a *non-real* eigenvalue. He fixed boundary conditions for the Green kernel depending on a beautiful geometric classification of the situation at the boundary points which we will now describe. We will essentially follow Weyl's argument, but in order to postpone choosing bases, we prefer to use the language of projective space rather than refer to affine coordinates as Weyl did in [29], p. 226. The reader may consult the appendix for a quick review of the description of circles in one dimensional complex projective space in terms of Hermitian forms of signature type $(1,1)$.

Returning to the singular Sturm–Liouville problem, we make the following observation about real boundary data at a point $x \in \,]\,a,b\,[\,$.

Lemma 4.1 *Let* $\lambda \in \mathbb{C}$ *and let* $f \in \mathcal{E}_\lambda \backslash \{0\}$. *Then the following assertions are equivalent.*

(a) $\exists\, \xi \in \mathbb{R}^2 \backslash \{0\} :\ \langle \varepsilon_x(f)\,,\,\xi \rangle = 0$;

(b) $[\varepsilon_x(f)] \in \mathbb{P}^1(\mathbb{R})$;

(c) $[f,f]_x = 0$.

Proof As $[f,f]_x = [\varepsilon_x(f),\varepsilon_x(f)]$, these are basically assertions about $\mathbb{C}^2$, which are readily checked. $\square$

It follows that the zero set

$$C_{\lambda,x} := \{ f \in \mathcal{E}_\lambda \mid [f,f]_x = 0 \}$$

defines a circle in the projective space $\mathbb{P}(\mathcal{E}_\lambda)$. Indeed, let $\bar{\varepsilon}_x : \mathbb{P}(\mathcal{E}_\lambda) \to \mathbb{P}^1(\mathbb{C})$ be the projective isomorphism induced by the evaluation map (2.5), then $\bar{\varepsilon}_x(C_{\lambda,x}) = \mathbb{P}^1(\mathbb{R})$.

The following important observation is made in Weyl's paper [29], Satz 1.

Proposition 4.1 *Let* $\lambda \in \mathbb{C} \backslash \mathbb{R}$. *Then the circle* $C_{\lambda,x}$ *in* $\mathbb{P}(\mathcal{E}_\lambda)$ *depends on* $x \in \,]\,a,b\,[$ *in a continuous and strictly monotonic fashion. Moreover, if* $x \to b$ *then* $C_{\lambda,x}$ *tends to either a circle or a point. A similar statement holds for* $x \to a$.

We shall denote by $C_{\lambda,b}$ the limit of the set $C_{\lambda,x}$ for $x \to b$. The notation $C_{\lambda,a}$ is introduced in a similar fashion. The proof of the above result is both elegant and simple.

Proof We fix a point $c \in \,]\,a,b\,[\,$. For $x \in \,]\,c,b\,[$ we define the Hermitian inner product $\langle\, \cdot\, ,\, \cdot\, \rangle_x$ on $\mathcal{E}_\lambda$ by $\langle f\,,\,g \rangle_x = \langle f\,,\,g \rangle_{[c,x]}$. It follows from (2.2) that

$$[f,f]_x = [f,f]_c + 2i\,\mathrm{Im}\,(\lambda)\langle f\,,\,f \rangle_x,$$

for $f \in \mathcal{E}_\lambda$ and $x \in \,]\,c,b\,[\,$. Without loss of generality, let $\mathrm{Im}\,(\lambda) > 0$. Then it follows that $x \mapsto -i[f,f]_x$ is a real valued, strictly increasing continuous function. All results follow from this. $\square$

In his paper [29], Weyl uses a basis $f_1, f_2 \in \mathcal{E}_\lambda$ such that $\varepsilon_c(f_2), \varepsilon_c(f_1)$ is the standard basis of $\mathbb{C}^2$. Then $[f_1, f_2]_c = 1$. In the affine chart determined by f_1, f_2 the circle $C_{\lambda, c}$ equals the real line. The circles $C_{\lambda, x}$ therefore form a decreasing family of circles which are either all contained in the upper half plane or in the lower half plane. The form $i[\,\cdot\,,\,\cdot\,]_x$ is with respect to the basis f_1, f_2 given by the Hermitian matrix $H_{kl} = i[f_k, f_l]_x$. It follows that the center of $C_{\lambda, x}$ is given by $i[f_1, f_2]_x / (-i[f_1, f_1]_x)$, see (11.2). If $\operatorname{Im} \lambda > 0$ then the denominator of this expression is positive for $t > c$ whereas the numerator has limit i for $x \downarrow c$. It follows that in the affine coordinate z parametrizing $f_2 + z f_1$ the circles $C_{\lambda, x}$ lie in the upper half plane. Likewise, for $\operatorname{Im} \lambda < 0$ all circles lie in the lower half plane.

The limit of $C_{\lambda, x}$ as x tends to one of the boundary points is closely related to the L^2-behavior of functions from $\mathcal{E}_\lambda$ at that boundary point.

Lemma 4.2 *Let $\lambda \in \mathbb{C} \setminus \mathbb{R}$, and let $f \in \mathcal{E}_\lambda \setminus \{0\}$ be such that $\mathbb{C}f \in C_{\lambda, b}$. Then $f \in L^2([c, b[)$ for all $c \in]a, b[$.*

Proof We may fix a basis f_1, f_2 of $\mathcal{E}_\lambda$ such that in the associated affine chart, $C_{\lambda, c}$ corresponds to the real line. Then for every $x \neq c$ the circle $C_{\lambda, x}$ is entirely contained in the associated affine chart. There exists a sequence of points $x_n \in]c, b[$ and $F_n \in C_{\lambda, x_n}$ such that $x_n \to b$ and $F_n \to F := \mathbb{C}f$.

We agree to write $f_z = z f_1 + f_2$. Then there exist unique $z_n \in \mathbb{C}$ such that $F_n = \mathbb{C}f_{z_n}$. Now z_n converges to a point z_∞ and $F = \mathbb{C}f_{z_\infty}$. For $m < n$ we have

$$\langle f_{z_n}, f_{z_n} \rangle_{x_m} \leq \langle f_{z_n}, f_{z_n} \rangle_{x_n} = -(\lambda - \bar{\lambda})^{-1} [f_{z_n}, f_{z_n}]_c.$$

The expression on the right-hand side has a limit L for $n \to \infty$. It follows that

$$\langle f_{z_\infty}, f_{z_\infty} \rangle_{x_m} \leq L.$$

This is valid for any m. Taking the limit for $x_m \to b$ we conclude that $f_{z_\infty} \in L^2([c, b[)$. $\square$

Lemma 4.3 *Let $\lambda \in \mathbb{C} \setminus \mathbb{R}$ and assume that $\mathcal{E}_\lambda|_{[c, b[} \subset L^2([c, b[)$.*

(a) *The Hermitian form $h_{\lambda, x} := i[\,\cdot\,,\,\cdot\,]_x|_{\mathcal{E}_\lambda}$ has a limit $h_{\lambda, b} = i[\,\cdot\,,\,\cdot\,]_{\lambda, b}$, for $x \to b$.*

(b) *The form $h_{\lambda, b}$ is Hermitian and non-degenerate of signature $(1, 1)$.*

(c) *The limit set $C_{\lambda, b}$ is the circle given by $[f, f]_{\lambda, b} = 0$.*

(d) *In the space* $\text{Hom}(\bar{\mathcal{E}}_\lambda^*, \mathcal{E}_\lambda)$ *the inverse* $h_{\lambda,x}^{-1}$ *converges to* $h_{\lambda,b}^{-1}$ *as* $x \to b$.

Proof (a) From (2.2) it follows by taking the limit for $x \to b$ that

$$[f,g]_{\lambda,b} = \lim_{x \to b} [f,g]_x = [f,g]_c + (\lambda - \bar{\lambda})\langle f, g \rangle_{[c,b]}$$

for all $f, g \in \mathcal{E}_\lambda$. This establishes the existence of the limit $h_{\lambda,b}$. As $h_{\lambda,x}$ is a Hermitian form for every $x \in]a, b[$, the limit is Hermitian as well.

(b, d) Fix a basis f_1, f_2 of $\mathcal{E}_\lambda$ and write $f(z) := z_1 f_1 + z_2 f_2$, for $z \in \mathbb{C}^2$. For $x \in [c, b]$ we define the Hermitian matrix H_x by $i[f(z), f(w)]_x = \langle H_x z, w \rangle$. Then $H_x \to H_b$ as $x \to b$. We will finish the proof by showing that $\det H_x$ is a constant function of $x \in [c, b[$, so that $\det H_b = \det H_c < 0$ and moreover $H_x^{-1} \to H_b^{-1}$.

Write $\varepsilon_x(f)$ for the linear endomorphism of $\mathbb{C}^2$ given by $z \mapsto \varepsilon_x(f(z))$. Then

$$[f(z), f(w)]_x = \langle J \varepsilon_x(f)z, \varepsilon_x(f)w \rangle = \langle \varepsilon_x(f)^* J \varepsilon_x(f)z, w \rangle,$$

so that $H_x = i \varepsilon_x(f)^* J \varepsilon_x(f)$. It follows that $\det H_x = -|\det \varepsilon_x(f)|^2$. By a straightforward calculation one sees that $\det \varepsilon_x(f) = [f_1, \bar{f}_2]_x$. Now $L\bar{f}_2 = \bar{\lambda} f_2$, so that from (2.2) it follows that $[f_1, \bar{f}_2]_x - [f_1, \bar{f}_2]_c = 0$ for all $x \in [c, b[$. Hence $\det \varepsilon_x(f) = \det \varepsilon_c(f)$ for all $x \geq c$.

Finally, the proof of (c) is straightforward. □

Combining Lemmas 4.2 and 4.3 we obtain the following corollary.

Corollary 4.4 *Let* $\lambda \in \mathbb{C} \setminus \mathbb{R}$. *Then precisely one of the following statements is valid.*

(a) *The limit set* $C_{\lambda,b}$ *is a circle. For any* $c \in]a, b[$ *the space* $\mathcal{E}_\lambda|_{[c,b[}$ *is contained in* $L^2([c, b[)$.

(b) *The limit set* $C_{\lambda,b}$ *consists of a single point. For any* $c \in]a, b[$ *the intersection of* $\mathcal{E}_\lambda|_{[c,b[}$ *with* $L^2([c, b[)$ *is one dimensional.*

At a later stage in his paper, [29], Satz 5, Weyl used spectral considerations to conclude that if (a) holds for a particular non-real eigenvalue $\lambda \in \mathbb{C}$, then $\mathcal{E}_\lambda|_{[c,b[}$ consists of square integrable functions for any eigenvalue λ. We will return to this in the next section, see Lemma 6.2. It follows that the validity of (a), and hence the validity of the alternative (b), is independent of the particular choice of the non-real eigenvalue λ.

If (a) holds, the operator L is said to be of *limit circle type* at b ('Grenzkreistypus'), and if (b) holds, L is said to be of *limit point type*

at b ('Grenzpunkttypus'). With obvious modifications, similar results
and terminology apply to the other boundary point, a. We note that
a regular Sturm–Liouville operator is of the limit circle type at both
boundary points.

Weyl observed that for each boundary point, the type of L determines
whether boundary conditions should be imposed or not. Indeed, if L is of
the limit point type at the boundary point, then no boundary condition
is needed there. On the other hand, if L is of the limit circle type
at a boundary point, then a boundary condition is required to ensure
self-adjointness. Following Weyl, we shall now describe how boundary
conditions can be imposed in the limit circle case.

The idea is to fix a non-real eigenvalue $\lambda \in \mathbb{C}$ and to construct a Green
function for the operator $L - \lambda I$. Weyl did this for the particular value
$\lambda = i$, but observed that the method works for any choice of non-real
λ, see [29], text above Satz 5. In the mentioned paper Weyl considers
the case $a = 0, b = \infty$, and L regular at a, but the method works in
general. In what follows, our treatment will deviate from Weyl's with
regard to technical details. However, in spirit we will stay close to his
original method.

We define $\mathcal{D}$ to be the space of functions $f \in C^1(\,]a, b\,[\,)$ such that
f' is locally absolutely continuous (so that Lf is locally integrable).
Moreover, we define $\mathcal{D}_b$ to be the subspace of functions $f \in \mathcal{D}$ such
that both f and Lf are square integrable on $[c, b\,[$ for some (hence any)
$c \in \,]a, b\,[$. The subspace $\mathcal{D}_a$ is defined in a similar fashion.

Given two functions $f, g \in \mathcal{D}_b$, it follows by application of (2.2) that

$$[f, g]_b := \lim_{x \to b} [f, g]_x$$

exists. If $\chi \in \mathcal{D}_b$, then we denote by $\mathcal{D}_b(\chi)$ the space of functions $f \in \mathcal{D}_b$
such that $[f, \chi]_b = 0$. We now select a non-zero function $\varphi_{b,\lambda} \in \mathcal{E}_\lambda$ such
that the associated point $\mathbb{C}\varphi_{b,\lambda} \in \mathbb{P}(\mathcal{E}_\lambda)$ belongs to the limit set $C_{\lambda,b}$. It
is possible to characterize the function $\varphi_{b,\lambda}$ by its limit behavior towards
b.

Lemma 4.5

   (a) *If L is of limit point type at b, then $\mathcal{E}_\lambda \cap \mathcal{D}_b = \mathbb{C}\varphi_{b,\lambda}$.*
   (b) *If L is of limit circle type at b, then there exists a function $\chi_b \in \mathcal{D}_b$
        such that $\mathcal{E}_\lambda \cap \mathcal{D}_b(\chi_b) = \mathbb{C}\varphi_{b,\lambda}$.*

Proof (a) follows from Corollary 4.4. For (b), assume that L is of limit circle type at b. Then $\mathcal{E}_\lambda \subset \mathcal{D}_b$. Take $\chi_b = \varphi_{b,\lambda}$. Then the space on the left-hand side of the equality equals the space of $f \in \mathcal{E}_\lambda$ with $[f, \chi_b]_c = 0$. The latter space is one dimensional since $[\cdot, \cdot]_b$ is non-degenerate on $\mathcal{E}_\lambda$. On the other hand, $\varphi_{b,\lambda}$ belongs to it, by Lemma 4.3, and the result follows.   □

To make the treatment as uniform as possible, we agree to always use the dummy boundary datum $\chi_b = 0$ in case L is of limit point type at b. In the limit circle case we select χ_b as in Lemma 4.5 (b). Then we always have

$$\mathcal{E}_\lambda \cap \mathcal{D}_b(\chi_b) = \mathbb{C}\varphi_{b,\lambda}.$$

We follow the similar convention for a choice of boundary datum $\chi_a \in \mathcal{D}_a$, so that $\mathcal{D}_a(\chi_a) \cap \mathcal{E}_\lambda$ is a line representing a point of the limit set $C_{\lambda,a}$. Moreover, we choose a non-zero eigenfunction $\varphi_{a,\lambda}$ spanning this line. Before proceeding we observe that it follows from Proposition 4.1 that

$$C_{\lambda,a} \cap C_{\lambda,b} = \emptyset.$$

This implies that $\varphi_{a,\lambda}$ and $\varphi_{b,\lambda}$ form a basis of $\mathcal{E}_\lambda$. The above choices having been made, we put

$$\mathcal{D}_\chi = \mathcal{D}_a(\chi_a) \cap \mathcal{D}_b(\chi_b).$$

Then $\mathcal{D}_\chi$ is a subspace of $L^2(\,]a, b[\,)$. It contains $C_c^2(\,]a, b[\,)$, hence is dense. Moreover, it follows from the above that

$$\mathcal{D}_\chi \cap \mathcal{E}_\lambda = 0.$$

Still under the assumption that $\lambda \in \mathbb{C} \setminus \mathbb{R}$, we now consider the differential equation

$$(L - \lambda)f = g \tag{4.1}$$

where g is a given square integrable function on $]a, b[$. The equation may be rewritten as a first order equation for the $\mathbb{C}^2$-valued function $\varepsilon(f)$. The matrix with columns $\varepsilon(\varphi_{a,\lambda})$ and $\varepsilon(\varphi_{b,\lambda})$ is a fundamental matrix for this system. By variation of the constant one finds a function $f \in \mathcal{D}$, satisfying (4.1). If g has compact support then f can be uniquely fixed by imposing the boundary conditions

$$\lim_{x \to a} [f, \chi_a]_x = 0, \qquad \lim_{x \to b} [f, \chi_b]_x = 0.$$

This function is expressed in terms of g by means of an integral operator,

$$f(t) = \mathcal{G}_\lambda g(t) := \int_a^b G_\lambda(t,\tau)\, g(\tau)\, d\tau, \qquad (4.2)$$

whose integral kernel, called the Green function, is given by

$$G_\lambda(t,\tau) = w(\lambda)^{-1}\, \varphi_{a,\lambda}(t)\varphi_{b,\lambda}(\tau), \qquad (t \le \tau), \qquad (4.3)$$

and $G_\lambda(t,\tau) = G_\lambda(\tau,t)$ for $t \ge \tau$. Here $w(\lambda)$ is the Wronskian, defined by

$$w(\lambda) = [\varphi_{b,\lambda}, \overline{\varphi}_{a,\lambda}]_c,$$

for a fixed $c \in \,]\,a,b\,[$; note that the expression on the right-hand side is independent of c, by (2.2). It is an easy matter to show that $\mathcal{G}_\lambda$ is well defined on $L^2(\,]\,a,b\,[)$, with values in $\mathcal{D}$. Moreover, $(L - \lambda I)\mathcal{G}_\lambda = I$. Finally, $\mathcal{G}_\lambda$ maps functions with compact support into $\mathcal{D}_\chi$.

At this point Weyl essentially proves the following result. He specializes to $\lambda = i$ and splits G_λ into real and imaginary part, but the crucial idea is to approximate the Green kernel by Green kernels associated to a regular Sturm–Liouville problem on smaller compact intervals, where the spectral decomposition of Theorem 3.1 is applied.

Theorem 4.2 *The Green operator $\mathcal{G}_\lambda$ is a bounded linear endomorphism of $L^2(\,]\,a,b\,[)$, with operator norm at most $|\mathrm{Im}\,\lambda|^{-1}$.*

Proof For $z \in \mathbb{C}$ we consider the eigenfunction $\varphi_b^z = \varphi_{b,\lambda} + z\varphi_{a,\lambda}$. As $\mathbb{C}\varphi_{b,\lambda}$ is contained in the limit set $C_{\lambda,b}$, there exists a sequence of points $b_n \in \,]\,a,b\,[$ and $z_n \in \mathbb{C}$ such that $b_n \nearrow b$, $z_n \to 0$ and $\varphi_b^n := \varphi_b^{z_n}$ represents a point of the circle C_{λ,b_n}. Similarly, there is a sequence of points $a_n \in \,]\,a,b\,[$, $w_n \in \mathbb{C}$ such that $a_n \searrow a$, $w_n \to 0$ and $\varphi_a^n := \varphi_{a,\lambda} + w_n\varphi_{b,\lambda}$ represents a point of C_{λ,a_n}. Define G_λ^n as in (4.3), but with $\varphi_{a,\lambda}$ and $\varphi_{b,\lambda}$ replaced by φ_a^n and φ_b^n respectively. Then it is readily seen that $G_\lambda^n \to G_\lambda$, locally uniformly on $]\,a,b\,[\times]\,a,b\,[$. Apply the spectral decomposition associated with the regular Sturm–Liouville problem for L on $[a_n, b_n]$ with boundary data φ_a^n and φ_b^n. Then the operator $\mathcal{G}_\lambda^n : L^2([a_n,b_n]) \to L^2([a_n,b_n])$ satisfies $(L-\lambda) \circ \mathcal{G}_\lambda^n = I$, hence diagonalizes with eigenvalues $(\nu - \lambda)^{-1}$, $\nu \in \mathbb{R}$. All of these eigenvalues have length at most $|\mathrm{Im}\,\lambda|^{-1}$, so that $\|\mathcal{G}_\lambda^n\| \le |\mathrm{Im}\,\lambda|^{-1}$. It now follows by taking limits that for all $f, g \in C_c(\,]\,a,b\,[)$,

$$|\langle f, \mathcal{G}_\lambda g\rangle| = \lim_{n\to\infty} |\langle f, \mathcal{G}_\lambda^n g\rangle\| \le |\mathrm{Im}\,\lambda|.$$

This implies the result. □

Corollary 4.6 *The operator $\mathcal{G}_\lambda$ is a bounded linear endomorphism of the space $L^2(\,]\,a,b\,[\,)$ with image equal to $\mathcal{D}_\chi$. Moreover, $\mathcal{G}_\lambda$ is a two-sided inverse to the operator $L - \lambda I : \mathcal{D}_\chi \to L^2(\,]\,a,b\,[\,)$.*

Proof It is easy to check that $\mathcal{G}_\lambda$ is continuous as a map $L^2(\,]\,a,b\,[\,) \to C^1(\,]\,a,b\,[\,)$. Using (2.2) and Theorem 4.2 it is then easy to check that $g \mapsto [\mathcal{G}_\lambda g, \chi_b]_b$ is continuous on $L^2(\,]\,a,b\,[\,)$. As this functional vanishes on functions with compact support, it follows that $\mathcal{G}_\lambda$ maps $L^2(\,]\,a,b\,[\,)$ into $\mathcal{D}_b(\chi_b)$. By a similar argument at the other boundary point we conclude that $\mathcal{G}_\lambda$ maps into $\mathcal{D}_\chi$.

We observed already that $(L - \lambda I)\mathcal{G}_\lambda = I$ on $L^2(\,]\,a,b\,[\,)$. It follows that $(L-\lambda I) \circ [\mathcal{G}_\lambda(L-\lambda I) - I] = 0$ on $\mathcal{D}_\chi$. As $\mathcal{G}_\lambda$ maps into $\mathcal{D}_\chi$, on which $L - \lambda I$ is injective, it follows that $\mathcal{G}_\lambda(L - \lambda I) - I$ on $\mathcal{D}_\chi$. All assertions follow. □

Looking at Weyl's result from a modern perspective, it is now possible to show that the densely defined operator L with domain $\mathcal{D}_\chi$ is self-adjoint. To prepare for this, we need a better understanding of the boundary conditions.

In what follows we will assume that L is of limit circle type at b, so that $\mathcal{E}_\lambda \subset \mathcal{D}_b$. For $x \in \,]\,a,b\,[$, the map $\varepsilon_x : \mathcal{E}_\lambda \to \mathbb{C}^2$ is a linear isomorphism. We define the map $\beta_{\lambda,x} : \mathcal{D}_b \to \mathcal{E}_\lambda$ by $\varepsilon_x \circ \beta_x(f) = \varepsilon_x(f)$, for $f \in \mathcal{D}_b$. Then $\beta_{\lambda,x}$ may be viewed as a projection onto $\mathcal{E}_\lambda$. For $f, g \in \mathcal{D}_b$,

$$[\beta_{\lambda,x}(f), \beta_{\lambda,x}(g)]_x = [f,g]_x. \tag{4.4}$$

This implies that

$$[\beta_{\lambda,x}(f), \cdot\,]_x = [f, \cdot\,]_x \quad \text{on } \mathcal{E}_\lambda. \tag{4.5}$$

As the form $[\,\cdot\,,\,\cdot\,]_b$ is non-degenerate on $\mathcal{E}_\lambda$, we may define a linear map $\beta_{\lambda,b} : \mathcal{D}_b \to \mathcal{E}_\lambda$ by (4.5) with $x = b$. Then again $\beta_{\lambda,b} = I$ on $\mathcal{E}_\lambda$, so that $\beta_{\lambda,b}$ may be viewed as a projection onto $\mathcal{E}_\lambda$.

Lemma 4.7 *Let $f \in \mathcal{D}_b$. Then $\beta_{\lambda,x}(f) \to \beta_{\lambda,b}(f)$ in $\mathcal{E}_\lambda$, as $x \to b$.*

Proof Let γ_x denote the sesquilinear form $[\,\cdot\,,\,\cdot\,]_x$ on $\mathcal{E}_\lambda$, for $a < x \le b$. Then for $x \to b$, we have the limit behavior $\gamma_x \to \gamma_b$ in $\mathrm{Hom}(\mathcal{E}_\lambda, \overline{\mathcal{E}}_\lambda^*)$ and $\gamma_x^{-1} \to \gamma_b^{-1}$ in the space $\mathrm{Hom}(\overline{\mathcal{E}}_\lambda^*, \mathcal{E}_\lambda)$, see Lemma 4.3.

From (4.5) we deduce that $\gamma_x(\beta_{\lambda,x}(f)) = [f, \cdot]_x$, for all $f \in \mathcal{D}_b$. It follows that $\gamma_x(\beta_{\lambda,x}(f)) \to [f, \cdot]_b = \gamma_b(\beta_{\lambda,b})(f)$, hence

$$\beta_{\lambda,x}(f) = \gamma_x^{-1}\gamma_x\beta_{\lambda,x}(f) \to \beta_{\lambda,b}(f),$$

for $x \to b$. $\qquad\qquad\square$

Corollary 4.8 *For all $f, g \in \mathcal{D}_b$ we have $[f, g]_b = [\beta_{\lambda,b}(f), \beta_{\lambda,b}(g)]_b$.*

Proof This follows from (4.4) by passing to the limit for $x \to b$. $\qquad\square$

The following immediate corollary clarifies the nature of the boundary datum χ_b.

Corollary 4.9 *Let $\chi_b \in \mathcal{D}_b$. Then $\mathcal{D}_b(\chi_b)$ depends on χ_b through its image $\beta_{\lambda,b}(\chi_b)$ in $\mathcal{E}_\lambda$.*

It follows that in the present setting (L of limit circle type at b), the equality of Lemma 4.5 (b) is equivalent to $\mathbb{C}\beta_{\lambda,b}(\chi_b) = \mathbb{C}\varphi_{b,\lambda}$. In other words, let $\widetilde{C}_{\lambda,b}$ denote the preimage in $\mathcal{E}_\lambda \setminus \{0\}$ of the limit circle $C_{\lambda,b}$. Then functions from $\beta_{\lambda,b}^{-1}(\widetilde{C}_{\lambda,b})$ provide appropriate boundary data at b.

The following result is needed to determine the adjoint of the Green operator $\mathcal{G}_\lambda$. As $\mathcal{E}_\lambda \subset \mathcal{D}_b$ it follows that $\mathcal{E}_{\bar\lambda} = \overline{\mathcal{E}_\lambda}$ is contained in $\mathcal{D}_b$ as well. We put $\varphi_{b,\bar\lambda} := \overline{\varphi}_{b,\lambda}$; then the above definitions are valid with $\bar\lambda$ instead of λ. It is immediate from the definitions that

$$\beta_{\lambda,b}(\bar f) = \overline{\beta_{\bar\lambda,b}(f)}, \qquad (f \in \mathcal{D}_b). \qquad (4.6)$$

Lemma 4.10

(a) $\beta_{\bar\lambda,b} \circ \beta_{\lambda,b} = \beta_{\bar\lambda,b}$;

(b) $\beta_{\bar\lambda,b}(\varphi_{b,\lambda}) = c\overline{\varphi}_{b,\lambda}$, *with c a non-zero complex scalar;*

(c) $\beta_{\lambda,b}(\bar\chi_b) = \bar c\,\varphi_{b,\lambda}$;

(d) $\mathcal{D}_b(\overline{\chi}_b) = \mathcal{D}_b(\chi_b)$.

Proof (a) We check that $\beta_{\bar\lambda,x} = \beta_{\bar\lambda,x}\beta_{\lambda,x}$ by applying ε_x on the left. Now use Lemma 4.7.

(b) Let ψ be an eigenfunction in $\mathcal{E}_{\bar\lambda}$. Then $[\varphi_{\lambda,b}, \psi]_x$ is constant as a function of x, by (2.2). Hence, $[\beta_{\bar\lambda,b}(\varphi_{\lambda,b}), \psi]_b = [\varphi_{\lambda,b}, \psi]_x$. It follows that $\beta_{\bar\lambda,b}(\varphi_{\lambda,b})$ is non-zero, whereas $[\beta_{\bar\lambda,b}(\varphi_{\lambda,b}), \varphi_{\bar\lambda,b}]_b = [\varphi_{\lambda,b}, \varphi_{\bar\lambda,x}]_x = 0$. The assertion follows.

(c) Applying (4.6), (a) and (b) we obtain: $\overline{\beta_{\lambda,b}(\bar\chi_b)} = \beta_{\bar\lambda,b}(\chi_b) =$

$\beta_{\bar{\lambda},b}\beta_{\lambda,b}(\chi_b) = \beta_{\bar{\lambda}}(\varphi_{b,\lambda}) = c\overline{\varphi}_{b,\lambda}$. Finally, (d) is an immediate consequence of (c). $\qquad\square$

We now return to the situation of a general singular Sturm–Liouville operator L.

Theorem 4.3 *The operator L with domain $\mathcal{D}_\chi$ is self-adjoint.*

Proof From Lemma 4.10 it follows that $\mathcal{D}_\chi = \mathcal{D}_{\bar{\chi}}$. The adjoint $\mathcal{G}^*$ of the operator $\mathcal{G} = \mathcal{G}_\lambda$ has integral kernel $G_\lambda^*(t,\tau) := \overline{G_\lambda(\tau,t)}$. This is precisely the Green kernel associated with the eigenvalue $\bar{\lambda}$ and the boundary data $\bar{\chi}_a, \bar{\chi}_b$. As its image $\mathcal{D}_{\bar{\chi}}$ equals $\mathcal{D}_\chi$, it follows that $\mathcal{G}^*$ is the two-sided inverse of the bijection $L - \bar{\lambda}I : \mathcal{D}_\chi \to L^2(\,]\,a,b\,[\,)$. These facts imply that the adjoint L^* equals L. $\qquad\square$

At this point one can prove the following generalization of Theorem 3.1, due to Weyl, [29], Satz 4. Let $\sigma(L,\chi)$ be the set of $\lambda \in \mathbb{C}$ for which $\mathcal{E}_\lambda \cap \mathcal{D}_\chi \neq 0$.

Theorem 4.4 *(Weyl 1910) Let L be of limit circle type at both end points. Then $\sigma(L,\chi)$ is a discrete subset of $\mathbb{R}$, without accumulation points. Moreover, $L^2(\,]\,a,b\,[\,)$ is the orthogonal direct sum of the spaces $\mathcal{E}_\lambda \cap \mathcal{D}_\chi$.*

Proof Weyl proved this by using the Green operator $\mathcal{G}$ corresponding to the eigenvalue i. Let G_2 be the imaginary part of its kernel. Then G_2 is real valued, symmetric and square integrable, hence admits a diagonalization. In today's terminology, the associated integral operator $\mathcal{G}_2$, which equals $(2i)^{-1}[\mathcal{G} - \mathcal{G}^*]$, is self-adjoint and Hilbert-Schmidt, hence compact. All its eigenspaces are finite dimensional, and contained in $\mathcal{D}_\chi$, since $\mathcal{D}_\chi = \mathcal{D}_{\bar{\chi}}$. Moreover, each of them is invariant under the symmetric operator L. $\qquad\square$

The regular case may be viewed as a special case of the above. Indeed, if ξ_a, ξ_b are the boundary data of Theorem 3.1, let $\mu \in \mathbb{C}\backslash\mathbb{R}$ be arbitrary, and for $x = a,b$, let χ_x be the constant function with value $J\xi_x$. Then $\mathcal{E}_{\lambda,\xi} = \mathcal{E}_\lambda \cap \mathcal{D}_\chi$, for all $\lambda \in \mathbb{C}$.

5 Weyl's spectral theorem

Using Green's function G_λ for non-real λ and his earlier work on singular integral equations, [28], Weyl was able to establish the existence of a spectral decomposition of $L^2(\,]\,a,b\,[\,)$ in terms of eigenfunctions of L with real eigenvalue. In [29] he considers the case $a=0, b=\infty$, and assumes that L is regular (hence of circle limit type) at 0. Let $\xi_a \in \mathbb{R}^2 \setminus \{0\}$ be a boundary datum at a, and fix a unit vector $\eta \in \mathbb{R}^2$ perpendicular to ξ. Let χ_0 be the constant function with value η and let χ_∞ be a boundary datum at ∞ (if L is of limit point type at ∞, we take the dummy boundary datum $\chi_\infty = 0$). For each $\lambda \in \mathbb{C}$ let φ_λ be the unique eigenfunction of L with eigenvalue λ and $\varphi_\lambda(a) = \eta$. Then according to Weyl, [29], Satz 5,7, there exists a right-continuous monotonically increasing function ρ such that each function $f \in C^2(\,]\,0,\infty\,[\,) \cap \mathcal{D}_\chi$ admits a decomposition of the form

$$f(x) = \int_{\mathbb{R}} \varphi_\lambda(x)\, dF(\lambda) \qquad (5.1)$$

with uniformly and absolutely converging integral; here $dF(\lambda)$ is a regular Borel measure, defined by

$$dF(\Delta) = \int_0^\infty f(t) \int_\Delta \varphi_\lambda(t)\, d\rho(\lambda)\, dt. \qquad (5.2)$$

In the above, $d\rho$ denotes the regular Borel measure determined by the formula $d\rho(\,]\,\mu,\nu\,]) = \rho(\nu) - \rho(\mu)$, for all $\mu < \nu$.

Actually, Weyl's original formulation was different and involved a discrete and a continuous part. His formulation follows from the one above by the observation that ρ admits a unique decomposition $\rho = \rho_d + \rho_c$ with ρ_c a continuous monotonically increasing function with $\rho_c(0) = 0$, and with ρ_d a right-continuous monotonically increasing function which is constant on each interval where it is continuous.

In case L is of the limit circle type at infinity, the decomposition is discrete by Theorem 4.4, so that $\rho_c = 0$, so that the above gives rise to a discrete decomposition. In case L is of the limit point type at ∞, the decomposition is of mixed discrete and continuous type.

It has now become customary to write

$$dF(\lambda) = \mathcal{F}f(\lambda)\, d\rho(\lambda), \qquad \mathcal{F}f(\lambda) = \int_0^\infty f(t)\, \varphi_\lambda(t)\, dt, \qquad (5.3)$$

with the interpretation that the integral converges as an integral with values in $L^2(\mathbb{R}, d\rho)$.

We will call ρ the spectral function associated with the operator L, the boundary data χ_0, χ_∞, and the choice of eigenfunctions φ_λ. In [29], Weyl also addressed the natural problem to determine its continuous part ρ_c.

Theorem 5.1 *(Weyl 1910) Assume L is a Sturm–Liouville operator of the form* (2.1) *on* $[0, \infty[$, *regular at 0. Assume moreover that the coefficients p and q satisfy the conditions*

(a) $\lim_{t \to \infty} t|p(t) - 1| = 0$, $\lim_{t \to \infty} t\, q(t) = 0$,

(b) $\int_0^\infty t|p(t) - 1|\, dt < \infty$, $\int_0^\infty t|q(t)| < \infty$.

Then L is of the limit point type at ∞. Let ξ_a, η, χ_a and φ_λ be defined as above and let ρ be the associated spectral function. Then the support of $d\rho_d$ is finite and contained in the open negative real half line $] - \infty, 0[$. The support of $d\rho_c$ is contained in the closed positive real half line $[0, \infty[$. There exist uniquely determined continuous functions $a, b :]0, \infty[\to \mathbb{R}$ such that

$$\varphi_\lambda(t) = a(\lambda)\cos(t\sqrt{\lambda}) + b(\lambda)\sin(t\sqrt{\lambda}) + o(t). \tag{5.4}$$

In terms of these coefficients, the spectral measure $d\rho_c$ is given by

$$d\rho_c(\lambda) = \frac{1}{a(\lambda)^2 + b(\lambda)^2} \frac{d\lambda}{\pi\sqrt{\lambda}}.$$

Here we note that by (5.4) and the condition on f, the integral (5.3) is absolutely convergent. If $p = 1$ and $q = 0$, then of course one has $a(\lambda) = \eta_1$ and $b(\lambda) = \eta_2$, and one retrieves the continuous measure $d\rho_c(\lambda) = (\pi\sqrt{\lambda})^{-1} d\lambda$.

Let $c(\sqrt{\lambda}) := \frac{1}{2}(a(\lambda) - ib(\lambda))$. Then

$$\varphi_\lambda(t) = c(\lambda)e^{it\sqrt{\lambda}} + \overline{c(\lambda)}e^{-it\sqrt{\lambda}} + o(t)$$

and the spectral measure is given by

$$d\rho(\lambda) = \frac{d\sqrt{\lambda}}{2\pi|c(\sqrt{\lambda})|^2} \tag{5.5}$$

We may view the operator L as a perturbation of the operator $-d^2/dt^2$. At infinity the eigenfunction φ_λ behaves asymptotically as a linear combination of the exponential eigenfunctions for the unperturbed problem, with amplitudes of equal modulus $|c(\sqrt{\lambda})|$. The spectral measure of the perturbed problem is obtained from the spectral measure of the unperturbed problem by dividing through $|c(\sqrt{\lambda})|^2$. As we will see later, this principle is omnipresent in the theory of harmonic analysis of non-compact Riemannian symmetric spaces, of non-compact real semisimple Lie groups, and of their common generalization, the so-called semisimple symmetric spaces.

6 Dependence on the eigenvalue parameter

In this section we will prove holomorphic dependence of the Green function G_λ on the parameter λ. This is not obvious from the definition (4.3). Indeed, in the limit circle case at b, the particular normalization of $\varphi_{b,\lambda}$ chosen only guarantees real analytic dependence on the parameter λ (this fairly easy result will not be needed in the sequel). In the limit point case, only the line $\mathbb{C}\varphi_{b,\lambda}$ does not depend on the choices made, but the dependence of $\varphi_{b,\lambda}$ on λ may be arbitrary. The following result suggests to look for differently normalized eigenfunctions, which do depend holomorphically on λ.

Lemma 6.1 *The Green kernel G_λ defined by (4.2) depends on $\varphi_{a,\lambda}$ and $\varphi_{b,\lambda}$ through their images in $\mathbb{P}(\mathcal{E}_\lambda)$.*

Proof This is caused by the division by the Wronskian $w(\lambda)=[\varphi_{b,\lambda},\overline{\varphi}_{a,\lambda}]$.
□

The following result follows by application of the method of variation of the constant as explained in [7], Thm. 2.1, p. 225. The assertion about holomorphic dependence is not given there, but follows by the same method of proof.

Lemma 6.2 *Let $a < c < b$ and assume that for some $\lambda_0 \in \mathbb{C}$ the eigenspace $\mathcal{E}_{\lambda_0}|_{[c,b[}$ is contained in $L^2([c,b[)$. Then for each eigenvalue $\lambda \in \mathbb{C}$ the associated eigenspace $\mathcal{E}_\lambda|_{[c,b[}$ is contained in $L^2([c,b[)$.*

Moreover, for each $c \in]a,b[$ and all $v \in \mathbb{C}^2$, the function $\lambda \mapsto s_c(\lambda, \cdot)v|_{[c,b[}$ (see (2.8)) is entire holomorphic as a function with values in $L^2([c,b[)$.

In the following, we assume that L is of limit circle type at b. From the text below (4.5) we recall the definition of the map $\beta_{\lambda,b} : \mathcal{D}_b \to \mathcal{E}_\lambda$, for every $\lambda \in \mathbb{C} \setminus \mathbb{R}$.

Lemma 6.3 *Let L be of the limit circle type at b. Then for all $\mu, \lambda \in \mathbb{C} \setminus \mathbb{R}$,*

(a) $\beta_{\mu,b} \circ \beta_{\lambda,b} = \beta_{\mu,b}$;

(b) *the restriction $\beta_{\mu,b}|_{\mathcal{E}_\lambda}$ is a linear isomorphism onto $\mathcal{E}_\mu$;*

(c) *the restriction $\beta_{\mu,b}|_{\mathcal{E}_\lambda}$ induces a projective isomorphism $\mathbb{P}(\mathcal{E}_\lambda) \to \mathbb{P}(\mathcal{E}_\mu)$, mapping the limit circle $C_{\lambda,b}$ onto the limit circle $C_{\mu,b}$.*

Proof Assertion (a) is proved in the same fashion as assertion (a) of Lemma 4.10. Since $[\,\cdot\,,\,\cdot\,]_b$ is non-degenerate on both $\mathcal{E}_\lambda$ and $\mathcal{E}_\mu$, assertion (b) follows by application of Corollary 4.8. Finally, (c) follows from the identity of Corollary 4.8, in view of Lemma 4.3. $\square$

The following result suggests the modification of the eigenfunctions in (4.2) that we are looking for.

Lemma 6.4 *Let L be of the limit circle type at b. Let $\chi_b \in \mathcal{D}_b$ and assume that for some $\mu \in \mathbb{C} \setminus \mathbb{R}$ the function $\beta_{\mu,b}(\chi_b)$ is non-zero and represents a point of the limit circle $C_{\mu,b}$. Then*

- (a) *for each $\lambda \in \mathbb{C} \setminus \mathbb{R}$ the function $\beta_{\lambda,b}(\chi)$ is a non-zero eigenfunction in $\mathcal{E}_\lambda$ which represents a point of the limit circle $C_{\lambda,b}$;*
- (b) *for each $c \in \,]a,b[$, the map $\lambda \mapsto [\varepsilon_c(\beta_{\lambda,b}(f))]$ is holomorphic from $\mathbb{C} \setminus \mathbb{R}$ to $\mathbb{P}^1(\mathbb{C}) \setminus \mathbb{P}^1(\mathbb{R})$.*

Proof By the first assertion of Lemma 6.3, $\beta_{\lambda,b}(\chi) = \beta_{\lambda,b}\beta_{\mu,b}(\chi) = \beta_{\lambda,b}(\varphi_{b,\mu})$. Assertion (a) follows by application of the remaining assertions of the mentioned lemma.

We now turn to (b). We will prove the holomorphy in a neighborhood of the fixed point $\lambda_0 \in \mathbb{C} \setminus \mathbb{R}$. As $\beta_{\lambda,b}(\chi) = \beta_{\lambda,b}(\beta_{\lambda_0,b}(\chi))$ we may as well assume that $\chi \in \mathcal{E}_{\lambda_0}$ and that $[\chi] \in C_{\lambda_0,b}$. Select a sequence x_n in $\,]a,b[$ converging to b, and for each n a point $p_n \in C_{\lambda_0,x_n}$ such that $p_n \to [\chi]$. There exist $\chi_n \in \mathcal{E}_{\lambda_0}$ such that $[\chi_n] = p_n$ and $\chi_n \to \chi$ in $\mathcal{E}_{\lambda_0}$. We define $\varphi_{\lambda,n} \in \mathcal{E}_\lambda$ by $\varepsilon_{x_n}\varphi_{n,\lambda} = \varepsilon_{x_n}\chi_n$. Then in the notation of (4.5), $\varphi_n(\lambda)$ equals $\beta_{\lambda,x_n}(\chi_n)$ and represents a point of C_{λ,x_n}. For each fixed λ the sequence $\beta_{\lambda,x_n}|_{\mathcal{E}_{\lambda_0}}$ in $\mathrm{Hom}(\mathcal{E}_{\lambda_0}, \mathcal{E}_\lambda)$ has limit $\beta_{\lambda,b}$. Hence $\varphi_n(\lambda) \to \beta_{\lambda,b}(\chi)$, pointwise in λ.

Let $c \in \,]a,b[$. Passing to a subsequence we may assume that $x_n > c$ for all $n \geq 1$. The map $\bar{\varepsilon}_c : \mathbb{P}(\mathcal{E}_\lambda) \to \mathbb{P}^1(\mathbb{C})$ maps the circle $C_{\lambda,c}$ onto $\mathbb{P}^1(\mathbb{R})$. Let Ω be a connected open neighborhood of λ_0. Then it follows by application of Proposition 4.1 that all circles $\bar{\varepsilon}_c(C_{\lambda,x_n})$ are contained in one particular connected component U of $\mathbb{P}^1(\mathbb{C}) \setminus \mathbb{P}^1(\mathbb{R})$. This implies that $\psi_n(\lambda) := \bar{\varepsilon}_c\beta_{x_n,\lambda}(f) \in U$ for every $\lambda \in \Omega$. By using an affine chart containing the compact closure of U we see that the sequence ψ_n has a subsequence converging locally uniformly to a holomorphic limit function $\psi : \Omega \to U$. By pointwise convergence, $\psi(\lambda) = [\varepsilon_c\beta_{\lambda,b}(\chi)]$, and (b) follows. $\square$

Corollary 6.5 *Let L be of the limit circle type at b and let $\chi_b \in \mathcal{D}_b$ be as in the above lemma. There exists a family of functions $\varphi_{\lambda,b} \in C^2(\,]\,a,b\,[\,)$ depending holomorphically on $\lambda \in \mathbb{C} \setminus \mathbb{R}$ such that for each $\lambda \in \mathbb{C} \setminus \mathbb{R}$*

(a) $\varphi_{b,\lambda} \in \mathcal{E}_\lambda \setminus \{0\}$;
(b) $\varphi_{b,\lambda}$ represents the point $[\beta_{\lambda,b}(\chi_b)]$ of the limit circle $C_{\lambda,b}$.

The following analogous result in the limit point case can be proved using a similar method, see [7], Thm. 2.3, p. 229, for details.

Lemma 6.6 *Let L be of the limit point type at b. Then there exists a family of functions $\varphi_{b,\lambda} \in C^2(\,]\,a,b\,[\,)$, depending holomorphically on the parameter $\lambda \in \mathbb{C} \setminus \mathbb{R}$, such that for each $\lambda \in \mathbb{C} \setminus \mathbb{R}$,*

(a) $\varphi_{b,\lambda} \in \mathcal{E}_\lambda \setminus \{0\}$;
(b) the function $\varphi_{b,\lambda}$ represents the limit point in $\mathbb{P}(\mathcal{E}_\lambda)$.

Let L be arbitrary again. We fix boundary data χ_a, χ_a as indicated in the previous section, so that $L : \mathcal{D}_\chi \to L^2(\,]\,a,b\,[\,)$ is self-adjoint. Accordingly, we fix holomorphic families of eigenfunctions $\varphi_{a,\lambda}, \varphi_{b,\lambda} \in \mathcal{E}_\lambda$ in the manner indicated in Corollary 6.5 and Lemma 6.6.

Finally, we define the Green function G_λ by means of the formula (4.3). The functions $\varphi_{a,\lambda}, \varphi_{b,\lambda}$ used here are renormalizations of those used in Section 4. By Lemma 6.1 this does not affect the definition of the Green kernel.

Corollary 6.7 *The Green kernel $G_\lambda \in C(\,]\,a,b\,[\,\times\,]\,a,b\,[\,)$ depends holomorphically on the parameter $\lambda \in \mathbb{C} \setminus \mathbb{R}$.*

This result of course realizes the resolvent $(L-\lambda I)^{-1}$ of the self-adjoint operator L with domain $\mathcal{D}_\chi$ explicitly as an integral operator with kernel depending holomorphically on λ.

7 A paper of Kodaira

For the general singular Sturm–Liouville problem, there exists a spectral decomposition similar to (5.1), but with a spectral matrix instead of the spectral function ρ. Weyl observed this in [30]. The spectral matrix was later determined by E.C. Titchmarsh who used involved direct computations using the calculus of residues, see [27].

Independently, K. Kodaira [23] rediscovered the result by a very elegant method, combining Weyl's construction of the Green function with the general spectral theory for self-adjoint unbounded operators on Hilbert space, as developed in the 1930's by J. von Neumann and M. Stone. Weyl was very content with this work of Kodaira, as becomes clear from the following quote from the Gibbs lecture delivered in 1948, [31], p. 124: 'The formula (7.5) was rediscovered by Kunihiko Kodaira (who of course had been cut off from our Western mathematical literature since the end of 1941); his construction of ρ and his proofs for (7.5) and the expansion formula [...], still unpublished, seem to clinch the issue. It is remarkable that forty years had to pass before such a thoroughly satisfactory direct treatment emerged; the fact is a reflection on the degree to which mathematicians during this period got absorbed in abstract generalizations and lost sight of their task of finishing up some of the more concrete problems of undeniable importance.'

We will now describe the spectral decomposition essentially as presented by Kodaira [23]. Fix boundary data χ_a and χ_b as in Theorem 4.3. We use the notation $\mathcal{H} := L^2(\,]a,b\,[\,)$. Then the operator L with domain $\mathcal{D}_\chi$ is a self-adjoint operator in the Hilbert space $\mathcal{H}$; it therefore has a spectral resolution dE.

To obtain a suitable parametrization of the space of eigenfunctions for L, fix $c \in \,]a,b\,[$ and recall that the map $\varepsilon_c : f \mapsto (f(c), p(c)f'(c))$ is a linear isomorphism from $\mathcal{E}_\lambda$ onto $\mathbb{C}^2$, for each $\lambda \in \mathbb{C}$. We define the function $s(\lambda) = s_c(\lambda,\,\cdot\,) : \,]a,b\,[\,\to \mathrm{Hom}(\mathbb{C}^2,\mathbb{C})$ as in (2.6). Then $\lambda \mapsto s(\lambda)$ may be viewed as an entire holomorphic map with values in $C^2(\,]a,b\,[\,)\otimes\mathrm{Hom}(\mathbb{C}^2,\mathbb{C}))$. Moreover, for each $\lambda \in \mathbb{C}$ the map $v \mapsto s(\lambda)v$ is a linear isomorphism from $\mathbb{C}^2$ onto $\mathcal{E}_\lambda$.

For $f \in C_c(\,]a,b\,[\,)$ and $\lambda \in \mathbb{R}$ we define the Fourier transform

$$\mathcal{F}f(\lambda) = \int_a^b s(\lambda,x)^* f(x)\,dx, \qquad (7.1)$$

where $s(\lambda,x)^* \in \mathrm{Hom}(\mathbb{C},\mathbb{C}^2)$ is the adjoint of $s(\lambda,x)$ with respect to the standard Hermitian inner products on $\mathbb{C}^2$ and $\mathbb{C}$.

By a spectral matrix we shall mean a function $P : \mathbb{R} \to \mathrm{End}(\mathbb{C}^2)$ with the following properties

(a) $P(x)^* = P(x)$, i.e., $P(x)$ is Hermitian with respect to the standard inner product, for all $x \in \mathbb{R}$;

(b) P is continuous from the right;

(c) $P(0) = 0$ and $P(y) - P(x)$ is positive semi-definite for all $x \le y$.

Associated with a spectral matrix as above there is a unique regular Borel measure dP on $\mathbb{R}$, with values in the space of positive semi-definite Hermitian endomorphisms of $\mathbb{C}^2$, such that $dP(\,]\mu,\nu]) = P(\nu) - P(\mu)$

for all $\mu \leq \nu$. Conversely, a measure with these properties comes from a unique spectral matrix P. Given a spectral matrix P, we define $\mathcal{M}_2 = \mathcal{M}_{2,P}$ to be the space of Borel measurable functions $\varphi : \mathbb{R} \to \mathbb{C}^2$ with

$$\langle \varphi, \varphi \rangle_P := \int_{\mathbb{R}} \langle \varphi(\nu), dP(\nu)\varphi(\nu) \rangle < \infty.$$

Moreover, we define $\mathfrak{H} = \mathfrak{H}_P$ to be the Hilbert space completion of the quotient $\mathcal{M}_2/\mathcal{M}_2^\perp$.

Let $T_\lambda := \varepsilon_c \circ \mathcal{G}_\lambda$. Then $T_\lambda : \mathcal{H} \to \mathbb{C}^2$ is a continuous linear map. We denote its adjoint by T_λ^*. Kodaira uses the elements $\gamma_1(\lambda), \gamma_2(\lambda)$ of $\mathcal{H}$ determined by $\mathrm{pr}_j \circ T_\lambda = \langle \cdot, \gamma_j(\lambda) \rangle$.

Theorem 7.1 *(Kodaira 1949) The spectral function P determined by*

$$dP(\nu) = |\nu - \lambda|^2 \, T_\lambda \circ dE(\nu) \circ T_\lambda^* \tag{7.2}$$

is independent of $\lambda \in \mathbb{C} \setminus \mathbb{R}$. Moreover, it has the following properties.

(a) The Fourier transform extends to an isometry from the Hilbert space $\mathcal{H} = L^2(\,]a,b\,[)$ onto the Hilbert space $\mathfrak{H} = \mathfrak{H}_P$.

(b) The spectral resolution $dE(\nu)$ of the self-adjoint operator L with domain $\mathcal{D}_\chi$ is given by

$$\mathcal{F} \circ dE(S) = 1_S \circ \mathcal{F},$$

for every Borel measurable set $S \subset \mathbb{R}$; here 1_S denotes the map induced by multiplication with the characteristic function of S.

For the proof of Theorem 7.1, which involves ideas of Weyl [29], we refer the reader to Kodaira's paper [23]. In addition to the above, Kodaira proves more precise statements about the nature of the convergence of the integrals in the associated inversion formula.

After having introduced the spectral matrix, Kodaira gives an ingenious short proof of an expression for the spectral matrix which had been found earlier by Titchmarsh. We observe that the $\mathbb{C}^2$-valued functions $F_a(\lambda) = \varepsilon_c(\varphi_{a,\lambda})$ and $F_b(\lambda) = \varepsilon_c(\varphi_{b,\lambda})$ are holomorphic functions of $\lambda \in \mathbb{C} \setminus \mathbb{R}$. The matrix $F(\lambda)$ with columns $F_a(\lambda)$ and $F_b(\lambda)$ is invertible for $\lambda \in \mathbb{C} \setminus \mathbb{R}$. By the above definitions,

$$\varphi_{a,\lambda}(t) = s(\lambda,t)F_a(\lambda), \qquad \varphi_{b,\lambda}(t) = s(\lambda,t)F_b(\lambda). \tag{7.3}$$

We now define the 2×2 matrix $M(\lambda)$, the so-called characteristic matrix, by $M(\lambda) = -(\det F)^{-1} F_a F_b^{\mathrm{T}}$, i.e.,

$$M(\lambda) = -\det F(\lambda)^{-1} \begin{pmatrix} F_{a1} F_{b1} & F_{a1} F_{b2} \\ F_{a2} F_{b1} & F_{a2} F_{b2} \end{pmatrix}_\lambda \qquad (7.4)$$

Actually, Kodaira uses the symmetric matrix $M(\lambda) - \frac{1}{2}J$, which has the same imaginary part. The matrix $M(\lambda)$ depends holomorphically on the parameter $\lambda \in \mathbb{C} \setminus \mathbb{R}$.

Theorem 7.2 *(Titchmarsh, Kodaira) The spectral matrix P is given by the following limit:*

$$P(\nu) = \lim_{\delta \downarrow 0} \lim_{\varepsilon \downarrow 0} \frac{1}{\pi} \int_{[\delta, \nu+\delta]+i\varepsilon} \operatorname{Im} M(\lambda) \, d\lambda. \qquad (7.5)$$

Proof Multiplying both sides of (7.2) with $|\nu - \lambda|^{-2}$ and integrating over $\mathbb{R}$, we find that

$$\int_{\mathbb{R}} |\nu - \lambda|^{-2} \, dP(\nu) = T_\lambda T_\lambda^*.$$

By a straightforward, but somewhat tedious calculation, using (2.2) and $[\varphi_{a,\lambda}, \varphi_{a,\lambda}]_a = [\varphi_{b,\lambda}, \varphi_{b,\lambda}]_b = 0$, it follows that

$$\operatorname{Im} \lambda \, T_\lambda T_\lambda^* = \frac{1}{2i} \frac{1}{|[F_b, \overline{F}_a]|^2} \left([F_a, F_a] F_b \overline{F}_b^{\mathrm{T}} - [F_b, F_b] F_a \overline{F}_a^{\mathrm{T}} \right).$$

This in turn implies that $\operatorname{Im} \lambda \cdot T_\lambda T_\lambda^* = \operatorname{Im} M(\lambda)$. Hence,

$$\int_{\mathbb{R}} |\nu - \lambda|^{-2} \operatorname{Im} \lambda \, dP(\nu) = \operatorname{Im} M(\lambda).$$

From this (7.5) follows by a straightforward argument. □

After this, Kodaira shows that the above result can be extended to a more general basis of eigenfunctions. A fundamental system for L is a linear map $s(\lambda) : \mathbb{C}^2 \to \mathcal{E}_\lambda$, depending entire holomorphically on $\lambda \in \mathbb{C}$ as a $C^2(\,]a, b\,[\,)$-valued function, such that the following conditions are fulfilled for all $\lambda \in \mathbb{C}$:

  (a) $\overline{s(\lambda)v} = s(\bar{\lambda})(\bar{v})$, $(v \in \mathbb{C}^2)$;
  (b) $\det(\varepsilon_x \circ s(\lambda)) = 1$, $(x \in \,]a, b\,[\,)$.

Put $s_j(\lambda) = s(\lambda)e_j$, then condition (b) means precisely that the Wronskian $[s_1(\lambda), s_2(\lambda)]_x$ equals 1. Write $\psi(\lambda) = \varepsilon_c \circ s(\lambda) \in \operatorname{End}(\mathbb{C}^2)$. Then it follows that ψ entire holomorphic, and that $\det \psi(\lambda) = 1$. Moreover,

$$s(\lambda, x) = s_c(\lambda, x)\psi(\lambda).$$

We may define the Fourier transform associated with s by the identity
(7.1). The associated spectral function P is expressed in terms of the
spectral matrix P_c for s_c by the equation

$$dP(\lambda) = \psi(\lambda)^{-1} \, dP_c(\lambda) \, \psi(\lambda)^{*-1}.$$

We define the matrix F for s by the identity (7.3). Then the associated
matrix M, defined by (7.4) is given by

$$M(\lambda) = \psi(\lambda)^{-1} \, M_c(\lambda) \, \psi(\lambda)^{\mathrm{T}-1}.$$

Kodaira shows that with these definitions, the identity (7.5) is still valid.

8 A special equation

In the second half of the paper [23], Kodaira applies the above results
to the time independent one dimensional Schrödinger operator

$$L = -\frac{d^2}{dt^2} + m(m+1)t^{-2} + V(t),$$

with $m \geq -\frac{1}{2}$ and $tV(t)$ a real valued real analytic function on an open
neighborhood of $[0, \infty\,[$, such that

$$tV(t) = \mathcal{O}(t^{-\varepsilon}), \quad \text{for } t \to \infty,$$

with $\varepsilon > 0$. Actually, Kodaira considers a more general problem with
weaker requirements both at infinity and zero, but we shall not need
this. It is in fact not clear that his condition on the behavior of V
at 0 is strong enough for the subsequent argument to be valid, as was
pointed out by [22], p. 206. Kodaira's argumentation, which we shall now
present, is valid under the hypotheses stated above, as they imply that
the eigenequation $Lf = \lambda f$ has a regular singularity at zero. Because
of this, the asymptotic behavior of the eigenfunctions towards zero is
completely understood. Indeed, the associated indicial equation has
solutions $m + 1$ and $-m$, where $m + 1 \geq -m$. Let c_0 be any non-zero
real constant. Then there exists a unique eigenfunction $s_1(\lambda) \in \mathcal{E}_\lambda$ such
that

$$s_1(\lambda, t) = c_0 \, t^{m+1} \varphi(\lambda, t)$$

with $\varphi(\lambda, \cdot)$ real analytic in an open neighborhood of 0 and $\varphi(\lambda, 0) = 1$.
It can be shown that $\varphi(\lambda, t)$ is entire holomorphic in λ and real valued for
real λ. Kodaira claims that there exists a second eigenfunction $s_2(\lambda) \in$
$\mathcal{E}_\lambda$, depending holomorphically on λ, such that s_1, s_2 form a fundamental

system fulfilling the requirements (a) and (b) stated below Theorem 7.2. Using the theory of second order differential equations with a regular singularity this can indeed be proved along the following lines.

If $k := (m+1) - (-m) = 2m + 1$ is strictly positive, there exists a second eigenfunction $s_2(\lambda) \in \mathcal{E}_\lambda$ with

$$s_2(\lambda, t) = -c_0^{-1}(2m+1)^{-1} t^{-m} + \mathcal{O}(t^{-m+\varepsilon}), \qquad (t \to 0).$$

If k is not an integer, this eigenfunction is unique. If k is an integer, then $s_2(\lambda)$ has a series expansion in terms of t^{-m+r} and $t^{m+1+s} \log t$, $(r, s \in \mathbb{N})$, and is uniquely determined by the requirement that the coefficient of $t^{-m+k} = t^{m+1}$ is zero. Finally, if $k = 0$, i.e., $m = -\frac{1}{2}$, then there exists a unique second eigenfunction $s_2(\lambda, t)$ with

$$s_2(\lambda, t) = c_0^{-1} t^{1/2} \log t + \mathcal{O}(t^{1/2+\varepsilon}), \qquad (t \to 0).$$

In all cases, by arguments involving monodromy for t around zero it can be shown that $s_2(\lambda, t)$ is entire holomorphic in λ and real valued for real λ. Finally, from the series expansions for these functions and their derivatives, it follows that the Wronskian $[s_1(\lambda), \overline{s_2(\lambda)}]_t$ behaves like $1 + \mathcal{O}(t^\varepsilon)$ for $t \to 0$. Since the Wronskian is constant, this implies that s_1, s_2 is a fundamental system.

From the asymptotic behavior of s_1, s_2 it is seen that at the boundary point 0, the operator L is of limit circle type if and only if $m > \frac{1}{2}$. It is of limit point type if $-\frac{1}{2} \leq m \leq \frac{1}{2}$. In the first case we fix the boundary datum $\chi_0 = s_1(0, \cdot)$ at 0 and in the second case we fix the (dummy) boundary datum $\chi_0 = 0$. In all cases s_1 is square integrable on $]0, 1]$, so that $\varphi_{0\lambda} = s_1(\lambda)$ and $F_0(\lambda) = (1, 0)^T$, in the notation of (7.3).

We now turn to the asymptotic behavior at ∞. Kodaira first shows that for every ν with $\mathrm{Im}\,\nu \geq 0$, $\nu \neq 0$, there is a unique solution Φ_ν to the equation $Lf = \nu^2 f$ such that

$$\Phi_\nu(t) \sim e^{i\nu t}, \qquad (t \to \infty),$$

the asymptotics being preserved if the expressions on both sides are differentiated once with respect to t. Moreover, both $\Phi_\nu(t)$ and $\Phi'_\nu(t)$ are continuous in (t, ν) and holomorphic in ν for $\mathrm{Im}\,\nu > 0$.

For $\mathrm{Im}\,\nu < 0$ the function $\Psi_\nu = \overline{\Phi_{\bar\nu}}$ belongs to $\mathcal{E}_{\nu^2}$ and $\Psi_\nu(t) \sim e^{-i\nu t}$ for $t \to \infty$. This shows that Ψ_ν is not square integrable towards infinity, so that L is of limit point type at infinity. We may therefore take

$$\varphi_{\infty\lambda} = \Phi_\nu, \qquad (\mathrm{Im}\,\lambda > 0,\ \mathrm{Im}\,\nu > 0,\ \nu^2 = \lambda).$$

It follows from the above that $\Phi_\nu(t) = a(\nu)\, s_1(\nu^2, t) + b(\nu)\, s_2(\nu^2, t)$, with

a, b continuous on $\operatorname{Im}\nu \geq 0$, $\nu \neq 0$, and holomorphic on $\operatorname{Im}\nu > 0$. We note that $F_\infty(\lambda) = (a(\nu), b(\nu))^{\mathrm{T}}$. Using the similar expression for $\Phi_{-\bar\nu}$ it follows that

$$\overline{a(\nu)} = a(-\bar\nu), \quad \overline{b(\nu)} = b(-\bar\nu). \tag{8.1}$$

If ν is real and non-zero, then Φ_ν and $\Phi_{-\nu}$ form a basis of $\mathcal{E}_{\nu^2}$ and from the asymptotic behavior of the (constant) Wronskian $[\Phi_\nu, \overline{\Phi}_{-\nu}]_t$ one reads off that

$$b(\nu)\,a(-\nu) - a(\nu)\,b(-\nu) = 2i\nu, \quad (\nu \in \mathbb{R} \setminus \{0\}). \tag{8.2}$$

From (8.1) and (8.2) it follows that

$$\operatorname{Im} a(\nu)\overline{b(\nu)} = -\nu, \quad (\nu \in \mathbb{R} \setminus \{0\}). \tag{8.3}$$

In particular, a and b do not vanish anywhere on $\mathbb{R} \setminus \{0\}$.

We can now determine the spectral matrix for this problem. Indeed, for $\operatorname{Im}\lambda > 0$ and $\operatorname{Im}\nu > 0, \nu^2 = \lambda$,

$$F(\lambda) = \begin{pmatrix} 1 & a(\nu) \\ 0 & b(\nu) \end{pmatrix},$$

so that

$$\operatorname{Im} M(\lambda) = -\operatorname{Im}\begin{pmatrix} a(\nu)b(\nu)^{-1} & 1 \\ 0 & 0 \end{pmatrix} = \begin{pmatrix} -\operatorname{Im}\frac{a(\nu)}{b(\nu)} & 0 \\ 0 & 0 \end{pmatrix}$$

by (7.4). From this we conclude that the spectral matrix $P(\lambda)$ has zero entries except for the one in the upper left corner, which we denote by $\rho(\lambda)$. The second component of $\mathcal{F}f$ now plays no role in the Plancherel formula. Indeed, define

$$\mathcal{F}_1 f(\lambda) = \int_0^\infty f(t)\, s_1(\lambda, t)\, dt,$$

then we have the following.

Corollary 8.1 *$\mathcal{F}_1$ extends to an isometry from the space $L^2(\,]0, \infty\,[\,)$ onto $L^2(\mathbb{R}, d\rho)$. The spectral function ρ is given by*

$$\rho(\lambda) = -\frac{1}{\pi}\lim_{\delta\downarrow 0}\lim_{\varepsilon\downarrow 0}\int_{[\delta,\lambda+\delta]+i\varepsilon} \operatorname{Im}\frac{a(\sqrt{\mu})}{b(\sqrt{\mu})}\, d\mu, \tag{8.4}$$

where the square root $\sqrt{\mu}$ with positive imaginary part should be taken.

Since a and b are holomorphic in the upper half plane, $a(\sqrt{\mu})\, b(\sqrt{\mu})^{-1}$ is meromorphic over the interval $]-\infty, 0[$, so that on $]-\infty, 0[$, the measure $d\rho$ is a countable sum of point measures. Indeed, let S be the (discrete) subset of zeros for a on the positive imaginary axis $i\,]\,0,\infty\,[\,$. Then

$$d\rho|_{]-\infty,0[} = \sum_{\sigma\in S} 2\,\mathrm{Res}_{\nu=\sigma}\,\frac{\nu\,a(\nu)}{b(\nu)}\cdot\delta_{[\sigma^2]}$$

On the other hand, for $\lambda > 0$, if $\mu \to \lambda$, then the integrand of (8.4) tends to $\mathrm{Im}\,a(\sqrt{\lambda})b(\sqrt{\lambda})^{-1}$, with local uniformity in λ. In view of (8.3) it now follows that

$$d\rho(\lambda)|_{]\,0,\infty\,[} = -\frac{1}{\pi}\mathrm{Im}\,\frac{a(\sqrt{\lambda})}{b(\sqrt{\lambda})} = \frac{1}{\pi}\frac{\sqrt{\lambda}\,d\lambda}{|b(\sqrt{\lambda})|^2} = \frac{2}{\pi}\frac{\lambda\,d\sqrt{\lambda}}{|b(\sqrt{\lambda})|^2}.$$

Finally, if $s_1(0)$ is not square integrable at infinity, then $\rho_0 := d\rho(\{0\}) = 0$. On the other hand, if it is, then $\rho_0 := d\rho(\{0\})$ equals the squared L^2-norm of $s_1(0)$.

Finally, since $s_1(0,\lambda)$ is real valued for λ real, whereas $\overline{\Phi_{-\nu}} = \Phi_\nu$ for real ν, there exists a real analytic function $c : \mathbb{R}\setminus\{0\} \to \mathbb{C}$ such that

$$s_1(0,\nu^2) = c(\nu)\Phi_\nu + \overline{c(-\nu)}\Phi_{-\nu}$$

for all $\nu \in \mathbb{R}\setminus\{0\}$. This gives rise to the equations

$$\begin{cases} a(\nu)c(\nu) + a(-\nu)\overline{c(-\nu)} &= 1 \\ b(\nu)c(\nu) + b(-\nu)\overline{c(-\nu)} &= 0. \end{cases}$$

Using (8.2) we now deduce that

$$c(\nu) = -b(\nu)/2i\nu, \qquad (\nu \in \mathbb{R}\setminus\{0\}). \tag{8.5}$$

Therefore,

$$d\rho(\lambda)|_{]\,0,\infty\,[} = \frac{1}{2\pi}\frac{d\sqrt{\lambda}}{|c(\sqrt{\lambda})|^2}.$$

We thus see that the principle formulated below (5.5) still holds in this setting.

9 Riemannian symmetric spaces

A Riemannian symmetric space is a connected Riemannian manifold X with the property that the local geodesic reflection at each point extends to a global isometry of X. Up to covering, each such space allows a decomposition into a product of three types of symmetric space. Those with zero sectional curvature (the Euclidean spaces), those with positive sectional curvature (among which the Euclidean spheres) and those with negative sectional curvature (among which the hyperbolic spaces). It follows from the work of E. Cartan, that the spaces of negative sectional curvature are precisely those given by $X = G/K$, where G is a connected real semisimple Lie group of non-compact type, with finite center, and where K is a maximal compact subgroup of G. The Killing form of G naturally induces a G-invariant Riemannian metric on G/K. The group K is the fixed point group of a Cartan involution θ of G; this involution induces the geodesic reflection in the origin $\bar{e} = eK$ of X.

A typical example of a symmetric space of this type is the space X of positive definite symmetric $n \times n$-matrices on which $G = \mathrm{SL}(n, \mathbb{R})$ acts by $(g, h) \mapsto ghg^{\mathrm{T}}$. The stabilizer of the identity matrix equals $\mathrm{SO}(n)$ and the associated Cartan involution $\theta : G \to G$ is given by $g \mapsto (g^{\mathrm{T}})^{-1}$. The geodesic reflection in the identity matrix I is given by $h \mapsto h^{-1}$.

In the general setting, the derivative of the Cartan involution at the identity element of G induces an involution θ_* of the Lie algebra $\mathfrak{g}$. The Lie algebra $\mathfrak{g}$ decomposes as a direct sum of vector spaces

$$\mathfrak{g} = \mathfrak{k} \oplus \mathfrak{p},$$

where $\mathfrak{k}$ and $\mathfrak{p}$ are the $+1$ and -1 eigenspaces of θ_*, respectively. It can be shown that the map

$$(X, k) \mapsto \exp Xk, \quad \mathfrak{p} \times K \to G \tag{9.1}$$

is an analytic diffeomorphism onto G. In particular, this implies that the exponential map induces a diffeomorphism $\mathrm{Exp} : X \mapsto \exp XK$, $\mathfrak{p} \to G/K$. Let $\mathfrak{a}$ be a subspace of $\mathfrak{p}$, maximal subject to the condition that it is abelian for the Lie bracket of $\mathfrak{g}$. Every other such subspace is K-conjugate to $\mathfrak{a}$. The dimension r of $\mathfrak{a}$ is called the rank of the symmetric space G/K.

In the example $G = \mathrm{SL}(n, \mathbb{R})$, the Lie algebra $\mathfrak{sl}(n, \mathbb{R})$ consists of all traceless $n \times n$-matrices, and θ_* is given by $X \mapsto -X^{\mathrm{T}}$. The Cartan decomposition (9.1) is given by the decomposition of a matrix in terms of a positive definite symmetric one times an orthogonal one. The algebra $\mathfrak{a}$ now consists of the traceless diagonal matrices, so that the rank of $\mathrm{SL}(n, \mathbb{R})/\mathrm{SO}(n)$ equals $n - 1$. For $n = 2$ the space is isomorphic to the hyperbolic upper half plane, equipped with the action of $\mathrm{SL}(2, \mathbb{R})$ through fractional linear transformations.

By a result of Harish-Chandra, the algebra $\mathbb{D}(G/K)$ of G-invariant linear partial differential operators on G/K is a polynomial algebra of rank r. More precisely, let M be the centralizer of $\mathfrak{a}$ in K, and let

$W := N_K(\mathfrak{a})/M$, the normalizer modulo the centralizer of $\mathfrak{a}$ in K. As a subgroup of $\mathrm{GL}(\mathfrak{a})$, this group is the reflection group associated with the roots of $\mathfrak{a}$ in $\mathfrak{g}$. It is therefore called the Weyl group of the pair $(\mathfrak{g}, \mathfrak{a})$. There exists a canonical isomorphism γ from $\mathbb{D}(G/K)$ onto $P(\mathfrak{a}_{\mathbb{C}}^*)^W$, the algebra of W-invariants in the polynomial algebra of the complexified dual space $\mathfrak{a}_{\mathbb{C}}^*$ (equipped with the dualized Weyl group action). By a result of C. Chevalley, the algebra $P(\mathfrak{a}_{\mathbb{C}}^*)^W$ is known to be polynomial of rank r.

In the example $\mathrm{SL}(n, \mathbb{R})$, the Weyl group is given by the natural action of the permutation group S_n on the space $\mathfrak{a}$ of traceless diagonal matrices. Here the algebra $P(\mathfrak{a}_{\mathbb{C}}^*)^W$ corresponds to the algebra of S_n-invariants in $\mathbb{C}[T_1, \ldots, T_n]/(T_1 + \cdots + T_n)$, which is of course well known to be a polynomial algebra of $n-1$ generators of its own right. We note that in the case of rank 1, the algebra $\mathbb{D}(G/K)$ consists of all polynomials in the Laplace-Beltrami operator.

In the papers [12],[13], Harish-Chandra created a beautiful theory of harmonic analysis for left K-invariant functions on the symmetric space G/K, culminating in a Plancherel formula for $L^2(G/K)^K$, the space of left-K-invariant functions on G/K, square integrable with respect to the Riemannian volume form. We will now give a brief outline of the main results.

For $\nu \in \mathfrak{a}_{\mathbb{C}}^*$, we consider the following system of simultaneous eigenequations on G/K :

$$Df = \gamma(D, i\nu)f, \qquad (D \in \mathbb{D}(G/K)). \tag{9.2}$$

For $r = 1$, this system is equivalent to a single eigenequation for the Laplace operator. Each eigenfunction is analytic, by ellipticity of the Laplace operator. The space of K-invariant functions satisfying (9.2) is one dimensional and spanned by the so-called elementary spherical function φ_ν, normalized by $\varphi_\nu(eK) = 1$. This function can be constructed as a matrix coefficient $x \mapsto \langle 1_K, \pi_\nu(x)1_K \rangle$, with 1_K a K-fixed vector in a suitable continuous representation of G in an infinite dimensional Hilbert space, obtained by the process of induction. By Weyl invariance of the polynomials $\gamma(D)$ it follows that $\varphi_{w\nu} = \varphi_\nu$, for all $w \in W$.

In terms of the elementary spherical functions one may define the so-called Fourier transform of a function $f \in C_c^\infty(G/K)^K$ by

$$\mathcal{F}_{G/K} f(\nu) = \int_{G/K} f(x)\,\varphi_{-\nu}(x)dx, \qquad (\nu \in \mathfrak{a}^*),$$

with dx the G-invariant volume measure on G/K.

By analyzing the system of differential equations (9.2) it is possible
to obtain rather detailed information on the asymptotic behavior of the
elementary spherical functions φ_ν towards infinity. It can be shown that
the map $K/M \times \mathfrak{a} \to G/K$, $(kM, X) \mapsto k \exp X K$ is surjective. For
obvious reasons, the associated decomposition

$$G/K = K \exp \mathfrak{a} \cdot \bar{e} \tag{9.3}$$

is called the polar decomposition of G/K. In it, the $\mathfrak{a}$-part of an element is uniquely determined modulo the action of W. Let $\mathfrak{a}^+$ be a
choice of positive Weyl chamber relative to W, then it follows that
$G/K = K \exp \overline{\mathfrak{a}^+} \cdot \bar{e}$ with uniquely determined $\overline{\mathfrak{a}^+}$-part. Moreover, the
map $K/M \times \mathfrak{a}^+ \to G/K$ is an analytic diffeomorphism onto an open
dense subset of G/K.

Accordingly, each elementary spherical function is completely determined by its restriction to $A^+ := \exp(\mathfrak{a}^+)$. Moreover, the restricted
function $\varphi_\nu|_{A^+}$ satisfies the system of equations arising from (9.2) by
taking radial parts with respect to the polar decomposition (9.3). Using
a characterization of $\mathbb{D}(G/K)$ in terms of the universal algebra $U(\mathfrak{g})$,
Harish-Chandra was able to analyze these radial differential equations
in great detail. This allowed him to show that, for generic $\nu \in \mathbb{C}$, the
behavior of the function φ_ν towards infinity is described by

$$\varphi_\nu(k \exp X) = \sum_{w \in W} c(w\nu) e^{(iw\nu - \rho)(X)} [1 + R_{w\nu}(X)], \tag{9.4}$$

for $k \in K$ and $X \in \mathfrak{a}^+$. Here $\rho \in \mathfrak{a}^*$ is half the sum of the positive roots,
counted with multiplicities, $c(\nu)$, the so-called c-function, is a certain
meromorphic function of the parameter $\nu \in \mathfrak{a}_\mathbb{C}^*$, and $R_\nu(X)$ is a certain
analytic function of X, depending meromorphically on the parameter ν.
Moreover, the asymptotic behavior of R_ν is described by

$$R_\nu(tX) = \mathcal{O}(e^{-tm(X)}), \quad (X \in \mathfrak{a}^+, t \to \infty), \tag{9.5}$$

with $m(X)$ a positive constant, depending on X in a locally uniform
way. Each of the summands in (9.4) is an eigenfunction of the radial
system of differential equations of its own right.

Theorem 9.1 *(Harish-Chandra's Plancherel formula). The function c
has no zeros on $\mathfrak{a}^*$. Moreover, let*

$$dm(\nu) := \frac{d\nu}{|c(\nu)|^2}, \tag{9.6}$$

with $d\nu$ a suitable normalization of Lebesgue measure on $\mathfrak{a}^*$ (see further *down). Then $dm(\nu)$ is Weyl-group invariant and the Fourier transform $\mathcal{F}_{G/K}$ extends to an isometry from $L^2(G/K)^K$ onto $L^2(\mathfrak{a}^*, dm(\nu))^W$.*

In footnote 3), p. 242, to the introduction of his paper [12], Harish-Chandra mentions: 'This is reminiscent of a result of Weyl [[29], p. 266] on ordinary differential equations.' It seems that Harish-Chandra was actually very inspired by Weyl's paper. In [5], p. 38, A. Borel writes: '[...] less obviously maybe, Weyl was also of help via his work on differential equations [29], which gave Harish-Chandra a crucial hint in his quest for an explicit form of the Plancherel measure. [...] It was the reading of [29] which suggested to Harish-Chandra that the measure should be the inverse of the square modulus of a function in λ describing the asymptotic behavior of the eigenfunctions [...] and I remember well from seminar lectures and conversations that he never lost sight of that principle, which is confirmed by his results in the general case as well.'

It is the purpose of the rest of this section to show that for the rank one case Theorem 9.1 is in fact a rather direct consequence of Kodaira's generalization of Weyl's result, described in Section 8.

Before we proceed it should be mentioned that in [12] and [13] Theorem 9.1 was completely proved for spaces of rank 1. Moreover, for these spaces the c-function was explicitly determined as a certain quotient of Gamma factors.

For spaces of arbitrary rank Theorem 9.1 was proved modulo two conjectures. The first of these concerned the injectivity of the Fourier transform and the second certain estimates for the c-function. The first conjecture was proved by Harish-Chandra himself, in his work on the so-called discrete series of representations for G, [14]. The validity of the second conjecture followed from the work of S. Gindikin and F. Karpelevic, [11], where a product decomposition of the c-function in terms of rank one c-functions was established. Simpler proofs of Theorem 9.1 were later found through the contributions of [20], [10], [26].

The precise normalization of the Lebesgue measure $d\nu$ may be given as follows. The polar decomposition (9.3) gives rise to an integral formula

$$\int_{G/K} f(x)\, dx = \int_K \int_{\mathfrak{a}^+} f(k \exp X) J(X)\, dX\, dk, \qquad (9.7)$$

with dk normalized Haar measure on K, J a suitable Jacobian, and dX suitably normalized Lebesgue measure on $\mathfrak{a}$. The Jacobian J and the measure dX are uniquely determined by the above formula and the requirement that $J(tX)$ behaves asymptotically as $e^{t2\rho(X)}$, for $X \in \mathfrak{a}^+$

and $t \to \infty$. Let $d\xi$ denote the dual Lebesgue measure on $\mathfrak{a}^*$. Then

$$d\nu = \frac{1}{|W|} \frac{d\xi}{(2\pi)^n},$$

with $|W|$ the number of elements of the Weyl group.

We now turn to the setting of a space of rank 1. A typical example of such a space is the n-dimensional hyperbolic space X_n, which may be realized as the submanifold of $\mathbb{R}^{n+1}$ given by the equation $x_1^2 - (x_2^2 + \cdots + x_n^2) = 1$, $x_1 > 0$. Its Riemannian metric is induced by the indefinite standard inner product of signature $(1, n)$ on $\mathbb{R}^{n+1}$. As a homogeneous space $X_n \simeq \mathrm{SO}(1,n)/\mathrm{SO}(n)$.

More generally, as $\mathfrak{a}$ is one dimensional, all roots in R are proportional. Let α be the simple root associated with the choice of positive chamber $\mathfrak{a}^+$. Then $-\alpha$ is a root as well, and possibly $\pm 2\alpha$ are roots as well. No other multiples of α occur. We fix the unique element $H \in \mathfrak{a}$ with $\alpha(H) = 1$.

Via the map $tH \mapsto t$ we identify $\mathfrak{a}$ with $\mathbb{R}$; likewise, via the map $t\alpha \mapsto t$ we identify $\mathfrak{a}^*$ with $\mathbb{R} \simeq \mathbb{R}^*$. Then $\mathfrak{a}^+ =]0, \infty[$. Rescaling the Riemannian metric if necessary we may as well assume that under these identifications, both dX and $d\xi$ correspond to the standard Lebesgue measure on $\mathbb{R}$.

Let m_1, m_2 denote the root multiplicities of $\alpha, 2\alpha$, i.e., m_j is the dimension of the eigenspace of $\mathrm{ad}(H)$ in $\mathfrak{g}$ with eigenvalue j. Then with the above identifications,

$$\rho = \frac{1}{2}(m_1 + 2m_2).$$

The Laplace operator Δ satisfies $\gamma(\Delta, i\nu) = (-\|\nu\|^2 - \|\rho\|^2)$ with $\|\cdot\|$ the norm on $\mathfrak{a}^*$ dual to the norm on $\mathfrak{a}$ induced by the Riemannian inner product on $\mathfrak{g}/\mathfrak{k} \simeq \mathfrak{p}$. Multiplying Δ with a suitable negative constant, we obtain an operator L_0 with $\gamma(L_0, i\nu) = \nu^2 + \rho^2$. Let $\widetilde{L} = L_0 - \rho^2$, then $\widetilde{L}\varphi_\nu = \nu^2 \varphi_\nu$.

The Jacobian J mentioned above is given by the formula

$$J(t) = (e^t - e^{-t})^{m_1} (e^{2t} - e^{-2t})^{m_2}.$$

Let $L := J^{1/2} \circ \mathrm{rad}\,(\widetilde{L}) \circ J^{-1/2}$ be the conjugate of the radial part of $\widetilde{L}$ by multiplication with $J^{1/2}$. Put

$$s(\nu, t) := J^{1/2}(t)\, \varphi_\nu(\exp tH). \tag{9.8}$$

Then the system of equations (9.2) is equivalent to the single eigenequation

$$Ls(\nu, \cdot) = \nu^2 s(\nu, \cdot)$$

By a straightforward calculation, see [10], p. 156, it follows that

$$L = -\frac{d^2}{dt^2} + q(t), \qquad q(t) = \frac{1}{2}J^{-1}\frac{d^2}{dt^2}J - \frac{1}{4}J^{-2}(\frac{d}{dt}J)^2 - \rho^2.$$

By using the Taylor series of $J(t)$ at 0 we see that there exists a real analytic function V on $\mathbb{R}$ such that

$$q(t) = m(m+1)t^{-2} + V(t), \qquad (t > 0),$$

where

$$m = \frac{1}{2}(m_1 + m_2) - 1 \geq -\frac{1}{2}.$$

On the other hand, at infinity, $J(t)$ equals $e^{2t\rho}$ times a power series in terms of powers of e^{-2t} with constant term 1. From this we see that $q(t) = \mathcal{O}(e^{-2t})$, so that $V(t) = \mathcal{O}(t^{-2})$ as $t \to \infty$. It follows that our operator L satisfies all requirements of Section 8.

We now observe that $\varphi_\nu(e) = 1$ and $J(t)^{1/2} \sim 2^\rho t^{m+1}$ $(t \to 0)$. Let $c_0 := 2^\rho$ and let $s_1(\lambda, t)$ be defined as in Section 8, for $\lambda \in \mathbb{C}$. Then it follows that $s(\nu, t) = s_1(\nu^2, t)$ for all $\nu \in \mathbb{C}$. Moreover, it follows from (9.4) that the function Φ_ν of Section 8 is given by

$$\Phi_\nu(t) = e^{-t\rho} J(t)^{1/2} c(\nu) e^{i\nu} (1 + R_\nu(t)).$$

In particular, it depends meromorphically on the parameter ν. From this it follows that the functions $\nu \mapsto a(\nu), b(\nu)$ are meromorphic. By analytic continuation it now follows that the identity (8.5) extends to an identity of meromorphic functions. From its explicitly known form as a quotient of Gamma factors, it follows that the function $\nu \mapsto c(\nu)$ has no zeros on $i\,]0, \infty[$. Moreover, it has a zero of order 1 at 0. Using (8.5) we now see that b has no zeros on $i[0, \infty]$, so that the spectral measure $d\rho$ has no discrete part. Hence,

$$d\rho(\lambda) = \frac{d\sqrt{\lambda}}{2\pi|c(\sqrt{\lambda})|^2}\Big|_{]0,\infty[}.$$

Let $\mathcal{F}_1$ be the Fourier transform defined in terms of $s_1(\lambda, t) = s(\sqrt{\lambda}, t)$, see Section 8. Then it follows by application of (9.7) and (9.8) that

$$\mathcal{F}_{G/K} f(\nu) = \mathcal{F}_1(J^{1/2} f \circ \exp|_{\mathfrak{a}^+})(\nu^2), \qquad (\nu \in \mathbb{R}), \qquad (9.9)$$

for every $f \in C_c(G/K)$.

By Corollary 8.1 the Fourier transform $\mathcal{F}_1$ is an isometry from the space $L^2(\mathfrak{a}^+, dX)$ onto the space $L^2(]0, \infty[, d\rho)$. Moreover, the map $f \mapsto J^{1/2} f \circ \exp|_{\mathfrak{a}^+}$ is an isometry from $L^2(G/K)^K$ onto $L^2(\mathfrak{a}^+, dX)$.

Finally, since $W = \{\pm I\}$, whereas the function $\nu \mapsto |c(\nu)|^2$ is even by (8.5) and (8.1), pull-back by the map $\nu \mapsto \nu^2$ defines an isometry

$$L^2(]0, \infty[, d\rho) \xrightarrow{\simeq} L^2(\mathfrak{a}^*, \frac{1}{2}\frac{d\xi}{2\pi|c(\nu)|^2})^W. \tag{9.10}$$

By (9.9) $\mathcal{F}_{G/K}$ is the composition of the three mentioned isometries. The assertion of Theorem 9.1 follows.

10 Analysis on groups and symmetric spaces

After his work on the Riemannian symmetric spaces, Harish-Chandra continued to work on a theory of harmonic analysis for real semisimple Lie groups in the 1960's. His objective was to obtain an explicit Plancherel decomposition for $L^2(G)$, the space of square integrable functions with respect to a fixed choice of (bi-invariant) Haar measure on G.

In the case of a compact group, the Plancherel formula is described in terms of representation theory and consists of the Peter–Weyl decomposition combined with the Schur-orthogonality relations.

In the more general case of a real semisimple Lie group, the situation is far more complicated. If G is simple and non-compact, then the non-trivial irreducible unitary representations of G are infinite dimensional. Moreover, there is a mixture of discrete and continuous spectrum.

An irreducible unitary representation is said to be of the discrete series if it contributes discretely to $L^2(G)$, i.e., it is embeddable as a closed invariant subspace for the left regular representation. Equivalently, this means that its matrix coefficients are square integrable. An irreducible unitary representation has a character, which is naturally defined as a conjugation invariant distribution on G. A deep theorem of Harish-Chandra in the beginning of the 1960's asserts that in fact all such characters are locally integrable. Moreover, they are analytic on the open dense subset of regular elements.

In [14] and [15], Harish-Chandra gave a complete classification of the discrete series. First of all, G has discrete series if and only if it has a compact Cartan subgroup. Moreover, the representations of the discrete series are completely determined by the restriction of their characters to this compact Cartan subgroup. Harish-Chandra achieved their classification and established a character formula on the compact Cartan which shows remarkable resemblance with Weyl's character formula.

In the early 1970's, Harish-Chandra, [16], [17], [18], completed his work on the Plancherel decomposition. The orthocomplement of the discrete part of $L^2(G)$ is decomposed in terms of representations of the so-called generalized principal series. These are induced representations of the form

$$\pi_{P,\xi,\lambda} = \text{Ind}_P^G(\xi \otimes e^{i\lambda} \otimes 1),$$

where P is a (cuspidal) parabolic subgroup of G, with a so-called Langlands decomposition $P = M_P A_P N_P$. Moreover, ξ is a discrete series representation of M_P and $e^{i\lambda}$ is a unitary character of the vectorial group A_P. The space $L^2(G)$ splits into a finite orthogonal direct sum of closed subspaces $L^2(G)_{[P]}$, each summand corresponding to an equivalence class of parabolic subgroups with K-conjugate A_P-part. Here G counts for a parabolic subgroup, and $L^2(G)_{[G]}$ denotes the discrete part of $L^2(G)$.

Each summand $L^2(G)_{[P]}$ decomposes discretely into a countable orthogonal direct sum of spaces $L^2(G)_{[P],\xi}$ parametrized by (equivalence classes of) discrete series representations of M_P. Finally, each of the spaces $L^2(G)_{[P],\xi}$ has a continuous decomposition parametrized by $\lambda \in \mathfrak{a}_P^*$. Harish-Chandra achieved this continuous decomposition by reduction to the space of functions transforming finitely under the action of the maximal compact subgroup K.

Let δ_L, δ_R be two irreducible representations of K and let $L^2(G)_{[P],\xi,\delta}$ be the part of $L^2(G)_{[P],\xi}$ consisting of bi-K-finite functions of left K-type δ_L and right K-type δ_R. The decomposition of this space is described in terms of *Eisenstein integrals*. These are essentially $K \times K$-finite matrix coefficients of type (δ_L, δ_R) of the induced representation involved. The Eisenstein integrals $E([P], \xi, \lambda, \psi)$ are functions on G which depend analytically on the parameter $\lambda \in \mathfrak{a}^*$. In addition, they depend linearly on a certain parameter ψ, which ranges over a certain finite dimensional Hilbert space $\mathcal{A}_2(M_P, \xi, \delta)$ of functions $M \times K \times K \to \mathbb{C}$. The Eisenstein integrals satisfy eigenequations coming from the bi-G-invariant differential operators on G. As in the previous section these equations can be analyzed in detail, and it can be shown that the integrals behave asymptotically like

$$E([P], \xi, \lambda, \psi)(k_1 \, m \exp X \, k_2)$$
$$\sim \sum_{w \in W(\mathfrak{a}_Q | \mathfrak{a}_P)} e^{(iw\lambda - \rho_Q)(X)} \, [c_{Q|P,\xi}(w, \lambda)\psi](m, k_1, k_2)$$

for $m \in M_Q$, $k_1, k_2 \in K$, and as X tends to infinity in $\mathfrak{a}_Q^+$; here Q is a parabolic subgroup in the same equivalence class as P and $W(\mathfrak{a}_Q|\mathfrak{a}_P)$ denotes the finite set of isomorphisms $\mathfrak{a}_P \to \mathfrak{a}_Q$ induced by the adjoint action of K. Each coefficient $c_{Q|P,\xi}(w, \lambda)$ is an isomorphism from the finite dimensional Hilbert space $\mathcal{A}_2(M_P, \xi, \delta)$ onto the similar space $\mathcal{A}_2(M_Q, w\xi, \delta)$. It can be shown that

$$c_{Q|P,\xi}(w, \lambda)^* c_{Q|P,\xi}(w, \lambda) = \eta(P, \xi, \lambda) \, I$$

with $\eta(P, \xi, \lambda)$ a strictly positive scalar, independent of Q, w, δ and depending real analytically on $\lambda \in \mathfrak{a}_P^*$. Finally, the measure for the Plancherel decomposition of $L^2(G)_{[P],\xi}$ is given by

$$\frac{d\lambda}{\eta(P, \xi, \lambda)}. \tag{10.1}$$

In this sense, Weyl's principle is valid for all continuous spectral parameters in the Plancherel decomposition for G.

In the 1980's and 1990's, much progress was made in harmonic analysis on general semisimple symmetric spaces. These are pseudo-Riemannian symmetric spaces of the form G/H, with G a real semisimple Lie group and H (an open subgroup of) the group of fixed points for an involution σ of G. This class of spaces contains both the Riemannian symmetric spaces and the semisimple groups. Indeed the group G is a homogeneous space for the action of $G \times G$ given by $(x, y) \cdot g = xgy^{-1}$. The stabilizer of the identity element e_G equals the diagonal H of $G \times G$, which is the group of fixed points for the involution $\sigma : (x, y) \mapsto (y, x)$. As a decomposition for the left times right regular action of $G \times G$ on $L^2(G)$ the Plancherel decomposition becomes multiplicity free. This is analogous to what happens for the Peter-Weyl decomposition for compact groups.

Another interesting class of semisimple symmetric spaces is formed by the pseudo-Riemannian hyperbolic spaces $\mathrm{SO}_e(p, q)/\mathrm{SO}_e(p-1, q)$, $p > 1$.

For general semisimple symmetric spaces, M. Flensted-Jensen, [9], gave the first construction of discrete series assuming the analogue of Harish-Chandra's rank condition. The full classification of the discrete series was then given by T. Oshima and T. Matsuki [25].

In [2], E.P. van den Ban and H. Schlichtkrull gave a description of the most continuous part of the Plancherel decomposition. Here, a new phenomenon is that the Plancherel decomposition may have finite multiplicities. Nevertheless, the multiplicities can be parametrized in such a way that Weyl's principle generalizes to this context. Then, P. Delorme, partly in collaboration with J. Carmona, determined the full Plancherel decomposition for G/H, [6], [8]. Around the same time this was also achieved by E.P. van den Ban and H. Schlichtkrull, [3],[4], with a completely different proof. In all these works, the appropriate analogue of (10.1) goes through. For more information, we refer the reader to the survey articles in [1].

Parallel to the developments sketched above, G. Heckman and E. Opdam [19] developed a theory of hypergeometric functions, generalizing the elementary spherical functions of the Riemannian symmetric spaces. For these spaces, the algebra of radial components of invariant differential operators is entirely determined by a root system and root multiplicities. The generalization is obtained by allowing these multiplicities to vary in a continuous fashion. In the associated Plancherel decomposition, established by Opdam, [24], Weyl's principle holds through the analogue of (9.6).

11 Appendix: circles in $\mathbb{P}^1(\mathbb{C})$

If V is a two dimensional complex linear space, then by $\mathbb{P}(V)$ we denote the 1-dimensional projective space of lines $\mathbb{C}v$, with $v \in V \setminus \{0\}$. In a natural way we will identify subsets of $\mathbb{P}(V)$ with $\mathbb{C}$-homogeneous subsets of V containing 0. In particular, the empty set is identified with $\{0\}$. The group $\mathrm{GL}(V)$ of invertible complex linear transformations of V naturally acts on $\mathbb{P}(V)$.

Let β be Hermitian form on V, i.e., $\beta : V \times V \to \mathbb{C}$ is linear in the first and conjugate linear in the second component, and $\beta(v, w) = \overline{\beta(w, v)}$ for all $v, w \in V$. By symmetry, $\beta(v, v) \in \mathbb{R}$ for all $v \in V$. We denote by $\mathcal{B}$ the space of Hermitian forms β on V for which the function $v \mapsto \beta(v, v)$ has image $\mathbb{R}$. Equivalently, this means that there exists a basis v_1, v_2 of V such that $\beta(v_1, v_1) = 1$ and $\beta(v_2, v_2) = -1$. It follows from this that the group $\mathrm{GL}(V)$ acts transitively on $\mathcal{B}$ by $g \cdot \beta(v, w) = \beta(g^{-1}v, g^{-1}w)$.

We note that for any Hermitian form β on V the map $v \mapsto \beta(v, \cdot)$ induces a linear map from V to the conjugate linear dual space $\overline{V}^*$. This map is an isomorphism if and only if β is non-degenerate. Let γ be any choice of positive definite Hermitian form on V. Then $H_\beta = \gamma^{-1} \circ \beta$ is a linear endomorphism of V; from $\beta(v, w) = \gamma(H_\beta v, w)$ for $v, w \in V$ we see that H_β is symmetric with respect to the inner product γ. The condition that $\beta \in \mathcal{B}$ is equivalent to the condition that H_β has both a strictly positive and a strictly negative eigenvalue, which in turn is equivalent to the condition that $\det H_\beta < 0$. For obvious reasons we will call $\mathcal{B}$ the space of Hermitian forms of signature $(1,1)$.

By a circle in $\mathbb{P}(V)$ we mean a set of the form

$$C_\beta := \{v \in V \mid \beta(v,v) = 0\}$$

with $\beta \in \mathcal{B}$. For $g \in \mathrm{GL}(V)$ we have $g(C_\beta) = C_{g\cdot\beta}$ so that the natural action of $\mathrm{GL}(V)$ on the collection of circles is transitive.

We now turn to the case of $\mathbb{C}^2$ equipped with the standard Hermitian inner product. Accordingly, any form $\beta \in \mathcal{B}$ is represented by a unique Hermitian matrix H of strictly negative determinant. We will use the standard embedding $\mathbb{C} \hookrightarrow \mathbb{P}^1(\mathbb{C}) := \mathbb{P}(\mathbb{C}^2)$ given by $z \mapsto \mathbb{C}(z,1)$. The complement of the image of this embedding consists of the single point $\infty_c := \mathbb{C}(1,0)$. The inverse map $\chi : \mathbb{P}^1(\mathbb{C}) \setminus \{\infty_c\} \to \mathbb{C}$ is called the standard affine chart. It is straightforwardly verified that ∞_c belongs to C_β if and only if the entry H_{11} equals zero. In this case the intersection of C_β with the standard affine chart is given by $2\,\mathrm{Re}(H_{21}z) = -H_{22}$, which is the straight line $-H_{21}^{-1}(\frac{1}{2}H_{22} + i\mathbb{R})$. In particular, the form

$$i[z,w] = i(z_1\overline{w}_2 - z_2\overline{w}_1) \tag{11.1}$$

is represented by the Hermitian matrix iJ (see (2.4)), and the associated circle in $\mathbb{P}^1(\mathbb{C})$ equals the closure $\mathbb{P}^1(\mathbb{R}) := \mathbb{C}\mathbb{R}^2 = \mathbb{R} \cup \{\infty_c\}$ of the real line.

In the remaining case the circle C_β is completely contained in the standard affine chart, and in the affine coordinate it equals a circle with respect to the standard Euclidean metric on $\mathbb{C} \simeq \mathbb{R}^2$. The radius r and the center α are given by

$$r^2 = -\frac{\det H}{|H_{11}|^2}, \qquad \alpha = -\frac{H_{12}}{H_{11}}. \tag{11.2}$$

The preimage under χ of the interior of this circle is the subset of $\mathbb{P}^1(\mathbb{C})$ given by the inequality

$$\mathrm{sign}(H_{11})\beta(z,z) < 0.$$

We note that all circles and straight lines in $\mathbb{C}$ are representable in the above fashion. In the standard affine coordinate, the action of the group $\mathrm{GL}(2,\mathbb{C})$ on $\mathbb{P}^1(\mathbb{C})$ is represented by the action through fractional linear transformations on $\mathbb{C}$. Accordingly, we retrieve the well-known fact that this action preserves the set of circles and straight lines.

More generally, let v_1, v_2 be a complex basis of V. Then the natural map $z \mapsto z_1v_1 + z_2v_2$ induces a diffeomorphism $\mathrm{v} : \mathbb{P}^1(\mathbb{C}) \to \mathbb{P}(V)$.

The map $\chi_v := \chi \circ v^{-1} : \mathbb{P}(V) \setminus \mathbb{C}v_1 \to \mathbb{C}$ is said to be the affine chart determined by v_1, v_2. Note that $z = \chi_v(\mathbb{C}(zv_1 + v_2))$, for $z \in \mathbb{C}$. The general linear group $GL(V)$ acts on the set of affine charts by $(g, \psi) \mapsto \psi \circ g^{-1}$, so that $g \cdot \chi_v = \chi_{gv}$. Clearly, the action is transitive. It follows that the transition map between any pair of affine charts is given by a fractional linear transformation.

From the above considerations it follows that a circle C in $\mathbb{P}(V)$ corresponds to a circle in the affine chart χ_v if and only if $\mathbb{C}v_1$ does not lie on C. Otherwise, the circle is represented by a straight line in χ_v.

References

[1] J.-P. Anker and B. Orsted, editors. *Lie theory*, volume 230 of *Progress in Mathematics*. Birkhäuser Boston Inc., Boston, MA, 2005. Harmonic analysis on symmetric spaces—general Plancherel theorems.

[2] E. P. van den Ban and H. Schlichtkrull. The most continuous part of the Plancherel decomposition for a reductive symmetric space. *Ann. of Math. (2)*, 145:267–364, 1997.

[3] E. P. van den Ban and H. Schlichtkrull. The Plancherel decomposition for a reductive symmetric space. I. Spherical functions. *Invent. Math.*, 161:453–566, 2005.

[4] E. P. van den Ban and H. Schlichtkrull. The Plancherel decomposition for a reductive symmetric space. II. Representation theory. *Invent. Math.*, 161:567–628, 2005.

[5] A. Borel. *Essays in the history of Lie groups and algebraic groups*, volume 21 of *History of Mathematics*. American Mathematical Society, Providence, RI, 2001.

[6] J. Carmona and P. Delorme. Transformation de Fourier sur l'espace de Schwartz d'un espace symétrique réductif. *Invent. Math.*, 134:59–99, 1998.

[7] E. A. Coddington and N. Levinson. *Theory of ordinary differential equations*. McGraw-Hill Book Company, Inc., New York-Toronto-London, 1955.

[8] P. Delorme. Formule de Plancherel pour les espaces symétriques réductifs. *Ann. of Math. (2)*, 147:417–452, 1998.

[9] M. Flensted-Jensen. Discrete series for semisimple symmetric spaces. *Ann. of Math. (2)*, 111:253–311, 1980.

[10] R. Gangolli. On the Plancherel formula and the Paley-Wiener theorem for spherical functions on semisimple Lie groups. *Ann. of Math. (2)*, 93:150–165, 1971.

[11] S. G. Gindikin and F. I. Karpelevič. Plancherel measure for symmetric Riemannian spaces of non-positive curvature. *Dokl. Akad. Nauk SSSR*, 145:252–255, 1962.

[12] Harish-Chandra. Spherical functions on a semisimple Lie group. I. *Amer. J. Math.*, 80:241–310, 1958.

[13] Harish-Chandra. Spherical functions on a semisimple Lie group. II. *Amer. J. Math.*, 80:553–613, 1958.

[14] Harish-Chandra. Discrete series for semisimple Lie groups. I. Construction of invariant eigendistributions. *Acta Math.*, 113:241–318, 1965.

[15] Harish-Chandra. Discrete series for semisimple Lie groups. II. Explicit determination of the characters. *Acta Math.*, 116:1–111, 1966.

[16] Harish-Chandra. Harmonic analysis on real reductive groups. I. The theory of the constant term. *J. Functional Analysis*, 19:104–204, 1975.

[17] Harish-Chandra. Harmonic analysis on real reductive groups. II. Wavepackets in the Schwartz space. *Invent. Math.*, 36:1–55, 1976.

[18] Harish-Chandra. Harmonic analysis on real reductive groups. III. The Maass-Selberg relations and the Plancherel formula. *Ann. of Math. (2)*, 104:117–201, 1976.

[19] G. J. Heckman and E. M. Opdam. Root systems and hypergeometric functions. I. *Compositio Math.*, 64:329–352, 1987.

[20] S. Helgason. An analogue of the Paley-Wiener theorem for the Fourier transform on certain symmetric spaces. *Math. Ann.*, 165:297–308, 1966.

[21] D. Hilbert. *Grundzüge einer allgemeinen Theorie der linearen Integralgleichungen.* Chelsea Publishing Company, New York, N.Y., 1953.

[22] I.S. Kac. The existence of spectral functions of generalized second order differential systems with boundary conditions at the singular end. *Amer. Math. Soc. Transl. (2)*, 62:204–262, 1964.

[23] K. Kodaira. The eigenvalue problem for ordinary differential equations of the second order and Heisenberg's theory of S-matrices. *Amer. J. Math.*, 71:921–945, 1949.

[24] E. M. Opdam. Harmonic analysis for certain representations of graded Hecke algebras. *Acta Math.*, 175:75–121, 1995.

[25] T. Oshima and T. Matsuki. A description of discrete series for semisimple symmetric spaces. In *Group representations and systems of differential equations (Tokyo, 1982)*, volume 4 of *Adv. Stud. Pure Math.*, pages 331–390. North-Holland, Amsterdam, 1984.

[26] J. Rosenberg. A quick proof of Harish-Chandra's Plancherel theorem for spherical functions on a semisimple Lie group. *Proc. Amer. Math. Soc.*, 63:143–149, 1977.

[27] E. C. Titchmarsh. *Eigenfunction Expansions Associated with Second-Order Differential Equations.* Oxford, at the Clarendon Press, 1946.

[28] H. Weyl. Singuläre Integralgleichungen. *Math. Ann.*, 66:273–324, 1908.

[29] H. Weyl. Über gewöhnliche Differentialgleichungen mit Singularitäten und die zugehörigen Entwicklungen willkürlicher Funktionen. *Math. Ann.*, 68:220–269, 1910.

[30] H. Weyl. Über gewöhnliche lineare Differentialgleichungen mit singulären Stellen und ihre Eigenfunktionen. (2. Note). *Göttinger Nachrichten*, pp. 442–467, 1910.

[31] H. Weyl. Ramifications, old and new, of the eigenvalue problem. *Bull. Amer. Math. Soc.*, 56:115–139, 1950.

3

Weyl's Work on
Singular Sturm–Liouville Operators

W.N. Everitt

School of Mathematics and Statistics
University of Birmingham
w.n.everitt@bham.ac.uk

H. Kalf

Mathematisches Institut
Universität München
hubert.kalf@mathematik.uni-muenchen.de

Wie doch ein einziger Reicher so viele Bettler in Nahrung
Setzt! Wenn die Könige baun, haben die Kärrner zu tun.

(Schiller, Kant und seine Ausleger)

Abstract

Up to the year 1910 there had been many significant mathematical contributions to the theory of linear ordinary differential, and of linear integral equations. Many of these advances were based on the original studies initiated by Sturm and Liouville commencing in 1829. In the closing years of the 19th century the work lead by the Göttingen school of mathematics gave a much needed overview of these significant and varied contributions to mathematical analysis.

The contributions of Hermann Weyl, in and around the year 1910, to the theory of Sturm-Liouville theory heralded the modern analytical and spectral study of boundary value problems. In particular the paper written for Mathematischen Annalen in 1910 stands today as a landmark not only in Sturm-Liouville theory, but in the development of mathematical analysis in the 20th century.

This paper discusses the work of Weyl, and indeed of the Göttingen school of mathematics, in introducing the now familiar terms of Sturm-Liouville theory; limit-point and limit-circle endpoint classifications; the point, continuous and essential spectra; singular eigenfunction expansions; and the interplay of these results with the development of quantum theory in physics.

1 Introduction

The investigation of second-order linear ordinary differential equations has a long and fascinating history, extending back to the middle of the 18th century. It shaped the concept of a function, led to Cantor's set theory and influenced the theories of measure and integration. It was essential to solving the initial-boundary-value problems for partial differential equations, for example the heat and wave equations, by separation of the variables. In particular, the study of the expansion of a function in terms of eigenfunctions of Sturm-Liouville equations affected the theory of integral equations and the early shape of functional analysis, and was in turn influenced by these disciplines.

When Hilbert in 1906 in the 4th of the six communications of the outline of a general theory of linear integral equations, introduced the new concept of a continuous spectrum (Streckenspektrum [25, Page 122]) it was natural to try to consider expansion theorems similar to the Fourier integral theorem, in the light of this new notion. The first mathematician to undertake this task was Hilb in his Erlangen Habilitationsschrift [22], who had spent the summer semester of 1906 in Göttingen, but it was three papers by Weyl, written in the years 1908 to 1910, that clinched the issue, see [55], [56] and [57]. The most detailed of these papers is [56]^1, which was submitted in 1909 and was the basis of his Habilitation the following year, and it is this memoir and its influence with which we are primarily concerned in this present paper. In [58] Weyl illustrated his theory by means of the example of the Fourier transform on the interval $(-\infty, \infty)$. The difficulty with this seemingly simplest possible case is that both endpoints of the interval are singular and that the multiplicity of the spectrum is two. Weyl had extended the results in [56] to this situation in [57], giving as interesting examples the Bessel and hypergeometric differential equations, the Hankel transform yielding the expansion theorem in the former case. (The Fourier expansion in an $L^2(\mathbb{R})$-setting is summarised by the Plancherel theorem. Replacements of the real line by spaces with a more complicated group structure, as by Harish-Chandra, are discussed by van den Ban in this present volume, see [3].)

In 1935 Weyl returned to his Habilitationsschift [56], albeit very briefly (see the references to [56] in Sections 2 and 7 below), when he considered in [59] a problem which is connected with the difference analogue of the

---

1 A translation into English of the first three of the four chapters can be found in [37].

Sturm-Liouville equation that had previously been treated by Hellinger [20]. Weyl's Gibbs lecture of 1948 [61] is a tour d'horizon of the diverse problems he pressed his mark upon, in particular putting the results of [56] into a historical perspective.

Extensive material on Weyl, biographical and otherwise, is mentioned on the web site of the MacTutor History of Mathematics, so that there is no need to give such details here in this present paper. The most detailed source of information on Hilb is to be found in [52]. Illuminating general accounts of the historical development of spectral theory are given in [5] and [10] (see also the epilogue in [26]). For the development of Sturm-Liouville theory we refer to the introduction in [1] and more specifically to [14]. The references [11, Pages 1581-92], [12, Chapter 2] and [13] are also of interest.

2 The Weyl circle method

For continuous $q : [0, \infty) \to \mathbb{R}$ and $p : [0, \infty) \to (0, \infty)$ Weyl in [56] considers the Sturm-Liouville differential expression

$$\mathfrak{L}(u) := -(pu')' + qu, \tag{2.1}$$

emphasising that in contrast to previous contributions (including his own paper [55]) the coefficient p need not be continuously differentiable, *i.e.*, he considers functions $u \in C^1([0, \infty))$ such that pu', which would much later be called the quasi-derivative of u, is again in $C^1([0, \infty))$. For such functions u, v the Lagrange identity reads

$$\int_0^x \left(\mathfrak{L}(u)\overline{v} - u\overline{\mathfrak{L}(v)} \right) = \int_0^x \left(u\overline{pv'} - pu'\overline{v} \right)' = [u, v](x) - [u, v](0), \tag{2.2}$$

where

$$[u, v](x) = u(x)\overline{(pv')(x)} - (pu')(x)\overline{v(x)}.$$

When the coefficient $p = 1$ the term $[u, \overline{v}]$ is the Wronskian of u and v.

It is natural to impose a boundary condition at 0 of the form

$$R_1(u) := u(0)\cos(\alpha) + (pu')(0)\sin(\alpha) = 0, \tag{2.3}$$

where $\alpha \in [0, \pi)$. The question is: What kind of boundary condition, if any, does one have to impose at infinity in order to obtain a properly formulated eigenvalue problem and an associated expansion formula?

For any $\lambda \in \mathbb{C}$ the equation[2]

$$\mathfrak{L}(u) = \lambda u \tag{2.4}$$

has a fundamental system of solutions $U_1(\cdot, \lambda), U_2(\cdot, \lambda)$ satisfying the initial conditions

$$\begin{array}{ll} U_1(0, \lambda) = 1 & (pU_1')(0, \lambda) = 0 \\ U_2(0, \lambda) = 0 & (pU_2')(0, \lambda) = 1. \end{array}$$

For fixed $x \geq 0$, the solutions $U_1(x, \cdot), U_2(x, \cdot)$ are entire functions on $\mathbb{C}$. Let

$$u_1 := U_1 \cos(\alpha) + U_2 \sin(\alpha) \text{ and } u_2 := -U_1 \sin(\alpha) + U_2 \cos(\alpha). \tag{2.5}$$

Then u_2 satisfies the boundary condition (2.3). Taking $\lambda \in \mathbb{C} \setminus \mathbb{R}$ and $\beta \in [0, \pi)$, Weyl asked: For which $w \in \mathbb{C}$ does $u_1 + wu_2$ satisfy a real boundary condition at $a > 0$

$$R_2(u) := u(a)\cos(\beta) + (pu')(a)\sin(\beta) = 0 \,? \tag{2.6}$$

Now $R_2(u_2) \neq 0$, because otherwise the non-real number λ would be an eigenvalue of the symmetric problem (2.3), (2.4), (2.6). Hence

$$w = -\frac{R_2(u_1)}{R_2(u_2)} = -\frac{u_1(a, \lambda)\cos(\beta) + (pu_1')(a, \lambda)\sin(\beta)}{u_2(a, \lambda)\cos(\beta) + (pu_2')(a, \lambda)\sin(\beta)}$$

or, defining $h := \cot(\beta)$,

$$w = l(h) := -\frac{Ah + B}{Ch + D}. \tag{2.7}$$

In view of

$$AD - BC = [u_1, u_2](0) = 1, \tag{2.8}$$

the fractional linear transformation (2.7) is non-degenerate. Since its denominator never vanishes, $l(\cdot)$ maps the extended real line onto a circle, $C_a(\lambda)$, the equation of which is straightforward to determine.

On account of (2.2) and (2.4) we have

$$2i \operatorname{Im}(\lambda) \int_0^a |u_1 + wu_2|^2 = [u_1 + wu_2, u_1 + wu_2](a) - [u_1 + wu_2, u_1 + wu_2](0) \tag{2.9}$$

for all $w \in \mathbb{C}$. Now

$$[u_1, u_1](0) = 0 = [u_2, u_2](0) \tag{2.10}$$

---

[2] There is no problem to replace λ by λk as long as the weight function $k : [0, \infty) \rightarrow \mathbb{R}$ is positive. The case that k may change sign is briefly considered in [56].

and (2.8) imply

$$[u_1 + wu_2, u_1 + wu_2](0) = -2i\,\mathrm{Im}(w).$$

Since the two terms

$$u_1(a,\lambda) + wu_2(a,\lambda) \text{ and } (pu_1')(a,\lambda) + w(pu_2')(a,\lambda)$$

are linearly dependent if and only if $w \in C_a(\lambda)$, we have

$$[u_1 + wu_2, u_1 + wu_2](a) = 0 \qquad (2.11)$$

if and only if $w \in C_a(\lambda)$ and so

$$C_a(\lambda) = \left\{ w \in \mathbb{C} : \int_0^a |u_1(x,\lambda) + wu_2(x,\lambda)|^2\,dx = \frac{\mathrm{Im}(w)}{\mathrm{Im}(\lambda)} \right\}, \qquad (2.12)$$

i.e., if $\mathrm{Im}(\lambda) > 0$ ($\mathrm{Im}(\lambda) < 0$) the set of $w \in \mathbb{C}$ for which $u_1 + wu_2$ satisfies a real boundary condition at a is a circle in the upper (lower) complex half-plane $\mathbb{C}_+$ ($\mathbb{C}_-$). With $w = 0$ certainly lying in the exterior of $C_a(\lambda)$, the interior of $C_a(\lambda)$, the disc $D_a(\lambda)$, is obtained by replacing " $=$ " with " $<$ " in (2.12). Hence

$$D_{a_2}(\lambda) \subsetneq D_{a_1}(\lambda) \text{ for } 0 < a_1 < a_2 < \infty.$$

As $a \to \infty$ the nesting circles $C_a(\lambda)$ shrink either to a circle (this is called the limit-circle case, LCC) or to a point (named the limit-point case, LPC) [56, Satz 1]. In any case it follows from (2.12) that for every $\lambda \in \mathbb{C} \setminus \mathbb{R}$ equation (2.4) has at least one non-trivial solution in $L^2(0,\infty)$, *viz.*, $u_1 + wu_2$ when $w \in C_\infty(\lambda)$ [56, Satz 2]. Jumping ahead of our story, the operator-theoretic version of this result is the following. If L is the maximal operator associated with (2.1) in $H := L^2(0,\infty)$ (see equation (7.1) below), the null space of $L - \lambda$ is non-trivial for every $\lambda \in \mathbb{C} \setminus \mathbb{R}$. The operator-theoretic proof is as follows. Suppose $N(L - \lambda) = \{0\}$ for some $\lambda \in \mathbb{C} \setminus \mathbb{R}$. Then (7.2) implies

$$\{0\} = N(L_0^* - \lambda) = R(L_0 - \overline{\lambda})^\perp.$$

This result is also true with λ replaced by $\overline{\lambda}$, because the coefficients of (2.1) are real-valued. As a consequence

$$R(L_0 - \lambda) = H = R(L_0 - \overline{\lambda}),$$

i.e., L_0 and so L are self-adjoint operators in H. However, L fails to be symmetric since

$$\langle L\tilde{u}_1, \tilde{u}_2 \rangle - \langle \tilde{u}_1, L\tilde{u}_2 \rangle = -[u_1, u_2](0) = -1$$

by (2.8), where $\tilde{u}_1, \tilde{u}_2$ are smooth functions with compact support which coincide with u_1, u_2 in a right neighbourhood of 0.

In order to determine the radius $r_a(\lambda)$ of $C_a(\lambda)$ we proceed as in [48, Section 2] and observe that the centre $z_a(\lambda)$ and $\infty = l(-D/C)$ are mirror points with respect to the circle. Therefore $l^{-1}(z_a(\lambda))$ and $l^{-1}(\infty)$ are mirror points with respect to the real line, *i.e.*,

$$\overline{l^{-1}(z_a(\lambda))} = l^{-1}(\infty) = -\frac{D}{C}$$

or

$$z_a(\lambda) = l(-\overline{D}/\overline{C}) = -\frac{A\overline{D} - B\overline{C}}{C\overline{D} - \overline{C}D}.$$

Since $l(0)$ is certainly a point on $C_a(\lambda)$, we have

$$\left\{ \begin{aligned} r_a(\lambda) &= |l(0) - z_a(\lambda)| &= \left| \frac{AD - BC}{C\overline{D} - \overline{C}D} \right| \\ &&= \frac{1}{|[u_1, u_2](a)|} \\ &&= \left(2|\mathrm{Im}(\lambda)| \int_0^a |u_2(x,\lambda)|^2 \, dx \right)^{-1}, \end{aligned} \right.$$

(2.13)

using (2.8) and (2.2) together with (2.10). Both in [56] and [61] Weyl himself determined $r_a(\lambda)$ by observing that the straight lines defined by the real and imaginary parts of $R_2(u_1) + wR_2(u_2)$ are orthogonal to each other, although he does use a fractional linear transformation in [61]. Rellich in his presentation of the Weyl theory, see [41] and [27], chose to avoid properties of fractional linear transformations altogether and was followed in this respect by Hellwig [21].

An important consequence of (2.13) is that u_2 is also of integrable square in the LCC. So all solutions of (2.4) then have this property. Next Weyl showed that if all solutions of (2.4) are in $L^2(0, \infty)$ for some λ_0 then this is also true for all other values of λ [56, Satz 5]. By the time he wrote [59] he regarded his original proof via an integral equation as unnatural. His argument in [59, Page 240 ff.], when adopted to the Sturm-Liouville case, amounts to applying the variation-of-constants formula to

$$(\mathfrak{L} - \lambda_0)u = (\lambda - \lambda_0)u =: f$$

(u is a solution of (2.4)), which is the standard proof used today ([41, Page 31 ff.], [27, Page 125 ff.], [21, Page 224 ff.]). Weyl found the issue important enough to append a note to Satz 5 when [56] and [59] were reprinted in 1955 in his Selected Papers. This note also went into his

Collected Papers, but unfortunately without adjusting the page references to the new situation.

The operator-theoretic background of this result is that the defect number

$$d_\lambda := \dim\{R(L_0 - \lambda)^\perp\}$$

of a symmetric operator in a Hilbert space is a constant in $\mathbb{C}_+$ and in $\mathbb{C}_-$ (in addition $d_{\overline{\lambda}} = d_\lambda$, since the coefficients of (2.1) are real-valued). Here we have the slightly more detailed information that $d_\lambda = 2$ for all $\lambda \in \mathbb{C}$ in the LCC. The LPC is of course also independent of $\lambda \in \mathbb{C}$ and in this case we have $d_\lambda = 1$ if $\lambda \in \mathbb{C} \setminus \mathbb{R}$.

3 Boundary conditions at infinity

Let $w \in C_\infty(\lambda)$ and $x \geq 0$. Abbreviate

$$u_w(x) := u_1(x, \lambda) + w u_2(x, \lambda), \tag{3.1}$$

and then define a Green's function by

$$G_w(x, y; \lambda) := \begin{cases} u_2(x, \lambda) u_w(y) & \text{if } 0 \leq x \leq y < \infty \\ u_2(y, \lambda) u_w(x) & \text{if } 0 \leq y < x < \infty. \end{cases} \tag{3.2}$$

For continuous $f \in L^2(0, \infty)$ let

$$v(x) := \int_0^\infty G_w(x, y; \lambda) f(y) \, dy = u_w(x) g(x) + u_2(x, \lambda) h_w(x),$$

where

$$g(x) := \int_0^x u_2(y, \lambda) f(y) \, dy \text{ and } h_w(x) := \int_x^\infty u_w(y) f(y) \, dy.$$

Let $\{a_i\}$ be a sequence of real numbers which tends to infinity and let $w \in C_\infty(\lambda)$. Take a sequence $\{w_i\}$ with $w_i \in C_{a_i}(\lambda)$ which converges to w in $\mathbb{C}$. Using (2.11) as well as (2.2) and (2.10), we see that

$$[v, u_{w_i}](a_i) = g(a_i)[u_{w_i}, u_{w_i}](a_i) + h_{w_i}(a_i)[u_2, u_{w_i}](a_i)$$
$$= h_{w_i}(a_i) 2i \operatorname{Im}(\lambda) \int_0^{a_i} u_2 \overline{u_{w_i}}.$$

In the LCC this last expression tends to zero as $i \to \infty$. From this result Weyl concludes in the LCC a one-parameter family of boundary conditions has to be imposed on the functions in

$$\vartheta := \left\{ u \in C^1([0, \infty)) : pu' \in C^1([0, \infty) \text{ and } u, \mathfrak{L}(u) \in L^2(0, \infty) \right\} \tag{3.3}$$

in addition to (2.3), viz.,

$$\lim_{a \to \infty} [u, u_1 + w u_2](a) = 0, \tag{3.4}$$

with w being any point on the circle $C_\infty(\lambda)$. In the LPC the Green's function (3.2) is completely determined by $w \in C_\infty(\lambda)$ and no boundary condition, in addition to (2.3), has to be imposed on the functions in ϑ given by (3.3); the choice of $\lambda \in \mathbb{C} \setminus \mathbb{R}$ is irrelevant. Using von Neumann's theory of defect numbers it can in fact be shown that in the LPC condition (3.4) is automatically satisfied for every $u \in \vartheta$ (see, e.g., [53, Satz 13.19a]).

Weyl then shows that the LPC occurs when

$$q(x) \geq -c \ (x \geq 0)$$

for some $c \geq 0$. Indeed, choose $\lambda := -(c + 1)$ and consider the function $u := U_1(\cdot, \lambda)$; then u is positive in a neighbourhood to the right of 0. At a first zero $x_0 > 0$ (zeros of non-trivial solutions of (2.4) are isolated) we would have $u'(x_0) < 0$, which is incompatible with

$$(pu')(x) = \int_0^x (pu')' = \int_0^x (q - \lambda)u \geq \int_0^x u \ (x \in [0, x_0]).$$

Hence u is strictly increasing and so not in $L^2(0, \infty)$. Weyl himself relies on a representation of u to yield this result, but arguments of the type just given are familiar in oscillation theory and are employed by him to prove the results mentioned at the beginning of Section 6 below.

The fact that there is no growth restriction on p is particularly striking in the light of the multidimensional analogue of (2.1),

$$\mathfrak{L}(u) := - \sum_{i,k=1}^{n} \partial_i (a_{ik} \partial_k) + q,$$

where the a_{ik} are the continuously differentiable entries of a strictly positive matrix function A. In general, $\mathfrak{L}$ does not have a unique self-adjoint realisation in $L^2(\mathbb{R}^n)$, even if q is absent, unless there is some growth restriction on the largest eigenvalue of A (see [8] and the literature cited therein).

4 The spectrum in the limit-circle case

In the LCC it follows from (3.2) that

$$\int_0^\infty \left(\int_0^\infty |G_w(x, y; \lambda)|^2 \, dy \right) dx < \infty$$

for all $w \in C_\infty(\lambda)$. As a consequence the resolvent of every operator S with domain (3.3) and boundary conditions (2.3) and (3.4), is a compact operator (completely continuous [vollstetig] in the terminology of the time; in fact it belongs to a very special class of compact operators, the Hilbert-Schmidt operators), so that Hilbert's results for such operators are applicable. The spectrum of S is purely discrete; the eigenvalues - they are all simple in our case - have no finite accumulation point. The point $+\infty$ is always an accumulation point, but $-\infty$ may be an accumulation point as well (this is not explicitly mentioned in [56], though). The corresponding eigenfunctions $\{\varphi_i\}$ form a complete orthonormal system in the sense that for $f \in D(S)$ we have

$$f = \sum_i \langle f, \varphi_i \rangle \varphi_i$$

where $\sum_i |\langle f, \varphi_i \rangle \varphi_i|$ is uniformly convergent on every compact subset of $[0, \infty)$ [56, Satz 4]. So the LCC is very similar to the regular case when $p > 0$ and q are continuous on a compact interval (in which case there are, however, at most finitely many negative eigenvalues). We add to these results that there is norm convergence if $f \in L^2(0, \infty)$.

5 Invariance of the essential spectrum

Using what is now called Weyl sequences or singular sequences, Weyl had shown in [54] that the union of the continuous spectrum and the accumulation points of the point spectrum[3] is invariant under compact perturbations. Assuming that there is LPC at ∞ (otherwise the essential spectrum is empty, as we have just seen), Weyl remarks that the difference of the two resolvents corresponding to a boundary condition of the form (2.3) with $\alpha \in [0, \pi)$ and $\gamma \in [0, \pi)$, $\gamma \neq \alpha$, is a compact operator and so the essential spectrum is independent of the boundary condition [56, Satz 8]. The simple proof is as follows. We give the functions in (2.5) and (3.1) the indices α and γ, denoting by w_α, w_γ the corresponding limit points. Since we are in the LPC, there is a number $C \neq 0$ such that

$$u_{w_\alpha}(x) = C u_{w_\gamma}(x) \ (x \geq 0).$$

It suffices to find $c_1, c_2 \in \mathbb{C}$ with

$$c_1 u_{2\alpha} - c_2 u_{2\gamma} = u_{w_\alpha}, \tag{5.1}$$

---

3 The name "essential spectrum" for this union is due to Wintner [62].

because then

$$c_1 C^{-1} u_{2\alpha}(x, \lambda) u_{w_\alpha}(y) - c_2 u_{2\gamma}(x, \lambda) u_{w_\gamma}(y) = u_{w_\alpha}(x) u_{w_\gamma}(y) \quad (x, y \geq 0).$$

We note that (5.1) is equivalent to

$$c_2 \sin(\gamma) - c_1 \sin(\alpha) = \cos(\alpha) - w_\alpha \sin(\alpha)$$
$$-c_2 \cos(\gamma) + c_1 \cos(\alpha) = \sin(\alpha) + w_\alpha \cos(\alpha),$$

which has a non-zero determinant. In operator-theoretic language Weyl's result can be rephrased as follows: Varying the boundary conditions of a Sturm-Liouville operator is a very special compact perturbation, *viz.*, a rank-one perturbation (see, *e.g.*, [53, Satz 10.17].

Weyl emphasises that he has shown in [54] that the continuous spectrum of a bounded (self-adjoint) operator is in general not invariant under compact perturbations and says, "At first glance it is plausible to conjecture that the continuous spectrum itself remains unaltered when the boundary condition is changed, but I am unable to confirm this". In 1957 Aronszajn, using the inverse spectral theory of Gel'fand-Levitan, produced a counter-example [2].

A more explicit and particularly amazing example is

$$p(x) = 1 \text{ and } q(x) = \cos\left(\sqrt{x}\right) \quad (x \geq 0)$$

for which the essential spectrum is $[-1, \infty)$. The absolutely continuous spectrum is $(1, \infty)$, and this part of the spectrum is independent of the boundary condition by virtue of a result of Aronszajn in [2]. For almost all $\alpha \in [0, \pi)$ there is a dense point spectrum in $[-1, 1]$, but for a dense set of angles in $[0, \alpha)$ the spectrum is singular continuous in $[-1, 1]$. Further explanations and references can be found in [9].

6 The spectrum in the limit-point case

Weyl shows that in the LPC there is an expansion formula which consists of two terms, a Fourier series which takes care of the (finite or infinite, possibly empty) set of eigenvalues and a Riemann-Stieltjes integral the integrand of which involves the function u_2 (which satisfies the boundary condition (2.3)) while the integrator, responsible for the continuous spectrum, is built up from solutions of (2.4) in a complicated way. The proof of this result [56, Satz 7], which Weyl regarded as the principal aim of his paper, is a veritable tour de force, using in particular, from his 1908 dissertation, his extension of Hilbert's theory to integral kernels on unbounded intervals, and from Hellinger's 1907 dissertation the concept of "eigendifferentials". A certain drawback of this form of the expansion theorem is mentioned in Section 8 below. However, Weyl's idea to relate the asymptotic behaviour of the solutions of (2.4) to the continuous part of the spectral function led him to significant results in situations similar to (6.1), (6.2) and (6.4) below.

Extending Sturm's classical results, Weyl proves that the spectrum is purely discrete if

$$\lim_{x \to \infty} q(x) = \infty.$$

More generally, if $c := \liminf_{x \to \infty} q(x)$ is finite, then the spectrum is purely discrete below c, the number of eigenvalues (they are all simple) being equal to the number of zeros in $(0, \infty)$ of the solution (unique up to a factor) of (2.3) and (2.4), with $\lambda = c$. Arranging the eigenvalues in a natural way, the eigenfunction belonging to the n-th eigenvalue has exactly $n - 1$ zeros in $(0, \infty)$ [56, Satz 9].

To describe his results concerning the continuous spectrum we assume for simplicity that $p = 1$ on $[0, \infty)$. Suppose

$$\int_0^\infty |q(x)| \, dx < \infty \tag{6.1}$$

or

$$\int_0^\infty x \, |q(x)| \, dx < \infty. \tag{6.2}$$

Weyl proves that there are at most finitely many negative eigenvalues when (6.2) holds. For $\lambda > 0$ he shows under condition (6.1) that the solutions of (2.4) behave asymptotically like $\sin(x\sqrt{\lambda})$ and $\cos(x\sqrt{\lambda})$ for large x and fixed λ and, also assuming (6.2), he is able to write the spectral function as

$$\rho(\lambda) = \frac{1}{\pi} \int_0^\lambda \frac{1}{\sqrt{s}[m_1^2(s) + m_2^2(s)]} \, ds \quad (\lambda > 0), \tag{6.3}$$

where the functions m_1 and m_2 are related to the asymptotic behaviour of the solution u_2. It was half a century later that Weyl's conditions (6.1) and (6.2) and his technique became prominent in the work of L.D. Fadeev on inverse scattering (see [33]).

In the case

$$p(x) = 1 \text{ and } q(x) = -x \quad (x \geq 0) \tag{6.4}$$

the solutions are Bessel functions of order $\pm 1/3$. From their known asymptotic behaviour Weyl derives the result that the spectral function is continuously differentiable on the whole real line.

7 Interplay with quantum mechanics

In his obituary for Hilbert, Weyl mused on the timeliness of discoveries and said [60, Page 645], "Most scientific discoveries are made when 'their time is fulfilled'; sometimes, but seldom, a genius lifts the veil decades earlier than could have been expected". Precisely this had happened with Weyl's papers [55], [56] and [57]; there was a full-fledged spectral analysis of one-dimensional or spherically symmetric quantum mechanical systems about 16 years before wave mechanics was discovered by Schrödinger.

This situation was acknowledged by Fues, Schrödinger's assistant in Zurich at the time, when he wrote [17, Page 295], "Prof. H. Weyl kindly provided me with papers of his own from which everything emerges that is necessary from the point of view of wave mechanics". It is well known that Schrödinger himself contacted his colleague in connection with his equation. There is an acknowledgement at the beginning of the first of his four communications "Quantisation as an eigenvalue problem", and in the fourth, referring to the difficulties the presence of the continuous spectrum causes, he writes, "The theory of such integral representations was developed by H. Weyl, however for ordinary differential equations only, but a generalisation to partial differential equations should probably be permitted." [46, Page 124]. In a footnote he adds, "I am indebted to Mr H. Weyl not only for these references to the literature[4], but also for his very valuable oral instructions concerning these not very easy things.".

These things were and are indeed not very easy and Kemble in his carefully written early book on quantum mechanics states that [56] is "a basic paper which unfortunately involves an elaborate mathematical technique and makes difficult reading for the non-specialist", [30, Page 163]. Since it suggests itself to regard the Fourier integral as a limiting case of a Fourier series, he outlined heuristically in four starred sections [30, Pages 165 to 173] a transition from the regular case to a general expansion theorem. Mathematically, this approximation by regular problems was realised around 1950 independently by Levitan,

---

4 Schrödinger mentions [55], [56] and two papers by Hilb - not the ones in our list of references.

Levinson and Yosida. The simplest account is probably that given in the book by Coddington and Levinson [7, Chapter 9].

The development of quantum mechanics had enormous repercussions on mathematics. In analysis, notably through the work of John von Neumann and Stone, it led to a theory of unbounded operators in an abstract Hilbert space, culminating in von Neumann's spectral theorem for self-adjoint operators. Weyl himself turned to the group-theoretical problems the new physical theory posed. This story has frequently been told, see [5] and [10]. Here we restrict ourselves to more technical details, in connection with Weyl's paper [56].

The abstract framework demands or is facilitated by the use of closed operators in a complete space.. This was why Stone, under now the common assumptions (I is an open interval)

$$p, q : I \to \mathbb{R} \text{ with } p^{-1}, q \in L^1_{\text{loc}}(I),$$

associated with (2.1) a maximal operator L in the Hilbert space $H :=$ $L^2(I)$ with domain

$$D(L) := \{u \in AC_{\text{loc}}(I) : pu' \in AC_{\text{loc}}(I) \text{ and } u, \mathfrak{L}(u) \in H\}, \qquad (7.1)$$

relaxing the requirement on a function to be continuously differentiable as in (3.3), to that of being locally absolutely continuous. In addition it is convenient to consider the smaller space

$$\vartheta_0 := \{u \in D(L) : \text{supp}(u) \subset I \text{ is compact}\}.$$

L_0, the closure of $\mathfrak{L} \upharpoonright \vartheta_0$, is called the minimal operator associated with (2.1), and Stone shows that

$$L_0^* = L \qquad (7.2)$$

[47, Pages 459 onwards]. Sometimes, as in questions of oscillation properties, it is necessary to demand $p > 0$ almost everywhere on I, but otherwise the analysis given in Sections 2 to 6 above is in no essential way affected by the new generalisation[5]. Stone shows that the self-adjoint realisations of (2.1) with separated boundary conditions are exactly those where $u \in D(L)$ is restricted by (3.4) ([47, Page 457 onwards]; more precisely he does this for slightly different spaces which are in one-to-one correspondence with (3.4) and which involve the function u_2 only). A

---

[5] Weyl wrote in [59, 649], "I think Hilbert was wise to keep within the bounds of continuous functions when there was no actual need for introducing Lebesgue's concepts."

direct proof of this result was later given by Rellich [41] and simplified by A. Schneider [45] (see also [21, Pages 239-243]).

The boundary conditions (3.4) (or those of Stone), while satisfactory from a theoretical point of view, are often not easy to apply to specific problems. In case p and q are such that the singular point of (2.4), here at ∞, is a regular singular point in the sense of complex analysis, Rellich [40] was able to replace (3.4) by a boundary condition that looks exactly like a regular boundary condition, but where $u(\infty)$ and $(pu')(\infty)$ (which in general do not exist) are replaced by "initial numbers" (see also [21, Chapter 15]). A modification of the Weyl circle method and of (3.4), which is inspired by the idea that problems which arise entirely within the realm of the real line should also be described by boundary conditions which involve real-valued functions only, is due to Mohr [34] (see also [29]). He illustrated the efficiency of his method in a number of papers; we only mention [35] here where he supplements (6.1) and (6.2) by conditions on p and the weight k which are more natural than those imposed by Weyl in [56] and [57].

Since the minimal operator has a one-parameter family of self-adjoint extensions in the LCC, it is natural to ask whether there is one extension which is distinguished from a mathematical or physical point of view. The first to raise and answer this question was Friedrichs [15] and [16]. He once said that all he used mathematically he obtained from Weyl - "except for Hilbert space and that I got from von Neumann" [39]. Friedrichs showed that such a distinguished extension exists when the minimal operator is bounded from below, but it took some time before a description of his extension was given by Rellich [42] in terms of a particular fundamental system of solutions of (2.4). In respect of later work on the Friedrichs extension generated by Sturm-Liouville differential expressions, see the papers of Rosenberger [44], and of Niessen and Zettl [37].

8 The expansion theorem and the m-function

Weyl's derivation of the expansion theorem was criticised by Hilb as "rather complicated" [23, Page 334] and his expansion formula as "not very transparent" [24, Page 1265]. He writes in italics [23, Page 334], "We want to show how easy the desired expansions can be obtained without this theory [Weyl's use of singular integral equations] directly from Cauchy's residue theorem". To do this, he needs to know that the Green's function is analytic on $\mathbb{C}_+$ in the spectral parameter. Let

$\lambda_0, \lambda \in \mathbb{C}_+$ and $w_0, w = m(\lambda)$ the corresponding Weyl limit points (we assume that there is LPC at ∞). Hilb then observes that Hilbert's resolvent identity [25, Page 140]

$$G_{w_0}(x, y; \lambda_0)$$
$$= G_w(x, y; \lambda) - (\lambda - \lambda_0) \int_0^\infty G_{w_0}(x, z; \lambda_0) G_w(z, y; \lambda) \, dz \ (x, y \geq 0)$$

can be viewed as an integral equation for G_w. This integral equation can be solved by means of a Neumann series, which is a power series in $\lambda - \lambda_0$. (The radius of convergence can be shown to be $\text{Im}(\lambda_0)$.) It follows in particular that w as a function of λ - which is what in the literature is called the Titchmarsh-Weyl, Weyl-Titchmarsh or even Weyl m-function - is analytic on $\mathbb{C}_+$ [23, Page 335]. (While Hellinger [20] applied the Vitali theorem to establish the analyticity of the m-coefficient for the difference analogue of (2.4), Weyl used Hilb's argument in [59, Page 244][6]; it occurs again in [61, Page 120].) Hilb then derived a formula which A. Kneser called the Fourier-Hilb integral representation, and which he evaluated in the Fourier and Bessel cases [31, Sections 51 and 52]. However, Weyl later remarked [61, Page 124] that Hilb did not carry the analysis "so far as to obtain the explicit construction of the differential $d\rho$ [the spectral measure]".

However, once it is known that the limit point $w = m(\lambda)$ is an analytic function, connection with the spectral function ρ can be made as follows. Performing the limit $a \to \infty$ in (2.12), it follows that

$$\text{Im}(m(\lambda)) > 0 \text{ if } \text{Im}(\lambda) > 0.$$

Such analytic functions (they are called "positive" by Weyl [59, Page 231], but today the names of Herglotz or Nevanlinna are more frequently attached to them) can be represented in the form

$$m(\lambda) = c_1 \lambda + c_2 + \int_{-\infty}^{\infty} \left(\frac{1}{s - \lambda} - \frac{s}{s^2 + 1} \right) d\rho(s) \qquad (8.1)$$

where $\rho : \mathbb{R} \to \mathbb{R}$ is non-decreasing, and satisfies the growth condition

$$\int_{-\infty}^{\infty} \frac{1}{1 + s^2} \, d\rho(s) < \infty;$$

then in (8.1)

$$c_1 := \lim_{t \to \infty} \frac{\text{Im}(m(it))}{t} \text{ and } c_2 := \text{Re}(m(i)).$$

---

6 We note in passing that this is Kodaira's reference for the analyticity of m [32, Theorem 1.2].

(The representation (8.1) is unique if ρ is required, say, to be continuous from the right and normalised by $\rho(0) = 0$.)

The study of analytic functions f on the unit disc which have a positive real part was initiated by Carathéodory in 1907. F. Riesz showed in 1911 that such functions can be represented (essentially uniquely) in the form

$$f(z) = \text{Im}(f(0)) + \int_0^{2\pi} \frac{\exp(it) + z}{\exp(it) - z} \, d\omega(t) \quad (|z| < 1) \qquad (8.2)$$

where ω is a non-decreasing function. In the same year a simplified proof was given by Herglotz, and so it is widely believed that the representation (8.2) originates with him (see, *e.g.*, [47, Page 571]). The representation (8.1) follows from (8.2) after a conformal mapping. Now the Stieltjes inversion formula applied to (8.1) yields

$$\rho(\lambda) = \lim_{\varepsilon \searrow 0} \frac{1}{\pi} \int_0^\lambda \text{Im}(m(s + i\varepsilon)) \, ds \quad (\lambda \in \mathbb{R}), \qquad (8.3)$$

which is the desired connection.

Formula (8.3) was first derived by Titchmarsh in the 5th of his series of eight papers on expansions in eigenfunctions ([49, (4.3)]; that his function k is indeed up to a factor π the spectral function ρ follows from [49, (5.1) to (5.3)]. He did not seem to be aware of formula (8.1) but he was certainly aware of the representation formula (8.2), because he cited Chapter VII, Section 2, of Nevanlinna's book [36], where (8.2) is proved and called the Poisson-Stieltjes representation. Titchmarsh, however, does not make the transformation to $\mathbb{C}_+$; rather he prefers to give a separate proof of (8.3), taking the Cauchy integral for the resolvent as his starting point[7]. In [48, Page 40] he had established the analyticity of the m-coefficient without knowing of Hilb's paper [23]. It is only the 2nd edition of his book [51, Page 188] which has a reference to [23]. In [50, Section 6] he proves that the spectral function is continuously differentiable on $(0, \infty)$ under the condition (6.1), showing that the integrand in (6.3) is $\text{Im}(m(s))$ (see also [51, Page 118]).

Formula (8.3) was derived independently, though slightly later, by Kodaira [32] in a way that Weyl found to be more direct, [61, Page 124] and [62, Page 168]. Stone had shown that the projection operators which form the spectral resolution of a general self-adjoint operator are, in our

---

7 Mary Cartwright concludes her obituary [6] of Titchmarsh in the following way. "It seems that his antipathy to geometry prevented him from using certain methods which would have led to the kind of simplification at which he aimed. His own simplifications paved the way for others to achieve further improvements, but it may be that, after all, it will be for his results that he will be remembered."

case, integral operators the kernels of which allow a representation as Riemann-Stieltjes integrals. Stone was completely aware of the fact that the link between the spectral function and the solutions of (2.4) was still missing [47, Page 530], and this link Kodaira provided by re-considering Weyl's proof of the expansion theorem in the light of Stone's analysis [32]. (Some care is needed when simplifying the expansion theorem in the case when two singular endpoints are present but the spectrum is nevertheless simple [28, Page 206].) A particularly lucid proof of the expansion theorem in Kodaira's spirit was recently given in Weidmann's book [53, Chapter 14.1]. A recent proof which is more in the spirit of Titchmarsh can be found in [4], in which the expansion theorem is proved for any interval I, any endpoint classifications and any separated boundary conditions, except LPC at both endpoints.

In 1987 Daphne Gilbert and D.B. Pearson identified the set of $\lambda \in \mathbb{R}$ where $\mathrm{Im}(m(\lambda))$ has a finite positive value as the minimal support of the absolutely continuous part of the measure generated by the spectral function ρ [19]. Using this, they proved the remarkable result that the spectrum of every self-adjoint realisation of (2.1) is absolutely continuous in an interval I if (2.4) has no subordinate solution for all $\lambda \in I$. A nontrivial solution u of (2.4) is called subordinate if there is a solution v of (2.4) such that

$$\lim_{a \to \infty} \left(\int_0^a |u|^2 \right) \left(\int_0^a |v|^2 \right)^{-1} = 0.$$

As a consequence, the absolutely continuous spectrum can be determined by finding those values of $\lambda \in \mathbb{R}$ for which (2.4) has two linearly independent solutions of "comparable size". A simplified version of the Gilbert-Pearson theory can be found in Weidmann's book [53, Chapter 14.5].

9 Aftermath

The years after 1950 saw a veritable surge of papers on specific Sturm-Liouville operators and even more specific problems concerning them (the bibliography in Zettl's book [64], far from claiming any completeness, lists 648 items), making Weyl's memoir [56] one of the most frequently cited, but, one may venture to guess, not one of the most frequently read papers in spectral analysis.

10 Acknowledgements

The second author thanks H.O. Cordes, Berkeley, for a number of discussions concerning the work of Hilb.

Bibliography

[1] W.O. Amrein, A.M. Hinz and D.B. Pearson. (Editors) *Sturm-Liouville Theory; Past and Present.* (Birkhäuser Verlag, Basel: 2005.)

[2] N. Aronszajn. On a problem of Weyl in the theory of singular Sturm-Liouville equations. *Amer. J. Math.* **79** (1957), 597-610.

[3] E.P. van den Ban. Weyl, eigenfunction expansions and analysis on noncompact symmetric spaces. (See contents of this present volume.)

[4] C. Bennewitz and W.N. Everitt. The Titchmarsh-Weyl eigenfunction expansion theorem for Sturm-Liouville differential equations. *Sturm-Liouville Theory; Past and Present*: Pages 137-171. (Birkhäuser Verlag, Basel: 2005; edited by W.O. Amrein, A.M. Hinz and D.B. Pearson.)

[5] G. Birkhoff and E. Kreyszig. The establishment of functional analysis. *Historia Math.* **11** (1984), 258-321.

[6] M.L. Cartwright. Edward Charles Titchmarsh. *J. London Math. Soc.* **39** (1964), 544-565.

[7] E.A. Coddington and N. Levinson. *Theory of ordinary differential equations.* (McGraw-Hill, New York: 1955.)

[8] E.B. Davies. L^1-properties of second-order elliptic operators. *Bull. London Math. Soc.* **17** (1985), 417-436.

[9] R. del Río. Boundary conditions and spectra of Sturm-Liouville operators. *Sturm-Liouville Theory; Past and Present*: Pages 217-235. (Birkhäuser Verlag, Basel: 2005; edited by W.O. Amrein, A.M. Hinz and D.B. Pearson.)

[10] J. Dieudonné. *History of functional analysis.* (North-Holland Mathematical Studies, Amsterdam: 1981.)

[11] N. Dunford and J.T. Schwartz. *Linear operators*: **II**. (Interscience, New York: 1963.)

[12] M.S.P. Eastham and H. Kalf. *Schrödinger-type operators with continuous spectra.* (Research Notes in Mathematics **65**; Pitman, London: 1982.)

[13] W.N. Everitt. A personal history of the m-coefficient. *J. Comput. Appl. Math.* **171** (2004), 185-197.

[14] W.N. Everitt. Charles Sturm and the development of Sturm-Liouville theory in the years 1900 to 1950. *Sturm-Liouville Theory; Past and Present*: Pages 45-74. (Birkhäuser Verlag, Basel: 2005; edited by W.O. Amrein, A.M. Hinz and D.B. Pearson.)

[15] K. Friedrichs. Spektraltheorie halbbeschränkter Operatoren und Anwendung auf die Spektralzerlegung von Differentialoperatoren: **I, II**. *Math. Ann.* **109** (1933/34), 465-487, 685-713. (Berichtigung: *Math. Ann.* **110** (1934/35), 777-779.)

[16] K. Friedrichs. Über die ausgezeichnete Randbedingung in der Spektraltheorie der halbbeschränkten gewöhnlichen Differentialoperatoren zweiter Ordnung. *Math. Ann.* **112** (1935/36), 1-23.

[17] E. Fues. Zur Intensität der Bandenlinien und des Affinitätsspektrums zweiatomiger Moleküle. *Ann. Physik* (4) **81** (1926), 281-313.

[18] C. Fulton. Parametrizations of Titchmarsh's $m(\lambda)$-functions in the limit circle case. (Dissertation; RWTH Aachen, Germany:1973) *Trans. Amer. Math. Soc.* **229** (1977), 51–63.

[19] D.J. Gilbert and D.B. Pearson. On subordinacy and analysis of the spectrum of one-dimensional Schrödinger operators. *J. Math. Anal. Appl.* **128** (1987), 30-56.

[20] E. Hellinger. Zur Stieltjesschen Kettenbruchtheorie. *Math. Ann.* **86** (1922), 18-29.

[21] G. Hellwig. *Differential operators of mathematical physics; An introduction.* (Translated from the German by Birgitta Hellwig. Addison-Wesley Publishing Co., Reading, Mass.-London-Don Mills, Ont. 1967.)

[22] E. Hilb. Über Integraldarstellungen willkürlicher Funktionen. *Math. Ann.* **66** (1909), 1-66.

[23] E. Hilb. Über gewöhnliche Differentialgleichungen mit Singularitäten und die dazugehörigen Entwicklungen willkürlicher Funktionen. *Math. Ann.* **76** (1915), 333-339.

[24] E. Hilb and O. Szász. Allgemeine Reihenentwicklungen. *Enc. Math. Wiss.* IIC11, 1229-1276, 1922. (Teubner, Leipzig: 1923-27.)

[25] D. Hilbert. *Grundzüge einer allgemeinen Theorie der linearen Integralgleichungen.* (Teubner, Leipzig: 1912.)

[26] D. Hilbert und E. Schmidt. *Integralgleichungen und Gleichungen mit unendlich vielen Unbekannten.* Herausgegeben und mit einem Nachwort versehen von A. Pietsch. (Teubner Archiv zur Mathematik: 11, Leipzig: 1989.)

[27] K. Jörgens und F. Rellich. *Eigenwerttheorie gewönlicher Differentialgleichungen.*
Bearbeitet von J. Weidmann. (Springer, Berlin:1976.)

[28] I.S. Kac. The existence of spectral functions of generalized second-order differential systems with a boundary condition at the singular end. *Amer. Math. Soc. Transl.* (2) **62** (1967), 204-262.

[29] H. Kalf. Ernst Mohrs Version der Weylschen Theorie der Sturm-Liouville-Operatoren. *Sitz.-Ber. Berliner Math. Ges.* 221-234. (Jahrgänge 1988-1992.)

[30] E.C. Kemble. *The fundamental principles of quantum mechanics.* (McGraw-Hill, New York: 1937. Dover reprint: 1958.)

[31] A. Kneser. *Die Integralgleichungen und ihre Anwendungen in der mathematischen Physik.* (Vieweg, Braunschweig: 1922. 2 Aufl.)

[32] K. Kodaira. The eigenvalue problem for differential equations of the second order and Heisenberg's theory of S-matrices. *Amer. J. Math.* **71** (1949), 921-945.

[33] V.A. Marchenko. *Sturm-Liouville operators and applications.* Operator theory: Advances and applications **22**. (Birkhäuser, Basel:1986.)

[34] E. Mohr. Eine Bemerkung zur Weylschen Theorie vom Grenzkreis- und Grenzpunktfall. *Ann. Mat. Pura. Appl.* (4) **129** (1981), 161-199.

[35] E. Mohr. Ein Beitrag zur Weylschen Theorie vom Grenzpunktfall. *Ann. Mat. Pura. Appl.* (4) **132** (1982), 331-352.

[36] R. Nevanlinna. Eindeutige analytische Funktionen. *Grundlehren der Math. Wiss.* **46** (Springer, Berlin: 1936. 2. Aufl. 1953.)

[37] H.-D. Niessen and A. Zettl. Singular Sturm-Liouville problems: the Friedrichs extension and comparison of eigenvalues. *Proc. London Math. Soc. (3)* **64** (1992), 545-578.

82   *W.N. Everitt and H. Kalf*

[38] D. Race. Limit-point and limit-circle: 1910-1970. (M.Sc. thesis, University of Dundee, Scotland, UK: 1976.)

[39] C. Reid. The life of Kurt Otto Friedrichs in *Kurt Otto Friedrichs, Selecta* I, 11-22. (C.S. Morawetz, Editor. Birkhäuser, Boston: 1986.) Note: II contains references [15] and [16] given above.

[40] F. Rellich. Die zulässigen Randbedingungen bei den singulären Eigenwertproblemen der mathematischen Physik. (Gewöhnliche Differentialgleichungen zweiter Ordnung.) *Math. Z.* **49** (1943/44), 702-723.

[41] F. Rellich. Spectral theory of second-order ordinary differential equations. (Lectures delivered 1950-1951; New York University: 1953.)

[42] F. Rellich. Halbbeschränkte gewöhnliche Differentialoperatoren zweiter Ordnung. *Math. Ann.* **122** (1950/51), 343 -368.

[43] F. Rellich. Eigenwerttheorie partieller Differentialgleichungen. (Vorlesung gehalten im Wintersemester 1952/53 an der Universität Göttingen.)

[44] R. Rosenberger. A new characterization of the Friedrichs extension of semibounded Sturm-Liouville operators. *J. London Math. Soc.* (2) **31** (1985), 501-510.

[45] A. Schneider. Eine Bemerkung zum Weyl-Stoneschen Eigenwertproblem. *Arch. Math.* (Basel) **17** (1966), 352-358.

[46] E. Schrödinger. Quantisierung als Eigenwertproblem. (Vierte Mitteilung) *Ann. Physik* (4) **81** (1926), 109-139.

[47] M.S. Stone. *Linear transformations in Hilbert space and their applications to analysis.* Amer. Math. Soc. Colloquium Pub. **XV**. (New York: 1932.)

[48] E.C. Titchmarsh. On expansions in eigenfunctions (IV). *Quart. J. Math. Oxford.* **12** (1941), 33-50.

[49] E.C. Titchmarsh. On expansions in eigenfunctions (V). *Quart. J. Math. Oxford.* **12** (1941), 89-107.

[50] E.C. Titchmarsh. On expansions in eigenfunctions (VI). *Quart. J. Math. Oxford.* **12** (1941), 154-166.

[51] E.C. Titchmarsh. Eigenfunction expansions associated with second-order differential equations. I (Oxford University Press: 2nd edition:1962; 1st edition: 1946.)

[52] H-J. Vollrath. Emil Hilb (1882-1929). In P. Baumgart (Hrsg.), *Lebensbilder bedeutender Würzburger Professoren:* 320-338. Degener, Neustadt/Aisch 1995.

[53] J. Weidmann. *Lineare Operatoren in Hilberträumen.* Teil **I**: Grundlagen. Teil **II**: Anwendungen. (Teubner, Stuttgart: 2000: 2003.)

[54] H. Weyl. Über beschränkte quadratische Formen, deren Differenz vollstetig ist. *Rend. Circ. Mat. Palermo* **27** (1909), 373-392.

[55] H. Weyl. Über gewöhnliche lineare Differentialgleichungen mit singulären Stellen und ihre Eigenfunktionen. *Nachr. Kgl. Ges. Wiss. Göttingen Math-Phys. Kl.* (1909), 37-63.

[56] H. Weyl. Über gewöhnliche Differentialgleichungen mit Singularitäten und die zugehörigen Entwicklungen willkürlicher Funktionen. *Math. Ann.* **68** (1910), 220-269.

[57] H. Weyl. Über gewöhnliche lineare Differentialgleichungen mit singulären Stellen und ihre Eigenfunktionen. (2. Note) *Nachr. Kgl. Ges Wiss. Göttingen Math-Phys. Kl.* (1910), 442-467.

[58] H. Weyl. Zwei Bemerkungen über des Fouriersche Integraltheorem. *Jahresber. Deutsch. Math.-Verein.* **20** (1911), 129-141. (Berichtigung *ibid.* **20** (1911), 339.)

[59] H. Weyl. Über das Pick-Nevanlinnasche Interpolationsproblem und sein infinitesimales Analogon. *Ann. of Math.* (2) **36** (1935), 230-254.

[60] H. Weyl. David Hilbert and his mathematical work. *Bull. Amer. Math. Soc.* **50** (1944), 612-654.

[61] H. Weyl. Ramifications, old and new, of the eigenvalue problem. *Bull. Amer. Math. Soc.* **56** (1950), 115-139.

[62] H. Weyl. Address of the President of the Fields medal committee 1954. In: *Proc. Inter. Congress of Mathematicians*: **I**, (1954), 161-174. (Noordhoff, Groningen, North-Holland, Amsterdam: 1957.)

[63] A. Wintner. On the location of continuous spectra. *Amer. J. Math.* **70** (1948), 22-30.

[64] A. Zettl. *Sturm-Liouville theory.* Mathematical surveys and monographs, **121**. (Amer. Math. Soc.: 2005)

11 Remarks on the references

Weyl's papers [54] to [58] can be found in Bd. **1**, the paper [59] in Bd. **3**, and papers [60] to [62] in Bd. **4** of his "Gesammelte Abhandlungen"; (Springer, Berlin: 1968). The papers [56] and [59] were also included in "Selecta Hermann Weyl"; (Birkhäuser, Basel: 1956) which were published on the occasion of his 70th birthday. Weyl added a short remark to Satz 5 of [56], and it is with this addition that [56] was included in his collected papers.

4

From Weyl quantization to modern algebraic index theory

Markus J. Pflaum

Fachbereich Mathematik
Goethe-Universität Frankfurt/Main
pflaum@math.uni-frankfurt.de

1 Introduction

One of the most influencial contributions of HERMANN WEYL to mathematical physics has been his paper *Gruppentheorie und Quantenmechanik* [WE27] from 1927 and its extended version, the book [WE28] which was published a year later and carries the same title. The main topic of this part of HERMANN WEYL'S work is the mathematics of quantum mechanics. After the fundamental papers by HEISENBERG and SCHRÖDINGER on the foundations of quantum mechanics had appeared in the twenties of the last century this was the central question studied in mathematical physics at that time and which to a certain degree still is present in all attempts to construct mathematically rigorous theories unifying quantum mechanics and general realtivity.

In his article *Gruppentheorie und Quantenmechanik*, HERMANN WEYL essentially introduced two novel aspects to the mathematics of quantum mechanics, namely the following:

(i) The representation theory of (compact) Lie groups on Hilbert spaces was applied to mathematically determine atomic spectra.

(ii) A conceptually clear quantization method was proposed which associates quantum mechanical operators to classical observables which mathematically are represented by appropriate functions of the space and momentum variables. Nowadays, this quantization scheme is named after his inventor Weyl quantization.

In this paper I will elaborate only on the second aspect, since the representation theory of compact Lie groups has already been covered in detail in other contributions to these proceedings.

Interestingwise, other than the group theoretical part in WEYL'S article from 1927, his quantization method did not immediately find acceptance in the scientific community as the following part from a review by JOHN VON NEUMANN in Zentralblatt shows:

Sodann wird eine Zuordnungsvorschrift von Matrizen zu beliebigen klassischen Größen (d.h. Funktionen der Koordinaten und Impulse) vorgeschlagen. (Da sie indessen gewisse wesentliche Anforderungen, die an eine solche Zuordnung zu stellen sind – z.B. die Definität der Matrix für wesentlich nichtnegative Größen u.ä. – verletzt, hat sie sich, trotz ihres einfachen und eleganten Baues, nicht durchsetzen können.)

Only much later after the invention of pseudodifferential operators [Hö] and deformation quantization [BFFLS] the virtue and power of Weyl quantization became fully clear. As we will see in Section 4 of this article one can namely show by using the modern language of pseudodifferential operators that Weyl quantization satisfies the axioms of a deformation quantization á la [BFFLS] (cf. [PF98, NETS96]). My impression is that H. WEYL with his vision for a mathematically sound quantization scheme was quite ahead of his time. The following quote from the book [WE28] supports this impression:

Ich kann es nun einmal nicht lassen, in diesem Drama von Mathematik und Physik – die sich im Dunkeln befruchten, aber von Angesicht zu Angesicht so gerne einander verkennen und verleugnen – die Rolle des (wie ich genügsam erfuhr, oft unerwünschten) Boten zu spielen.

Generalizing Heisenberg's commutation relations, P. M. DIRAC proposed in his influential book [DI, §. 21] that a quantization map $\mathfrak{q}$ which associates to every classical observable a an element $\mathfrak{q}(a)$ of an algebra of quantum mechanical observables should satisfy the following commutation relation:

$$[\mathfrak{q}(a), \mathfrak{q}(b)] = i\hbar\mathfrak{q}(\{a, b\}), \qquad (1.1)$$

where $\hbar$ denotes Planck's constant divided by 2π, a, b are classical observables, and $\{-, -\}$ is the Poisson bracket. As Dirac noticed, the commutation relations (1.1) show that "classical mechanics may be regarded as the limiting case of quantum mechanics when $\hbar$ tends to zero" (cf. [DI, §. 21]). For physical reasons Dirac's quantization conditions are usually supplemented by the requirement that the algebra of quantum observables $\mathfrak{q}(a)$ acts irreducibly on a Hilbert space $\mathcal{H}$. This Hilbert space $\mathcal{H}$ or more precisely the corresponding projective space $\mathbb{P}\mathcal{H}$ of rays in $\mathcal{H}$ is then interpreted as the space of (pure) states of the quantum mechanical system.

Let me explain now from the point of view of a mathematician what one means by quantization. This can be seen most easily by the following diagram:

	classical mechanics		quantum mechanics
states	points x of a symplectic manifold M respectively probability distributions on M	$\longrightarrow$	positive normed linear functionals μ on a noncommutative C^*-algebra A
observables	elements a of the Poisson algebra $(\mathcal{C}^\infty(M), \{\,,\,\})$	$\longleftarrow$	(self-adjoint) elements a of the C^*-algebra A
measuring process	evaluation $(x, a) \mapsto a(x)$		evaluation $(\mu, a) \mapsto \mu(a)$

The arrow from left to right is given by quantization while the arrow in the opposite direction is given by a classical limit process.

In 1946 it has been observed by GROENEWOLD [GR] and later refined by VAN HOVE [HO] that for the algebra of (polynomial) observables on $\mathbb{R}^{2n}$ with its standard Poisson bracket a quantization map fulfilling Dirac's commutation relations Eq. (1.1) together with the irreducibility condition cannot exist. The theorems by GROENEWOLD–VAN HOVE were extended by GOTAY et al. [GoGrHu, Go] to more general symplectic manifolds. By all these no go results the question arises, what conditions a reasonable quantization theory should satisfy then.

Weyl's quantization scheme motivated the right answer to that problem. As it has been pointed out by BAYEN, FLATO, FRONSDAL, LICHNEROWICZ and STERNHEIMER in [BFFLS], one should regard quantization as a formal deformation of the algebra of classical observables on a symplectic manifold in the sense of GERSTENHABER [GE]. This means that Dirac's quantization condition is required to hold only up to higher order in $\hbar$. Weyl quantization satisfies this requirement and thus provides an important example of a deformation quantization.

The paper [BFFLS] initiated quite an amount of research on the existence and uniqueness of deformation quantizations. The most outstanding are probably the existence theorem for deformation quantizations over a symplectic manifold by DEWILDE–LECOMTE [DEWILE], the geometric and intuitive construction of star products in the symplectic case by FEDOSOV [FE94], and the result on the existence and the classification of deformation quantizations for Poisson manifolds by KONTSEVICH [KO]. For a detailed overview on this see for example [DIST].

2 Weyl's commutation relations

In his analysis of quantization WEYL started from the Heisenberg commutation relations

$$[P, Q] = -i\hbar, \tag{2.1}$$

where Q resp. P denotes the quantum mechanical space resp. momentum operator. WEYL showed that these relations cannot be realized by bounded operators on a Hilbert space. His idea was then to integrate the Heisenberg commutation relations which leads to the relations

$$V(s)\, U(t) = e^{-ist\hbar} U(t) V(s), \quad s, t \in \mathbb{R}, \tag{2.2}$$

where $V(s) = e^{isQ}$ is the unitary abelian group generated by Q, and $U(t) = e^{itP}$ the one generated by P. For the Schrödinger representation on $L^2(\mathbb{R})$ given by

$$Q\, u(x) = x\, u(x), \quad P\, u(x) = -i\hbar \frac{du}{dx}(x) \quad \text{for } u \in \mathcal{S}(\mathbb{R}) \text{ and } x \in \mathbb{R} \tag{2.3}$$

one knows that

$$\bigl(V(s)u\bigr)(x) = e^{isx} u(x) \quad \text{and} \quad \bigl(U(t)u\bigr)(x) = u(x + \hbar t). \tag{2.4}$$

One thus obtains the integrated Schrödinger representation which obviously satisfies the Weyl commutation relations (2.2). Up to unitary equivalence, the integrated Schrödinger representation is the only irreducible nontrivial representation of the Weyl commutation relations. Note that the Heisenberg commutation relations have more than just one equivalence class of irreducible representations by (necessarily unbounded) symmetric operators on a Hilbert space (see [SCHM]).

Using the integrated Schrödinger representation let us define now the following projective representation of $\mathbb{R}^2$:

$$W(s, t) = e^{-\frac{i}{2} st} U(t)\, V(s). \tag{2.5}$$

For $a \in \mathcal{S}(\mathbb{R}^2)$, the space of Schwarz test functions on $\mathbb{R}^2$, define its Weyl quantization $\mathfrak{q}_\mathrm{w}(a) : \mathcal{C}^\infty_\mathrm{cpt}(\mathbb{R}) \to \mathcal{C}^\infty(\mathbb{R})$ by

$$\langle v, \mathfrak{q}_\mathrm{w}(a)u \rangle := \int_{\mathbb{R}^2} \hat{a}(s,t)\langle v, W(s,t)u \rangle \, ds \, dt, \quad u, v \in \mathcal{C}^\infty_\mathrm{cpt}(\mathbb{R}), \quad (2.6)$$

where $\hat{a}$ denotes the Fourier transform of a. This is the original form of Weyl quantization. Let us rewrite it in a more convenient form by applying the transformation rule and Fourier transformation:

$$\begin{aligned}
\langle v, \mathfrak{q}_\mathrm{w}(a)u \rangle &= \int_{\mathbb{R}^2} \hat{a} \int_{\mathbb{R}} \overline{v}(x)\big(W(s,t)u\big)(x)\, dx\, ds\, dt \\
&= \int_{\mathbb{R}^3} \hat{a}(s,t)\overline{v}(x)\, e^{-\frac{i}{2}\, st\hbar}\big(U(t)V(s)u\big)(x)\, dx\, ds\, dt \\
&= \int_{\mathbb{R}^3} \hat{a}(s,t)\overline{v}(x)\, e^{-\frac{i}{2}\, st\hbar}\, e^{is(x+t\hbar)}\, u(x+t\hbar)\, dx\, ds\, dt \\
&= \frac{1}{\hbar} \int_{\mathbb{R}^3} \hat{a}(s, \tfrac{t}{\hbar})\overline{v}(x)e^{is(x+t/2)}\, u(x+t)\, dx\, ds\, dt \\
&= \frac{1}{2\pi\hbar} \int_{\mathbb{R}^3} a\big(x + \tfrac{t}{2}, \xi\big)\, \overline{v}(x)e^{-\frac{i}{\hbar}t\xi}u(x+t)\, dx\, d\xi\, dt \\
&= \frac{1}{2\pi\hbar} \int_{\mathbb{R}^3} \overline{v}(x)e^{-\frac{i}{\hbar}t\xi}a\big(\tfrac{x}{2} + (\tfrac{x+t}{2}, \xi)\big)\, u(x+t)\, dx\, d\xi\, dt \\
&= \frac{1}{2\pi\hbar} \int_{\mathbb{R}^3} \overline{v}(x)e^{\frac{i}{\hbar}(x-y)\xi}a\big(\tfrac{x+y}{2}, \xi\big)u(y)\, dx\, dy\, d\xi \\
&= \frac{1}{2\pi\hbar} \Big\langle v, \int_{\mathbb{R}^2} e^{\frac{i}{\hbar}(\bullet-y)\xi}a\big(\tfrac{\bullet+y}{2}, \xi\big)\, u(y)\, dy\, d\xi \Big\rangle,
\end{aligned}$$

hence

$$\big[\mathfrak{q}_\mathrm{w}(a)u\big](x) = \frac{1}{2\pi\hbar} \int_{\mathbb{R}^2} e^{\frac{i}{\hbar}(x-y)\xi}a\big(\tfrac{x+y}{2}, \xi\big)\, u(y)\, dy\, d\xi, \quad (2.7)$$

which is the form of the Weyl quantization as it usually can be found in the literature. As one checks immediatley, $\mathfrak{q}_\mathrm{w}(a)$ is a densely defined (in general unbounded) linear operator on $L^2(\mathbb{R})$ which is symmetric, in case a is a real-valued function. If one interprets the right hand side of Eq. (2.7) as an oscillatory integral (see [GRSJ]), then Eq. (2.7) defines even for symbols $a \in \mathrm{S}^\infty(\mathbb{R})$ (cf. Sec. 4) a quantized observable $\mathfrak{q}_\mathrm{w}(a)$ which by definition then is a pseudodifferential operator on $\mathbb{R}$. By a standard argument in pseudodifferential calculus one shows that

$$\mathfrak{q}_\mathrm{w}(a)\,\mathfrak{q}_\mathrm{w}(a) - \mathfrak{q}_\mathrm{w}(b)\,\mathfrak{q}_\mathrm{w}(b) = -i\hbar\mathfrak{q}_\mathrm{w}(\{a,b\}) + o(\hbar^2) \quad \text{for } a, b \in \mathrm{S}^\infty,$$

which means that Weyl quantization satisfies Dirac's quantization condition up to higher orders in $\hbar$ or in other words that the algebra of pseudodifferential operators is a deformation of the algebra of symbols in direction of the Poisson bracket. Let us now explain the concept of a deformation quantization in some more detail.

3 Deformation quantization

Definition 1 ([BFFLS]) *By a deformation quantization of a symplectic manifold (M, ω) one understands an associative and $\mathbb{C}[[\hbar]]$-bilinear product $\star$ on the space $\mathcal{A}^\hbar := \mathcal{C}^\infty(M)[[\hbar]]$ of formal power series in the (now formal) variable $\hbar$ and with coefficients in the space $\mathcal{C}^\infty(M)$ such that the following axioms hold true:*

(i) *There exist bidifferential operators c_k on M such that $a \star b = \sum_{k=0}^\infty c_k(a, b)\, \hbar^k$ for all $a, b \in \mathcal{C}^\infty(M)$ and such that c_0 is the commutative pointwise product of smooth functions on M.*

(ii) *One has $a \star 1 = 1 \star a = a$ for all $a \in \mathcal{C}^\infty(M)$.*

(iii) *The commutation relation*

$$[a, b]_\star = -i\hbar\{a, b\} + o(\hbar^2)$$

is satisfied for all $a, b \in \mathcal{C}^\infty(M)$, where $[a, b]_\star := a \star b - b \star a$.

The product $\star$ is also called a star-product on M.

Example 3.1 *Consider a finite dimensional symplectic vector space (V, ω), and let*

$$\{-, -\} : \mathcal{C}^\infty(V) \otimes \mathcal{C}^\infty(V) \to \mathcal{C}^\infty(V), \quad a \otimes b \mapsto \sum_{1 \leq i, j \leq \dim V} \Pi_{ij} \frac{\partial a}{\partial x_i} \frac{\partial b}{\partial x_j}$$

be its Poisson bracket, where $(x_i)_{1 \leq i \leq \dim V}$ denote some coordinates of V. Since the standard Poisson bivector $\Pi := \sum \Pi_{ij} \frac{\partial}{\partial x_i} \otimes \frac{\partial}{\partial x_j}$ is constant, the operator

$$\hat{\Pi} : \mathcal{C}^\infty(V) \otimes \mathcal{C}^\infty(V) \to \mathcal{C}^\infty(V) \otimes \mathcal{C}^\infty(V), \quad a \otimes b \mapsto \sum_{1 \leq i, j \leq 2n} \Pi_{ij} \frac{\partial a}{\partial x_i} \otimes \frac{\partial b}{\partial x_j}$$

is well-defined. Denoting by μ the pointwise product of functions, one can now put

$$\star : \mathcal{C}^\infty(V)[[\hbar]] \otimes \mathcal{C}^\infty(V)[[\hbar]] \to \mathcal{C}^\infty(V)[[\hbar]],$$

$$a \otimes b \mapsto \sum_{k \in \mathbb{N}} \frac{(-i\hbar)^k}{k!} \mu\big(\hat{\Pi}^k(a \otimes b)\big)$$

and thus obtains a star product on $\mathcal{C}^\infty(V)$, the so-called Weyl–Moyal-product. It is immediately checked that $(\mathcal{C}^\infty(V)[[\hbar]], \star)$ is a deformation qunatization in the above sense.

By construction, the Weyl–Moyal-product makes sense also on the space $\mathbb{W}V$ of formal power series in $\hbar$ with formal power series at the origin of V as coefficients. One calls the resulting algebra $(\mathbb{W}V, \star)$ the formal Weyl algebra of V, and obtains an epimorphism of algebras

$$(\mathcal{C}^\infty(V)[[\hbar]], \star) \to (\mathbb{W}V, \star)$$

given by formal power series expansion at the origin in each degree of $\hbar$.

The existence of a star product on an arbitrary symplectic manifold was an open mathematical problem for almost ten years after the article [BFFLS] had appeared, and was settled by independant methods in the papers [DEWILE] and [FE94]. Another ten years passed until the existence of star products on Poisson could be proved in [KO].

Let us briefly sketch the main idea of the proof by FEDOSOV [FE96], since his approach contains essential tools which lead to algebraic index theory. Consider a symplectic manifold (M, ω) and its tangent bundle TM. Since each of the tangent spaces $T_p M$, $p \in M$ is a symplectic vector space, one can form the bundle of formal Weyl algebras $\mathbb{W}M := \bigsqcup_{p \in M} \mathbb{W}T_p M$. Note that the bundle of formal Weyl agebras is well-defined because the symplectic group acts as automorphisms on the formal Weyl algebra. Now consider the bundle of forms $\Lambda^\bullet \mathbb{W}M := \mathbb{W}M \otimes \Lambda^\bullet M$. Its space of smooth sections $\Omega^\bullet \mathbb{W}(M)$ obviously is a noncommutative algebra with product denoted by $\bullet$. The fundamental observation by FEDOSOV was that for an appropriate flat graded derivation D with respect to the product $\bullet$ on $\Omega^\bullet \mathbb{W}(M)$ the subalgebra

$$\mathcal{W}_D(M) := \{s \in \Omega^0 \mathbb{W}(M) \mid Ds = 0\}$$

of flat sections is linearly isomorphic to the space of formal power series $\mathcal{C}^\infty(M)[[\hbar]]$. Via the resulting isomorphism $\mathfrak{q} : \mathcal{C}^\infty(M)[[\hbar]] \to \mathcal{W}_D(M)$ one can then push down the product on $\mathcal{W}_D(M)$ to $\mathcal{C}^\infty(M)[[\hbar]]$ and thus obtains a star product on M. The flat connection needed for this construction has the form

$$D = \nabla + [A, -],$$

where ∇ is a symplectic connection on M, i.e. $\nabla \omega = 0$, $[-, -]$ is the commutator with respect to the product $\bullet$, and $A \in \Omega^1 \mathbb{W}(M)$. The cohomology class of the curvature

$$\Omega := \nabla A + \frac{1}{2}[A, A]$$

(note that it is a formal power series in $\hbar$) classifies the star product $\star$ up to equivalence.

For the application of deformation quantization to index theory, the notion of a trace on a deformation quantization $\big(\mathcal{C}^\infty_{\mathrm{cpt}}(M)[[\hbar]], \star\big)$ is crucial. By a that one understands a linear functional $\mathrm{tr} : \mathcal{C}^\infty_{\mathrm{cpt}}(M)[[\hbar]] \to \mathbb{C}[[\hbar, \hbar^{-1}]]$ which vanishes on commutators. The following result provides essential information on the existence and uniqueness of such traces.

Proposition 3.1 ([NETS95, FE96]) *The space of traces on a deformation quantization $\big(\mathcal{C}^\infty_{\mathrm{cpt}}(M)[[\hbar]], \star\big)$ over a connected symplectic manifold M is one-dimensional.*

4 Pseudodifferential operators

Next we will set up Weyl quantization within the language of pseudodifferential operators. Before we come to the details of this let us recall some basics of that theory.

Let $U \subset \mathbb{R}^n$ be open. By a symbol on $U \times \mathbb{R}^N$ of order $m \in \mathbb{R}$ one understands a function $a \in \mathcal{C}^\infty(U \times \mathbb{R}^N)$ such that for every compact $K \subset U$ and all multi-indices $\alpha \in \mathbb{N}^n$ and $\beta \in \mathbb{N}^N$ there exists a $C_{K,\alpha,\beta} > 0$ such that

$$\left| \partial_x^\alpha \partial_\xi^\beta a(x, \xi) \right| \leq C_{K,\alpha,\beta} \left(1 + ||\xi|| \right)^{m - |\beta|} \quad \text{for all } (x, \xi) \in K \times \mathbb{R}^N. \quad (4.1)$$

The space of such symbols is denoted by $\mathrm{S}^m(U, \mathbb{R}^N)$. Obviously, one can easily generalize the notion of a symbol of order m to smooth maps $a : E \to \mathbb{R}$ defined on a vector bundle $E \to M$ by requiring Eq. (4.1) to hold locally in bundle charts. For every manifold X we denote the space of symbols of order m on the cotangent bundle T^*X by $\mathrm{S}^m(X)$. Moreover, one puts $\mathrm{S}^\infty(X) := \bigcup_{m \in \mathbb{R}} \mathrm{S}^m(X)$ and $\mathrm{S}^{-\infty}(X) := \bigcap_{m \in \mathbb{R}} \mathrm{S}^m(X)$.

A pseudodifferential operator over $U \subset \mathbb{R}^n$ now is a linear operator $A : \mathcal{C}^\infty_{\mathrm{cpt}}(U) \to \mathcal{C}^\infty(U)$ which can be represented as an oscillatory integral (cf. [GRSJ, Sec. 1])

$$Au(x) = \frac{1}{(2\pi)^n} \int_{\mathbb{R}^n} \int_U e^{i\langle x - y, \xi \rangle} a(x, y, \xi)\, u(y)\, dy\, d\xi, \qquad (4.2)$$

where $u \in \mathcal{C}^\infty_{\mathrm{cpt}}(U)$ and $a \in \mathrm{S}^m(U \times U, \mathbb{R}^n)$ for some $m \in \mathbb{R} \cup \{\pm\infty\}$.

The space of thus defined pseudodifferential operators of order m will be denoted by $\Psi^m(U)$. More generally, if X is a manifold, the space $\Psi^m(X)$ of pseudodifferential operators of order m on X consists of all linear operators $A : \mathcal{C}^\infty_{\text{cpt}}(X) \to \mathcal{C}^\infty(X)$ which can be written in the form

$$Au = A_0 u + \sum_{j \in J} \varphi_j \left(A_j \left((\varphi_j u) \circ x_j^{-1} \right) \right) \circ x_j,$$

where the x_j, $j \in J$ run through an atlas of X, $(\varphi_j)_{j \in J}$ is a locally finite smooth partition of unity subordinate to the domains of the charts φ_j, the A_j are pseudodifferential operators on $\mathbb{R}^{\dim X}$ of order m, and finally A_0 is a smoothing operator, which means that its Schwartz kernel is smooth. Let us restrict our considerations now to the space $\Psi^\infty_{\text{ps}}(X)$ of properly supported pseudodifferential which means of all pseudodifferential operators A such that the projections $\text{pr}_{1/2} : \text{supp}\,K_A \to X$ of the support of the Schwartz kernel of A on the first resp. second coordinate are proper maps. Since every properly supported pseudodifferential operator maps functions with compact support to functions with compact support, $\Psi^\infty_{\text{ps}}(X)$ turns out to be an algebra which, as we will see in the following, can be interpreted as a quantization of the symbol algebra on TX.

Let us provide some details. After the choice of a riemannian metric on X and fixing an ordering parameter $s \in [0, 1]$, define for every symbol $a \in S^m(X)$ and $\hbar \in \mathbb{R}^*$ a quantization $\mathsf{q}_s(a) : \mathcal{C}^\infty_{\text{cpt}}(X) \to \mathcal{C}^\infty_{\text{cpt}}(X)$ by

$$[\mathsf{q}_s(a)u](x) = \frac{1}{(2\pi\hbar)^n} \int_{T^*X} \chi(x,y)\, e^{\frac{i}{\hbar}\langle \exp_y^{-1}(x),\xi\rangle} a\big(\tau_{g_s(x,y),y}\xi\big) u(y)\, dy d\xi.$$
$$(4.3)$$

The ingredients of this formula are given as follows. As usual, exp denotes the exponential function with respect to the riemannian metric on X, and χ is a properly supported cut-off function around the diagonal of $X \times X$ such that $\exp_y^{-1}(x)$ is defined for all $(x, y) \in \text{supp}\,\chi$. By $g_s(x, y)$ we mean the s-midpoint between x and y, or in other words the point $\exp\big(s \cdot \exp_x^{-1}(y)\big)$. For x and y close enough we denote by $\tau_{x,y}$ the parallel transport in T^*X from $T^*_y X$ to $T^*_x X$ along the geodesic joining x and y. Finally, $dy\,d\xi$ stands for the Liouville volume element on the symplectic manifold T^*X. One checks immediately (cf. [PF98, VO]) that $\mathsf{q}_s(a)$ is a (properly supported) pseudodifferential operator of order m. In case $s = 0$ one calls it the standard order quantization of the symbol a, if $s = \frac{1}{2}$, one obtains Weyl quantization on the riemannian manifold X. The reader is invited to check that on the cotangent bundle of $\mathbb{R}$,

$q_{\frac{1}{2}}$ coincides with the Weyl quantization q_W from Eq. (2.7) (up to some negligible smoothing operator).

The quantization map q_s has a pseudoinverse, namely the symbol map $\sigma_s : \Psi^\infty_{\mathrm{ps}}(X) \to S^\infty(X)$ which is defined by

$$\sigma_s(A)(x, \xi) =$$
$$\int_{T_x X} \chi_s(x, v)\, e^{\frac{i}{\hbar}\langle v, \xi\rangle}\, K_A\big(\exp_x(-sv), \exp_x((1-s)v)\big)\, \rho_s(x, v) d\theta_x(v),$$

where K_A is the Schwartz kernel of the operator A, the cut-off function χ_s is defined by $\chi_s(x, v) := \chi\big(\exp_x(-sv), \exp_x((1-s)v)\big)$, θ_x is the euclidean volume element of $T_x X$ induced by the riemannian metric on X, and the metric factor ρ_s satisfies $\rho_s(x, v) = \rho\big(\exp_x(-sv), \exp_x((1-s)v)\big)$ with $\rho(x, \exp_x v)\theta_x = \big(\exp_x^* \mu\big)(v)$, and μ the riemannian volume element on X. Then one has

$$\sigma_s \circ q_s(a) - a \in S^{-\infty}(X) \quad \text{and} \quad q_s \circ \sigma_s(A) - A \in \Psi^{-\infty}_{\mathrm{ps}}(X) \qquad (4.4)$$

for all symbols a and pseudodifferential operators A on X, which shows that they are inverse to each other up to smoothing operators resp. symbols. Moreover, one can prove (cf. [PF98]) that

$$[q_s(a), q_s(b)] = -i\hbar\, q_s(\{a, b\}) + o(\hbar^2) \qquad (4.5)$$

for all symbols a, b. This means that each q_s and in particular Weyl quantization $q_W := q_{\frac{1}{2}}$ induces a deformation quantization of the cotangent bundle TX.

Another usefull feature of the quantization q_s is that it allows to compute the operator trace of $q_s(a)$ for every symbol a of order $m < \dim X$. According to [PF98, VO] this trace is given by

$$\mathrm{tr}\, q_s(a) = \frac{1}{(2\pi\hbar)^{\dim X}} \int_{T^*X} a\, \omega^{\dim X}, \qquad (4.6)$$

where ω denotes the canonical symplectic form on the cotangent bundle T^*X.

5 The algebraic index theorem

Let us first recall some basic notions from index theory of Fredholm operators. Assume to be given two Hilbert spaces $\mathcal{H}_1$, $\mathcal{H}_2$ and a Fredholm operator $F : \mathcal{H}_1 \to \mathcal{H}_2$, which means that F is a linear operator which has finite dimensional kernel and cokernel. Its index is then defined as the integer

$$\operatorname{ind} F := \dim \ker F - \dim \operatorname{coker} F. \tag{5.1}$$

The index has the following crucial properties:

- it is homotopy invariant, i.e.

$$\operatorname{ind} F(0) = \operatorname{ind} F(1)$$

for every continuous path $F : [0,1] \to \operatorname{Fred}(\mathcal{H}_1, \mathcal{H}_2)$ of Fredholm operators,
- it is additive with respect to composition, i.e.

$$\operatorname{ind}(F_1 \circ F_2) = \operatorname{ind} F_1 + \operatorname{ind} F_2$$

for two composable Fredholm operators F_1 and F_2, and finally
- the index is invariant under compact perturbations, i.e.

$$\operatorname{ind}(F + K) = \operatorname{ind} F$$

for every Fredholm operator and every compact operator K from $\mathcal{H}_1$ to $\mathcal{H}_2$.

Every Fredholm operator $F : \mathcal{H}_1 \to \mathcal{H}_2$ has a pseudoinverse $R : \mathcal{H}_2 \to \mathcal{H}_1$, which in other words means that $\operatorname{id}_{\mathcal{H}_1} - R \circ F$ and $\operatorname{id}_{\mathcal{H}_2} - R \circ F$ are both compact operators. One can even choose R such that these operators are of trace class. Then one can compute the index of F by the following formula (cf. [FE96]):

$$\operatorname{ind} F = \operatorname{tr}(\operatorname{id}_{\mathcal{H}_1} - R \circ F) - \operatorname{tr}(\operatorname{id}_{\mathcal{H}_2} - R \circ F). \tag{5.2}$$

In the theory of linear partial differential equations, Fredholm operators appear abundantly. Namely, if $E \to X$ is a (metric) vector bundle over a compact (riemannian) manifold X, and $D : \Gamma^\infty(E) \to \Gamma^\infty(E)$ an elliptic differential operator (which means that its principle symbol is invertible) then it induces a Fredholm operator between appropriate Sobolev completions of $\Gamma^\infty(E)$. In particular, D then has a finite index, and this index does not depend on the particular choice of a Sobolev completion. By the celebrated index theorem of ATIYAH–SINGER [ATSI], the index of D can be computed by topological data as follows:

$$\operatorname{ind} D = (-1)^{\dim X} \int_X \operatorname{Ch}(\sigma_{\mathrm{p}}(D)) \operatorname{td}(T_\mathbb{C} X), \tag{5.3}$$

where $\sigma_{\mathrm{p}}(D)$ denotes the principal symbol of the differential operator D, Ch its Chern character and $\operatorname{td}(T_\mathbb{C} X)$ is the Todd class of the complexified tangent bundle.

As it has been observed by FEDOSOV [FE96] and NEST–TSYGAN [NETS95], an "algebraic" version of this index theorem can be formulated and proved within the framework of deformation quantization. Recall that the the index of an elliptic operator can be computed by Eq. (5.2). If one interprets a deformation quantization as a kind of "formal pseudodifferential calculus", it makes sense to consider elliptic elements in the deformed algebra and define an algebraic index for these objects. More precisely, given a symplectic manifold M with a star product $\star$, one understands by an elliptic pair in $\mathcal{C}^\infty(M)[[\hbar]]$ a pair of projections P, Q in the matrix algebra over $(\mathcal{C}^\infty(M)[[\hbar]], \star)$ such that the difference $P - Q$ has compact support. This means in particular, that every elliptic pair (P, Q) determines an element $[P] - [Q]$ of the K-theory of the deformed algebra. The algebraic index of the K-theory class $[P] - [Q]$ of an elliptic pair is defined by

$$\mathrm{ind}_a\left([P] - [Q]\right) := \mathrm{tr}(P - Q), \tag{5.4}$$

where tr is the (up to normalization) unique trace on the matrix algebra over $\mathcal{C}^\infty(M)[[\hbar]]$ (cf. Prop. 3.1). Note that the index is indeed well-defined on the K-theory of $\mathcal{C}^\infty(M)[[\hbar]]$.

The space of equivalence classes $[P] - [Q]$ of elliptic pairs is isomorphic to the space of equivalence classes of elliptic quadruples. These objects were introduced by FEDOSOV [FE96] and are the natural generalizations of elliptic operators to star product algebras. More precisely, an elliptic quadruple is a quadruple $(D, F, \tilde{P}, \tilde{Q})$ of elements of the matrix algebra over $\mathcal{C}^\infty(M)[[\hbar]]$ such that the following holds:

(i) $\tilde{P}$ and $\tilde{Q}$ are projections.
(ii) The elements $\tilde{P} - D \star R$ and $\tilde{Q} - R \star D$ have both compact support.

The element D of an elliptic quadruple hereby generalizes an elliptic pseudodifferential operator on a closed manifold, and F can be interpreted as its quasi-inverse.

By Eq. (5.2), the following definition of the algebraic index of an (equivalence class of an) elliptic quadruple appears to be reasonable:

$$\mathrm{ind}_a\left([D, F, \tilde{P}, \tilde{Q}]\right) := \mathrm{tr}(\tilde{Q} - R \star D) - \mathrm{tr}(\tilde{P} - D \star R). \tag{5.5}$$

In fact, one can show that the thus defined algebraic index is compatible with the algebraic index on elliptic pairs under the mentioned isomorphism between equivalence classes of elliptic pairs and of elliptic quadruples.

There is another remark in order, here. For an elliptic pair (P, Q) the coefficients (P_0, Q_0) of order 0 in the expansion in powers of $\hbar$ are obviously projections in the matrix algebra over $\mathcal{C}^\infty(M)$, and the virtual bundle $[P_0] - [Q_0]$ defines a K-theory class of M. This map turns out to be an isomorphism between K-theories, which shows that K-theory is invariant under deformation (cf. [Ro], [FE96, Sec. 6.1]).

The main result of algebraic index theory is the theorem below. By application of this algebraic index theorem to cotangent bundles of compact riemannian manifolds and the deformation quantization induced by Weyl quantization one obtains another proof of the index formula by ATIYAH–SINGER.

Theorem 5.1 ([NETS95, FE96]) *Let M be a symplectic manifold with a deformation quantization $\star$. The algebraic index of an elliptic pair $[P] - [Q]$ is then given by*

$$\text{ind}_{\text{a}} \left([P] - [Q] \right) = \int_M \text{Ch} \left([P_0] - [Q_0] \right) \exp \left(-\frac{\Omega}{2\pi\hbar} \right) \hat{A}(M), \quad (5.6)$$

where $\hat{A}(M)$ denotes the $\hat{A}$-genus of M, and Ω the characteristic class of the star product on M

The proof of the algebraic index theorem is quite involving. In the approach by NEST–TSYGAN, methods from cyclic homology theory and Lie algebra cohomology are used intensively. We refer to the original literature for that.

6 The algebraic index theorem for orbifolds

The problem to generalize index theorems to spaces more general than (compact) manifolds has been an active area of mathematical research since many years. For orbifolds, a class of singular spaces which has attained much interest in geometry and mathematical physics, an algebraic index theorem can be proved.

In local charts, orbifolds are represented as quotients of manifolds by finite groups. Globally, and that is the approach we use in our setup, orbifolds can be presented as orbits of proper étale Lie groupoids G (see [Mo] for details). The concepts of a deformation quantization, of vector bundles, and of K-theory can all be generalized to orbifolds by requiring the objects (like a star product or a vector bundle) to be invariant on the representing proper étale groupoid. For example, the orbifold K-theory

$K^0_{\mathrm{orb}}(X)$ of an orbifold X consists of equivalence classes of equivariant virtual bundles on the representing groupoid.

A quite usefull object associated to an orbifold X is its inertia orbifold $\tilde{X}$. Locally, $\tilde{X}$ consists of all fixed point manifolds of the locally representing orbifold charts. The inertia orbifold carries a lot of information about the singularities of the orbifold. The connected components of the inertia orbifold $\tilde{X}$ are sometimes called the sectors of the orbifold X.

The orbifold case is different to the manifold case in particular by one important aspect. The dimension of traces on a deformation quantization on a symplectic orbifold is in general not one (even if X is connected), but given by the number of sectors [NePfPoTa]. This means that one has to single out a particular trace to define the algebraic index for a deformation quantization on an orbifold. Fortunately, there exists a kind of "universal" trace on an orbifold, which captures from each sector a normalized contribution. With that universal trace the following algebraic index formula can be proved.

Theorem 6.1 ([PfPoTa]) *Let M be a symplectic orbifold presented by a proper étale Lie groupoid G carrying a G-invariant symplectic form ω. Let $\star$ be a star product on M, and let E and F be G-vector bundles which are isomorphic outside a compact subset of M. Then the following formula holds for the index of $[E] - [F] \in K^0_{\mathrm{orb}}(M)$:*

$$\mathrm{tr}_* \big([E] - [F] \big) = \int_{\tilde{M}} \frac{1}{m} \frac{\mathrm{Ch}_\theta \left(\frac{R^E}{2\pi i} - \frac{R^F}{2\pi i} \right)}{\det \left(1 - \theta^{-1} \exp(-\frac{R^\perp}{2\pi i}) \right)} \hat{A} \big(\frac{R^\perp}{2\pi i} \big) \exp \left(-\frac{\iota^* \Omega}{2\pi i \hbar} \right),$$

$$(6.1)$$

where $\mathrm{tr} : \mathcal{C}^\infty_{\mathrm{cpt}}[[\hbar]] \rtimes \mathsf{G} \to \mathbb{C}[[\hbar, \hbar^{-1}]]$ *is the universal trace on the convolution algebra capturing from each sector one contribution, m is a locally constant combinatorial function measuring the order of the isotropy group, and Ω is the characteristic class of the deformation quantization. The symbol θ denotes the action of the local isotropy groups, and $\mathrm{Ch}_\theta \left(\frac{R^E}{2\pi i} - \frac{R^F}{2\pi i} \right)$ is the equivariant Chern character which à la Chern-Weil is determined by equivariant curvatures R^E and R^F. Finally $R^\perp$ denotes the curvature of the normal bundle of the local embedding of $\tilde{X}$ into X.*

Like in the manifold case one can construct a symbol calculus and Weyl quantization for orbifolds. Since Weyl quantization on an orbifold X defines a deformation quantization over the symplectic orbifold T^*X, one can then derive an analytic index formula from the algebraic index theorem for orbifolds. One then obtains the KAWASAKI index formula for orbifolds (see [KA] and [FA]).

Bibliography

[ATSI] ATIYAH, M.F. and I.M. SINGER: *The Index of Elliptic Operators I.* Ann. Math. **87**, 484–530 (1968).

[BFFLS] BAYEN, F., M. FLATO, C. FRONSDAL, A. LICHNEROWICZ, and D. STERNHEIMER: *Deformation theory and quantization, I and II.* Ann. Phys. **111** (1978), 61–151.

[DI] DIRAC, P.A.M.: *The Principles of Quantum Mechanics.* 4th ed. Oxford, Clarendon Press, 1947.

[DIST] DITO, G., and STERNHEIMER, D.: *Deformation quantization: genesis, developments and metamorphoses.* in "Deformation quantization" (Strasbourg, 2001), 9–54, IRMA Lect. Math. Theor. Phys., 1, de Gruyter, Berlin, 2002.

[FA] FARSI, C.: *K-theoretical index theorems for orbifolds,* Quart. J. Math. Oxford Ser. (2) **43**, no. 170, 183–200 (1992).

[FE94] FEDOSOV, B.: *A simple geometrical construction of deformation quantization,* J. Diff. Geom. (1994).

[FE96] FEDOSOV, B.: *Deformation Quantization and Index Theory,* Akademie Verlag, 1996.

[GE] GERSTENHABER, M.; *On the deformations of rings and algebras,* Ann. of Math. **79**, 59–103 (1964).

[GO] GOTAY, M.J.: *Obstructions to quantization.* in "Mechanics: from theory to computation. Essays in honor of Juan-Carlos Simo." Papers invited by Journal of Nonlinear Science editors. Springer, New York. 171–216 (2000).

[GOGRHU] GOTAY, M. J., H. GRUNDLING and HURST, C.A.: *A Groenewold-Van Hove theorem for $S\mathfrak{sp}2$.* Trans. Am. Math. Soc. **348**, no. 4, 1579–1597 (1996).

[GRSJ] GRIGIS A., and J. SJÖSTRAND: *Microlocal Analysis for Differential Operators.,* London Mathematical Society Lecture note series, vol. **196**, Cambridge University Press, 1994.

[GR] GROENEWOLD, H.J.: *On the principles of elementary quantum mechanics.* Physics **12**, 405–460 (1946).

[HÖ] HÖRMANDER, L.: *Pseudodifferential operators* Comm. Pure Appl. Math. **18**, 501–517 (1965).

[HO] VAN HOVE, L.: *Sur le problème des relations entres les transformations unitaires de la mécanique quantique et les transformation canoniques de la mécanique classique.* Acad. Roy. Belgique Bull. Cl. Sci. (5) **37**, 610–620 (1951).

[KA] KAWASAKI, T.: *The index of elliptic operators over V-manifolds,* Nagoya Math. J. **84**, 135-157 (1981).

[KO] KONTSEVICH, M.: *Deformation quantization of Poisson manifolds, I,* arXiv:q-alg/9709040 (1997).

[MO] MOERDIJK, I.: *Orbifolds as groupoids: an introduction,* Adem, A. (ed.) et al., Orbifolds in mathematics and physics (Madison, WI, 2001), Amer. Math. Soc., Contemp. Math. **310**, 205–222 (2002).

[NePfPoTa] NEUMAIER, N., M. PFLAUM, H. POSTHUMA and X. TANG: *Homology of of formal deformations of proper étale Lie groupoids*, Journal f. die reine und angewandte Mathematik **593** (2006).

[NeTs95] NEST, R., and B. TSYGAN: *Algebraic index theorem*, Comm. Math. Phys **172**, 223–262 (1995).

[NeTs96] NEST, R., and B. TSYGAN: *Formal versus analytic index theorems*, Intern. Math. Research Notes **11**, 557–564 (1996).

[Pf98] PFLAUM, M.J.: *A deformation-theoretical approach to Weyl quantization on riemannnian manifolds*, Lett. Math. Physics **45** 277–294 (1998).

[PfPoTa] PFLAUM, M., H. POSTHUMA and X. TANG: *An algebraic index theorem for orbifolds*. Adv. Math. **210**, 83–121 (2007).

[Ro] ROSENBERG, J.: *Behavior of K-theory under quantization*, in "Operator Algebras and Quantum Field Theory", eds. S. Doplicher, R. Longo, J. E. Roberts, and L. Zsido, International Press, 404–415 (1997).

[Schm] SCHMÜDGEN, K.: *On the Heisenberg commutation relation. II*. Publ. Res. Inst. Math. Sci. 19, no. 2, 601–671 (1983).

[Vo] VORNOV, T.: *Quantization of forms on the cotangent bundle*. Comm. Math. Phys. **205**, no. 2, 315–336 (1999).

[We27] WEYL, H.: *Quantenmechanik und Gruppentheorie*. Z. f. Physik **46**, 1–46 (1927).

[We28] WEYL, H.: *Quantenmechanik und Gruppentheorie*. S. Hirzel Verlag, Leipzig, (1928).

[deWiLe] DE WILDE, M., and P. LECOMTE: *Formal deformations of the Poisson Lie algebra of symplectic manifold and star-products. Existence, equivalence, derivations*, in "Deformation Theory of Algebras and Structures and Applications" (Dordrecht) (M. Hazewinkel and M. Gerstenhaber, eds.), Kluwer Acad. Pub., 1988, pp. 897–960.

5

Sharp spectral inequalities for the Heisenberg Laplacian

A. M. Hansson

Department of Mathematics, KTH
anhan@math.kth.se

A. Laptev

Department of Mathematics
Imperial College, London
a.laptev@imperial.ac.uk & laptev@math.kth.se

ABSTRACT We obtain sharp inequalities for the spectrum of the Heisenberg Laplacian with Dirichlet boundary conditions in $2N+1$-dimensional domains of finite measure. We also give another proof of an inequality obtained by Strichartz.

1 Introduction

The aim of this paper is to prove uniform inequalities for a class of differential operators with discrete spectrum. In particular, we give a simple proof of a previously known result by Strichartz [24] for hypoelliptic operators.

Let X_1 and X_2 be two vector fields in $\mathbb{R}^3$, expressed in coordinates $x = (x_1, x_2, x_3)$ as

$$X_1 := \partial_{x_1} + \frac{1}{2}x_2\partial_{x_3}, \qquad X_2 := \partial_{x_2} - \frac{1}{2}x_1\partial_{x_3}. \qquad (1.1)$$

The Lie algebra of left-invariant vector fields on the first Heisenberg group $\mathbf{H}^1$ is spanned by X_1, X_2 and $X_3 := \partial_{x_3}$. $\mathbf{H}^1$ may be described as the set $\mathbb{R}^2 \times \mathbb{R}$ equipped with the group law

$$(x_1, x_2, x_3)(y_1, y_2, y_3) = \left(x_1 + y_1, \ x_2 + y_2, \ x_3 + y_3 + \frac{x_1 y_2 - x_2 y_1}{2} \right).$$

These vector fields satisfy the canonical commutation relation

$$[X_1, X_2] = -X_3.$$

The quadratic form

$$a[u] := \int_{\mathbb{R}^3} \left(|X_1 u(x)|^2 + |X_2 u(x)|^2 \right) dx, \qquad u \in H^1(\mathbb{R}^3), \qquad (1.2)$$

defines a self-adjoint operator

$$A := X_1^* X_1 + X_2^* X_2 = -X_1^2 - X_2^2$$

from the Kolmogorov-Hörmander class [8] of second-order hypoelliptic operators. A, usually referred to as the *Heisenberg Laplacian*, is non-elliptic because it is associated to a degenerate metric on $\mathbf{H}^1$, see [26]. The operators

$$Z := \frac{X_1 + iX_2}{\sqrt{2}} \quad \text{and} \quad \bar{Z} := \frac{X_1^* - iX_2^*}{\sqrt{2}}$$

are related to creation and annihilation operators; their corresponding energy operator $Z\bar{Z} + \bar{Z}Z$ coincides with A.

Let $\Omega \subset \mathbb{R}^3$ be a domain of finite measure. The closure of the form (1.2), defined on the class of functions $C_0^\infty(\Omega)$, gives us a self-adjoint operator A_Ω which is identified with the operator A with Dirichlet boundary conditions on the boundary $\partial\Omega$. It is well known that the spectrum of this operator is discrete and the corresponding eigenvalues $(\lambda_k)_{k=1}^\infty$ accumulate at ∞.

In a later section we will also study the elliptic operator

$$B := -X_1^2 - X_2^2 - X_3^2 \quad \text{in } L^2(\mathbb{R}^3),$$

which commutes with A and hence acts on each of its eigenspaces. By the approach used to prove the trace inequality for A we will obtain an estimate which turns out to be identical to the classical Weyl-type asymptotics.

2 Hypoelliptic case: main results

The aim of this article is to obtain a sharp uniform spectral inequality of Berezin-Li-Yau type for the trace $\operatorname{Tr}\varphi_\lambda(A_\Omega)$. Let

$$\varphi_\lambda(t) = (\lambda - t)_+ = \begin{cases} \lambda - t, & \lambda > t, \\ 0, & \lambda \leq t. \end{cases}$$

Theorem 2.1 *Let $\Omega \subset \mathbb{R}^3$ be a domain of finite measure and let A_Ω be the operator A with Dirichlet boundary conditions in Ω. Then the spectrum of A_Ω is discrete and*

$$\operatorname{Tr}\varphi_\lambda(A_\Omega) = \sum_k (\lambda - \lambda_k)_+ \leq \frac{1}{96}\,|\Omega|\,\lambda^3. \tag{2.1}$$

Remark 2.2 *The operator A is unitarily equivalent to a two-dimensional Laplacian with constant magnetic field, see (6.1). Note, however, that (2.1) cannot be considered as the (semi-classical) phase-volume estimate, which is infinity in the case of the Heisenberg Laplacian with Dirichlet boundary conditions.*

Remark 2.3 *For the Dirichlet Laplacian with constant magnetic field the uniform inequalities for the eigenvalues have been obtained in [6] and [7].*

This result implies several corollaries.

Corollary 2.1 *Let $N(\lambda)$ be the counting function of A_Ω,*

$$N(\lambda) := \{k : \lambda_k < \lambda\}. \tag{2.2}$$

Then under the conditions of Theorem 2.1 we have

$$N(\lambda) \le \frac{9}{2^7}\, |\Omega|\, \lambda^2.$$

Remark 2.4 *Uniform spectral inequalities for the number of eigenvalues below $\lambda > 0$ for Dirichlet Laplacian were obtained by Pólya [21] for tiling domains, in [4] and [5] for bounded domains and in [16], [M] and [23] for domain of finite Lebesgue measure. The sharp constant in this inequality remains unknown.*

Applying the Aizenmann-Lieb principle [1] for $\gamma > 1$ and an argument close in spirit [7] for $\gamma < 1$, we obtain

Corollary 2.2 *Under the conditions of Theorem 2.1 and for any $\gamma > 0$, the eigenvalues of the Dirichlet Heisenberg Laplacian satisfy the Lieb-Thirring inequality*

$$\sum_k (\lambda - \lambda_k)_+^\gamma \le K_\gamma\, |\Omega|\lambda^{\gamma+2}$$

with

$$K_\gamma = \begin{cases} \dfrac{9}{32} \dfrac{\gamma^\gamma}{(\gamma+2)^{\gamma+2}}, & 0 < \gamma \le 1, \\ \dfrac{1}{16} \dfrac{1}{(\gamma+1)(\gamma+2)}, & 1 \le \gamma. \end{cases} \tag{2.3}$$

Remark 2.5 *Note that $\lim_{\gamma \downarrow 0} K_\gamma$ is the constant in Corollary 2.1. However, taking this as the initial value and applying the technique we used to* *estimate K_γ 'upwards' would have given $2^{-6}9(\gamma+1)^{-1}(\gamma+2)^{-1}$, $\gamma < 1$, which decreases at a slower rate.*

In the case where $\partial\Omega$ is smooth, it has been obtained in [19] that the following spectral asymptotic formula holds true:

$$\lim_{\lambda\to\infty} \lambda^{-2}N(\lambda) = \int_\Omega \gamma(x)dx.$$

Here γ is a continuous and strictly positive function, which is given rather implicitly. Theorem 2.1 and Corollaries 2.1, 2.2 allow us to extend this statement, with an explicit right-hand side, to arbitrary domains of finite measure with non-smooth boundaries. Approximating the domain by cylinders and arguing by the variational principle, we obtain the necessary converse inequality, which confirms the asymptotic sharpness of the constant in Theorem 2.1 and Strichartz's Theorem 6.1 [24].

Corollary 2.3 *Under the conditions of Theorem 2.1, for any $\gamma \geq 1$,*

$$\lim_{\lambda\to\infty} \lambda^{-\gamma-2} \sum_k (\lambda - \lambda_k)_+^\gamma = K_\gamma |\Omega|$$

with K_γ as in (2.3).

By a standard Tauberian argument we also have

Corollary 2.4 *Under the conditions of Theorem 2.1, for $0 \leq \gamma < 1$,*

$$\lim_{\lambda\to\infty} \lambda^{-\gamma-2} \sum_k (\lambda - \lambda_k)_+^\gamma = \frac{\gamma+3}{\gamma+1} K_{\gamma+1} |\Omega|$$

with K_γ as in (2.3).

Another consequence of Theorem 2.1 follows immediately from the duality

$$f(x) \leq g(x), \; x \geq 0 \qquad \Leftrightarrow \qquad g^\star(p) \leq f^\star(p), \; p \geq 0,$$

where the Legendre transform of a convex, non-negative function f is given by $f^\star(p) = \sup_{x\geq 0}(px - f(x))$, $p \geq 0$. Transforming both sides of (2.1) (cf. [14]) we obtain

Corollary 2.5 *Under the conditions of Theorem 2.1 the eigenvalues of the Dirichlet Heisenberg Laplacian satisfy the Li-Yau inequality*

$$\sum_{k=1}^n \lambda_k \geq \frac{8\sqrt{2}}{3} |\Omega|^{-1/2} n^{3/2}.$$

This result constitutes the announced connection with [24]; more precisely, the same inequality follows from combining Theorems 4.2 and 6.1 therein.

3 Elliptic case: main results

Now consider the elliptic operator $B = -X_1^2 - X_2^2 - X_3^2$ in $L^2(\mathbb{R}^3)$. For $\Omega \subset \mathbb{R}^3$ an open set of finite measure we define as before the Dirichlet operator $B_\Omega := P_\Omega B P_\Omega$. In some contrast with Theorem 2.1 we prove

Theorem 3.1 *Let $\Omega \subset \mathbb{R}^3$ be a domain of finite measure and let B_Ω be the operator B with Dirichlet boundary conditions in Ω. Then the spectrum of B_Ω is discrete and*

$$\mathrm{tr}\varphi_\lambda(B_\Omega) = \sum_k (\lambda - \lambda_k)_+ \leq \frac{1}{15\pi^2} |\Omega| \lambda^{5/2}.$$

Remark 3.2 *The latter expression coincides which the semi-classical constant appearing in the phase-volume Weyl-type asymptotics. Indeed,*

$$\lim_{\lambda \to \infty} \lambda^{-5/2} \mathrm{tr}\varphi_\lambda(A_\Omega)$$

$$= \frac{1}{(2\pi)^3} \lim_{\lambda \to \infty} \lambda^{-5/2} \int_\Omega \int_{\mathbb{R}^3} (\lambda - a(x,\xi))_+ \, d\xi \, dx = \frac{1}{15\pi^2} |\Omega|,$$

where $a(x,\xi) = \left(\xi_1 + \frac{1}{2}x_2\xi_3\right)^2 + \left(\xi_2 - \frac{1}{2}x_1\xi_3\right)^2 + \xi_3^2$.

4 Hypoelliptic case: results obtained by heat-kernel techniques

So far we have restricted our study of the Dirichlet Heisenberg Laplacian A_Ω to the case where Ω is a domain of finite measure. Put differently, we have studied the Schrödinger-type operator $A - V$ for *potential wells*, i.e., with V being infinite outside and constant inside Ω. However, using heat-kernel methods we obtain the following upper bound on the counting function without this restriction on V.

Theorem 4.1 *Let $N(A - V)$ be the number of negative eigenvalues of $A - V$. Then for any $V \in L^2(\mathbb{R}^3)$.*

$$N(A - V) \leq C \int_{\mathbb{R}^3} V(x)_+^2 \, dx$$

with

$$C = \min_{a>0} \frac{1}{64a} \frac{1}{e^{-a} + a \operatorname{Ei}(-a)} \geq 0.09429.$$

Note that this result is not an improvement in the case of potential wells; then already Corollary 2.1 gives us the constant $9/2^8 \approx 0.03516$. However, the asymptotics on the moments of the eigenvalues can be obtained via heat-kernel estimates in conjuction with Karamata's well-known Tauberian theorem. [10]

Theorem 4.2 *Let (λ_j) be a non-decreasing sequence of non-negative numbers such that $C = \lim_{t \to 0} t^\alpha \sum_{j=0}^{\infty} e^{-t\lambda_j}$ exists for some $\alpha > 0$. Then for any $f \in C([0,1])$,*

$$\lim_{t \to 0} t^\alpha \sum_{j=0}^{\infty} f(e^{-t\lambda_j}) e^{-t\lambda_j} = \frac{C}{\Gamma(\alpha)} \int_0^\infty f(e^{-t}) t^{\alpha-1} e^{-t} dt.$$

In Section 9, in order to prove Theorem 4.1 we derive the equality

$$e^{-tA}(x,x) = \frac{1}{32t^2},$$

which implies

$$\operatorname{tr} e^{-tA_\Omega} \leq \frac{1}{32t^2} |\Omega|. \tag{4.1}$$

If the converse inequality also holds – which is still to be proved – so ! that

$$\lim_{t \to \infty} t^{-2} \operatorname{tr} e^{-tA_\Omega} = \frac{1}{32} |\Omega|,$$

then this implies, by Theorem 4.2 with $f(s) = s^{-1} \chi_{(1/e,1]}$,

$$\lim_{\lambda \to \infty} \lambda^{-2} N(\lambda) = \frac{1}{32\Gamma(3)} |\Omega|.$$

This gives us the asymptotic counterpart of Corollary 2.1, namely

Corollary 4.1 *Let $N(\lambda)$ be the counting function of A_Ω, see (2.2). Then under the conditions of Theorem 2.1 we have*

$$\lim_{t \to \infty} \lambda^{-2} N(\lambda) = \frac{1}{64} |\Omega|.$$

If we apply Theorem 4.2 with $f(s) = \lambda s^{-1}(1 + \log s) \chi_{(1/e,1]}$ instead (still assuming (4.1) is true), we get

Corollary 4.2 *Under the conditions of Theorem 2.1 we have*

$$\lim_{\lambda \to \infty} \lambda^{-3} \sum_k (\lambda - \lambda_k)_+ = \frac{1}{192} |\Omega|.$$

As we pointed out after Corollary 2.2, this is a better constant than what one would have got by straightforward Aizenmann-Lieb extrapolation.

5 Hypoelliptic case: generalisation to $\mathbb{R}^{2N+1}$

The Lie algebra on the Nth Heisenberg group $\mathbf{H}^N$ is spanned by vector fields $X_1, X_2, \ldots, X_{2N+1}$ in $\mathbb{R}^{2N+1}$. If we let

$$X_{2k-1} := \partial_{x_{2k-1}} + \frac{1}{2} x_{2k} \partial_{x_{2N+1}}, \qquad X_{2k} := \partial_{x_{2k}} - \frac{1}{2} x_{2k-1} \partial_{x_{2N+1}}$$

for $1 \leq k \leq N$ and $X_{2N+1} := \partial_{x_{2N+1}}$, then the only non-zero commutator is $[X_{2k-1}, X_{2k}] = -X_{2N+1}$.

This is the $2N + 1$-dimensional version of Theorem 2.1:

Theorem 5.1 *Let $\Omega \subset \mathbb{R}^{2N+1}$ be a domain of finite measure and let $A_{N,\Omega}$, $N \geq 1$, be the operator corresponding to the differential expression*

$$- \sum_{k=1}^{2N} X_k^2$$

defined in a self-adjoint way in $L^2(\mathbb{R}^2)$ with Dirichlet boundary conditions. The spectrum of $A_{N,\Omega}$ is discrete and

$$\mathrm{tr}\varphi_\lambda(A_{N,\Omega}) \leq \frac{2c_N}{(2\pi)^{N+1}(N+1)(N+2)} |\Omega| \lambda^{N+2}$$

with

$$c_N := \sum_{n_1,\ldots,n_N \geq 0} \frac{1}{(2(n_1 + \ldots + n_N) + N)^{N+1}}. \tag{5.1}$$

Remark 5.2 *The multiple sum (5.1) can be expressed as a single sum,*

$$c_N = \sum_{k=0}^{\infty} \binom{N+k-1}{k} \frac{1}{(2k+N)^{N+1}}.$$

Unaware of a closed expression for this number we give approximate values for the first constants: $c_2 \approx 2.055 \cdot 10^{-1}$, $c_3 \approx 2.737 \cdot 10^{-2}$, $c_4 \approx 2.929 \cdot 10^{-3}$, $c_5 \approx 2.601 \cdot 10^{-4}$, $c_6 \approx 1.969 \cdot 10^{-5}$.

Without hardly modifying the proofs at all, we can extend Corollaries 2.1 and 2.2 into

Corollary 5.1 *Let $N(\lambda)$ be the counting function of $A_{N,\Omega}$, see (2.2). Then under the conditions of Theorem 5.1 we have*

$$N(\lambda) \leq \frac{2c_N (N+2)^{N+1}}{(2\pi)^{N+1}(N+1)^{N+2}} |\Omega| \lambda^{N+1}.$$

Corollary 5.2 *Under the conditions of Theorem 5.1 and for any $\gamma > 0$ the eigenvalues of $A_{N,\Omega}$ satisfy the Lieb-Thirring inequality*

$$\sum_k (\lambda - \lambda_k)_+^\gamma \leq K_{N,\gamma} |\Omega| \lambda^{N+\gamma+1}$$

with

$$K_{N,\gamma} = \begin{cases} \dfrac{2c_N (N+2)^{N+1}}{(2\pi)^{N+1}(N+1)} \dfrac{\gamma^\gamma}{(\gamma+N+1)^{\gamma+N+1}}, & 0 < \gamma \leq 1, \\ \dfrac{2c_N}{(2\pi)^{N+1}(N+1)(N+2)} \dfrac{6}{(\gamma+1)(\gamma+2)}, & 1 \leq \gamma. \end{cases}$$

6 Proof of Theorems 2.1 and 3.1

Denote by $\mathcal{F}_3$ the partial Fourier transform

$$\mathcal{F}_3 u(x', \xi_3) = (2\pi)^{-1/2} \int_{-\infty}^{\infty} e^{-ix_3\xi_3} u(x', x_3)\, dx_3, \qquad x' = (x_1, x_2).$$

Then,

$$\mathcal{F}_3 A \mathcal{F}_3^* = \left(i\partial_{x_1} - \frac{1}{2} x_2 \xi_3 \right)^2 + \left(i\partial_{x_2} + \frac{1}{2} x_1 \xi_3 \right)^2 = (i\nabla_{x'} + \xi_3 \mathbf{A}(x'))^2, \tag{6.1}$$

where $\mathbf{A}(x') = \frac{1}{2}(-x_2, x_1)$. This is a Laplacian in x' with constant magnetic field ξ_3, on which the Landau levels $\mu_n(\xi_3)$ depend:

$$\mu_n(\xi_3) := |\xi_3|(2n+1), \qquad n \in \mathbb{N}_0 = \{0, 1, 2, \ldots\}.$$

Note that $\mathcal{F}_3 A \mathcal{F}_3^*$ acts as a multiplication operator with respect to the variable ξ_3. By the Spectral theorem,

$$\mathcal{F}_3 A \mathcal{F}_3^* = \sum_{n=0}^{\infty} \mu_n(\xi_3) \Pi_{\xi_3, n}, \qquad \Pi_{\xi_3, n} = \Pi'_{\xi_3, n} \otimes I_{L^2(\mathbb{R})},$$

where the projector $\Pi'_{\xi_3,n}$ has an explicit representation as an integral kernel depending on x' (see, e.g., [9] or [22]), which is constant on the diagonal:

$$\Pi'_{\xi_3,n}(x',x') = \frac{|\xi_3|}{2\pi}. \tag{6.2}$$

Let $P_\Omega : L^2(\mathbb{R}^3) \to L^2(\Omega)$ be the operator of multiplication by χ_Ω, the characteristic function of Ω. The operator A_Ω, which is A with Dirichlet boundary conditions in Ω, can be identified with the operator $P_\Omega A P_\Omega$. By the Berezin-Lieb inequality (see [2], [3], [15] and also [13]), we find

$$\begin{aligned}
\mathrm{tr}\varphi_\lambda(A_\Omega) &\leq \mathrm{tr} P_\Omega \varphi_\lambda(A) P_\Omega \\
&= \frac{1}{2\pi} \int_\Omega \int_{-\infty}^\infty \sum_{n=0}^\infty \varphi_\lambda(\mu_n(\xi_3)) \Pi_{\xi_3,n}(x',x') d\xi_3 dx \\
&= \frac{1}{2\pi^2} |\Omega| \sum_{n=0}^\infty \int_0^\infty (\lambda - \xi_3(2n+1))_+ \xi_3 d\xi_3.
\end{aligned}$$

Substituting $s = \xi_3(2n+1)$ we have

$$\begin{aligned}
\mathrm{tr}\varphi_\lambda(A_\Omega) &\leq \frac{1}{2\pi^2} |\Omega| \sum_{n=0}^\infty \frac{1}{(2n+1)^2} \int_0^\infty (\lambda - s)_+ s \, ds \\
&= \frac{1}{2\pi^2} |\Omega| \frac{\pi^2}{8} \frac{\lambda^3}{6} = \frac{1}{96} |\Omega| \lambda^3.
\end{aligned}$$

This completes the proof of Theorem 2.1.

Similarly to (6.1) we have

$$\begin{aligned}
\mathcal{F}_3 B \mathcal{F}_3^* &= \mathcal{F}_3 (-(\partial_{x_1} + x_2\partial_{x_3})^2 - (\partial_{x_2} - x_1\partial_{x_3})^2 - \partial_{x_3}^2) \mathcal{F}_3^* \\
&= (i\partial_{x_1} + x_2\xi_3)^2 + (i\partial_{x_2} - x_1\xi_3)^2 + \xi_3^2 \\
&= (i\nabla_{x'} + \xi_3 \mathbf{A}(x'))^2 + \xi_3.
\end{aligned}$$

The eigenvalues of this operator are again functions of ξ_3,

$$\nu_n(\xi_3) := |\xi_3|(2n+1) + |\xi_3|^2, \qquad n \in \mathbb{N}_0.$$

Applying the Berezin-Lieb inequality and recycling some of the computations above we find

$$\mathrm{tr}\varphi_\lambda(B_\Omega) \leq \mathrm{tr}P_\Omega\varphi_\lambda(B)P_\Omega$$

$$= \frac{1}{(2\pi)^2}|\Omega|\int_{-\infty}^{\infty}|\xi_3|\sum_{k=0}^{\infty}(\lambda - |\xi_3|(2k+1) - |\xi_3|^2)_+ \, d\xi_3$$

$$\leq \frac{1}{(2\pi)^2}|\Omega|\int_{-\infty}^{\infty}\int_0^{\infty}|\xi_3|(\lambda - 2|\xi_3|t - |\xi_3|^2)_+ \, dt \, d\xi_3$$

$$= \frac{1}{2^3\pi^2}|\Omega|\int_{-\infty}^{\infty}\int_0^{\infty}(\lambda - t - s^2)_+ \, dt \, ds$$

$$= \frac{1}{2^4\pi^2}|\Omega|\int_{-\infty}^{\infty}(\lambda - s^2)_+^2 \, ds = \frac{1}{15\pi^2}|\Omega|\lambda^{5/2}$$

as claimed.

7 Proof of Corollary 2.1

In order to prove Corollary 2.1 we take up an idea from [12]. Obviously, for any $\tau > 0$,

$$N(\lambda) \leq \frac{1}{\tau\lambda}\sum_k((1+\tau)\lambda - \lambda_k)_+ \leq \frac{1}{96}|\Omega|\lambda^2\frac{(1+\tau)^3}{\tau}. \qquad (7.1)$$

The minimum value of $(1+\tau)^3\tau^{-1}$ is reached at $\tau = 1/2$. Substituting this value into (7.1) we obtain Corollary 2.1.

8 Proof of Corollary 2.2

The first part of the proof, concerning $0 < \gamma < 1$, is based on the following result obtained in [7]

Lemma 8.1 *Let* $0 \leq \gamma < 1$ *and* $\lambda < \mu$. *Then for all* $E \geq 0$,

$$(\lambda - E)_+^\gamma \leq C(\gamma)(\mu - \lambda)^{\gamma-1}(\mu - E)_+$$

with $C(\gamma) := \gamma^\gamma(1-\gamma)^{1-\gamma}$.

The proof is elementary. Combining this inequality with Theorem 2.1 we have, for $0 < \gamma < 1$ and any $\mu > \lambda$,

$$\sum_k(\lambda - \lambda_k)_+^\gamma \leq C(\gamma)(\mu - \lambda)^{\gamma-1}\sum_k(\mu - \lambda_k)_+$$

$$\leq C(\gamma)(\mu - \lambda)^{\gamma-1}\frac{1}{96}|\Omega|\mu^3 \leq \frac{9}{32}\frac{\gamma^\gamma}{(\gamma+2)^{\gamma+2}}|\Omega|\lambda^{\gamma+2}$$

(choose $\mu = 3\lambda/(\gamma + 2)$).

To determine K_γ for $\gamma > 1$ we will use the identity

$$B(\gamma - 1, 2)(\lambda - t)_+^\gamma = \int_0^\infty (\lambda - t - \mu)_+ \mu^{\gamma - 2} d\mu, \qquad \gamma > 1,$$

B being the beta function. Writing $\sum(\lambda - \lambda_k)_+^\gamma = \int(\lambda - t)_+^\gamma dN(t)$ and applying Fubini's theorem,

$$B(\gamma - 1, 2)\lambda^{-\gamma-2} \sum_k (\lambda - \lambda_k)_+^\gamma = \lambda^{-\gamma-2} \int_0^\infty \mu^{\gamma-2} \sum_k (\lambda - \mu - \lambda_k)_+ d\mu$$

$$\leq K_1 |\Omega| \int_0^1 s^3 (1-s)^{\gamma-2} ds$$

$$= K_1 |\Omega| B(4, \gamma - 1)$$

by Theorem 2.1. Taking the supremum of the left-hand side we obtain, by the definition of K_γ,

$$K_\gamma \leq K_1 \frac{B(4, \gamma - 1)}{B(2, \gamma - 1)} = \frac{6}{(\gamma + 1)(\gamma + 2)} K_1 = \frac{1}{16(\gamma + 1)(\gamma + 2)}, \qquad \gamma > 1.$$

9 Proof of Theorem 4.1

In this section we assume (U, μ) to be a σ-finite measure space. Any selfadjoint non-negative operator H in $L^2(U, \mu)$ generates a contractive semigroup $\{e^{-tH}\}_{t \in \mathbb{R}_+}$. We will assume that the corresponding integral kernels $e^{-tH}(x, y) \geq 0$ a.e. in $\mathbb{R}_+ \times U \times U$. Put

$$M_H(t) := \|e^{-tH/2}\|_{L^2 \to L^\infty}^2$$

and suppose this quantity is bounded for all $t > 0$, $M_H(t) = \mathcal{O}(t^\alpha)$, $\alpha > 0$ at zero and integrable at infinity. Moreover, let $G(s)$ be a function on $\mathbb{R}_+$, polynomially growing at infinity and such that $G(s)/s$ is integrable at $s = 0$. We associate to any such G the function

$$g(\sigma) := \int_0^\infty \frac{G(s)}{s} e^{-s/\sigma} ds.$$

(Note that $g(1/\sigma)$ is the Laplace transform of $G(s)/s$.) We shall apply the following bound on the counting function, which is occasionally referred to as Lieb's Formula (see [16] and [25]).

Theorem 9.1 *Let H be an operator satisfying the hypotheses above and let $N(H - V)$ be the number of negative eigenvalues of $H - V$. For any*

$V \in L^2(U, \mu)$ *and any admissible* G,

$$N(H - V) \le \frac{1}{g(1)} \int_0^\infty \frac{dt}{t} \int_U M_H(t) G(tV(x)) \mu(dx). \qquad (9.1)$$

When M_H is a homogeneous function, this statement can be simplified in the following way. Writing $M_H(t) = K/t^{\nu/2}$ (and for ease of notation dx instead of $\mu(dx)$) and applying Fubini's theorem to the right-hand side, we find

$$\frac{K}{g(1)} \int_U \left(\int_0^\infty \frac{G(tV(x))}{t^{\nu/2+1}} dt \right) dx = \frac{K}{g(1)} \int_0^\infty \frac{G(s)}{s^{\nu/2+1}} ds \int_U V(x)^{\nu/2} dx.$$

In this case, (9.1) apparently is a (ν-dimensional) Cwikel-Lieb Rozenblyum inequality, whose constant only depends on the power ν.

It is easy to check that the operator A in $L^2(\mathbb{R}^3)$ fits into this framework and so $N(A - V)$ can be estimated using Theorem 9.1. Similarly to the proof of Theorem 2.1 we calculate, for $t > 0$, $x \in \mathbb{R}^3$,

$$e^{-tA}(x, x) = (\mathcal{F}_3^* e^{-t\mathcal{F}_3 A \mathcal{F}_3^*} \mathcal{F}_3)(x, x)$$

$$= (2\pi)^{-1} \int_{-\infty}^\infty \sum_{n=0}^\infty e^{-\mu_n(\xi_3)t} \Pi_{\xi_3,n}(x, x) d\xi_3$$

$$= \pi^{-2} \sum_{n=0}^\infty \int_0^\infty \xi_3 e^{-2\xi_3(2n+1)t} d\xi_3$$

$$= \frac{1}{(2\pi t)^2} \sum_{n=0}^\infty \frac{1}{(2n+1)^2} \int_0^\infty s e^{-s} ds = \frac{1}{32 t^2}.$$

This is exactly $M_A(t) = \operatorname{ess\,sup}_{x \in \mathbb{R}^3} e^{-tA}(x, x)$, so $\nu = 4$. One can argue (see [16]) that the optimal G is of the form $G(s) = (s - a)_+$, $a > 0$. Clearly, since

$$\int_0^\infty \frac{G(s)}{s^{\nu/2+1}} ds = \frac{1}{2a} \quad \text{and} \quad g(1) = \int_a^\infty \left(1 - \frac{a}{s} \right) e^{-s} ds = e^{-a} + a \operatorname{Ei}(-a),$$

the best constant is the minimum of $[64a(e^{-a} + a \operatorname{Ei}(-a))]^{-1}$, as stated in Theorem 4.1.

10 Proof of Theorem 5.1

Extending our previous definitions we let

$$\mathcal{F}_{2N+1} u(x', \xi_{2N+1}) = (2\pi)^{-1/2} \int_{-\infty}^\infty e^{-ix_{2N+1}\xi_{2N+1}} u(x', x_{2N+1}) dx_{2N+1},$$

$x' = (x_1, \ldots, x_{2N})$ and, for future convenience, $x'_k = (x_{2k-1}, x_{2k})$. Then,

$$\mathcal{F}_{2N+1} A_N \mathcal{F}_{2N+1}^* = -\sum_{k=1}^{N} \mathcal{F}_{2N+1}(X_{2k-1}^2 + X_{2k}^2)\mathcal{F}_{2N+1}^*$$

$$= \sum_{k=1}^{N}(i\nabla_{x'_k} + \xi_{2N+1}\mathbf{A}(x'_k))^2,$$

where $\mathbf{A}$ remains as in Section 6. The eigenvalues of this operator are

$$\mu_{\mathbf{n}}(\xi_{2N+1}) := |\xi_{2N+1}|(2|\mathbf{n}| + N), \qquad \mathbf{n} \in \mathbb{N}_0^N,$$

where $|\mathbf{n}| := n_1 + \ldots + n_N$. By the Spectral theorem,

$$\mathcal{F}_{2N+1} A_N \mathcal{F}_{2N+1}^* = \sum_{\mathbf{n}\in\mathbb{N}_0^N} \mu_{\mathbf{n}}(\xi_{2N+1})\Pi_{\xi_{2N+1},\mathbf{n}},$$

where $\Pi_{\xi_{2N+1},\mathbf{n}} = \Pi'_{\xi_{2N+1},n_1} \otimes \ldots \otimes \Pi'_{\xi_{2N+1},n_N} \otimes I_{L^2(\mathbb{R})}$ with Π' as in Section 6. Clearly,

$$\Pi_{\xi_{2N+1},\mathbf{n}}(x',x') = \frac{|\xi_{2N+1}|^N}{(2\pi)^N},$$

so that

$$\mathrm{tr}\varphi_\lambda(A_{N,\Omega}) \leq \mathrm{tr} P_\Omega \varphi_\lambda(A_N) P_\Omega$$

$$= \frac{1}{2\pi} \int_\Omega \int_{-\infty}^{\infty} \sum_{\mathbf{n}\in\mathbb{N}_0^N} \varphi_\lambda(\mu_{\mathbf{n}}(\xi_{2N+1}))\Pi_{\xi_{2N+1},\mathbf{n}}(x',x')d\xi_{2N+1}dx$$

$$= \frac{2}{(2\pi)^{N+1}}|\Omega| \sum_{\mathbf{n}\in\mathbb{N}_0^N} \int_0^{\infty} (\lambda - \xi_{2N+1}(2|\mathbf{n}| + N))_+ \xi_{2N+1}^N d\xi_{2N+1}$$

$$= \frac{2}{(2\pi)^{N+1}}|\Omega| \sum_{\mathbf{n}\in\mathbb{N}_0^N} \frac{1}{(2|\mathbf{n}| + N)^{N+1}} \int_0^{\infty} (\lambda - s)_+ s^N ds$$

$$= \frac{2}{(2\pi)^{N+1}}|\Omega|c_N \frac{\lambda^{N+2}}{(N+1)(N+2)}$$

with c_N as in (5.1).

11 Proof of Corollary 2.3

The asymptotics stated in the corollary will follow from Corollary 2.2 if we can prove the converse bound

$$\sum_k (\lambda - \lambda_k)_+^\gamma \geq K_\gamma |\Omega| \lambda^{\gamma+2} + o(\lambda^{\gamma+2}), \qquad \lambda \to \infty, \ \gamma \geq 1. \quad (11.1)$$

We first assume the domain to be a cylinder, $\Omega = \Omega' \times (0, L_3)$. By Theorem 2.1 the spectrum of A_Ω consists of eigenvalues λ_k of finite multiplicity. If we denote the corresponding $L^2(\Omega)$-normalised eigenfunctions by ω_k then by Plancherel's theorem $\|\mathcal{F}_3 \omega_k\|^2_{L^2(\Omega' \times \mathbb{R})} = 2\pi$. As explained in the proof of Theorem 2.1, the eigenstate of $\mathcal{F}_3 A \mathcal{F}_3^*$ corresponding to $\mu_n(\xi_3)$ has infinite dimensionality and is spanned by the eigenfunctions $(\tau_{\xi_3,m,n})_{m=-n}^\infty$, which we assume to be normalised in $L^2(\Omega)$. Evidently, the projection kernel has the form

$$\Pi'_{\xi_3,n}(x', y') = \sum_{m=-n}^\infty \tau_{\xi_3,m,n}(x')\overline{\tau_{\xi_3,m,n}(y')}.$$

We finally introduce a cut-off function $\chi_\varepsilon \in C_0^\infty(\Omega)$ such that $\chi_\varepsilon(x) = 1$ if $\mathrm{dist}(x, \partial\Omega) \geq \varepsilon$ for $\varepsilon > 0$, and will use $\chi_\varepsilon(x)e^{ix_3\xi_3}\tau_{\xi_3,m,n}(x')$ as an approximate eigenfunction.

Setting $\varphi(t) := (\lambda - t)_+^\gamma$ we have

$$\sum_k \varphi(\lambda_k) = \sum_k \varphi(\lambda_k)\|\omega_k\|_2^2$$

$$= \frac{1}{2\pi} \sum_k \varphi_\lambda(\lambda_k) \sum_{m,n} \int_{-\infty}^\infty |(\omega_k, e^{ix_3\xi_3}\tau_{\xi_3,m,n})|^2 d\xi_3$$

$$\geq \frac{1}{2\pi} \sum_k \varphi(\lambda_k) \sum_{m,n} \int_{-\infty}^\infty |(\omega_k, \chi_\varepsilon e^{ix_3\xi_3}\tau_{\xi_3,m,n})|^2 d\xi_3$$

$$= \frac{1}{2\pi} \int_{-\infty}^\infty \sum_{m,n} \int_0^\infty \varphi(\nu)(dE_\nu \chi_\varepsilon \tau_{\xi_3,m,n}, \chi_\varepsilon \tau_{\xi_3,m,n}) d\xi_3, \quad (11.2)$$

where E_ν is the spectral measure of $\mathcal{F}_3 A_\Omega \mathcal{F}_3^*$. However,

$$\sum_m \int_0^\infty (dE_\nu \chi_\varepsilon \tau_{\xi_3,m,n}, \chi_\varepsilon \tau_{\xi_3,m,n})$$

$$= \sum_m \|\chi_\varepsilon \tau_{\xi_3,m,n}\|_2^2 = \int_{\mathbb{R}^3} \chi_\varepsilon^2(x) \sum_m |\tau_{\xi_3,m,n}(x')|^2 dx$$

$$= \int_{\mathbb{R}^3} \chi_\varepsilon^2(x)\Pi'_{\xi_3,n}(x', x') dx = \frac{|\xi_3|}{2\pi} \int_{\mathbb{R}^3} \chi_\varepsilon^2(x) dx,$$

which implies

$$1 - C_1\varepsilon \leq \frac{2\pi}{|\xi_3||\Omega|} \sum_m \int_0^\infty (dE_\nu \chi_\varepsilon \tau_{\xi_3,m,n}, \chi_\varepsilon \tau_{\xi_3,m,n}) \leq 1 - c_1\varepsilon,$$

where $C_1 \geq c_1 > 0$ depend on Ω. The convexity of φ gives us, by

Jensen's inequality, the following lower bound to (11.2):

$$\frac{1-C_1\varepsilon}{(2\pi)^2}|\Omega|\int_{-\infty}^{\infty}\sum_n\varphi\left(\frac{1}{1-c_1\varepsilon}\sum_m\int_0^{\infty}\nu(dE_{\nu}\chi_{\varepsilon}\tau_{\xi_3,m,n},\chi_{\varepsilon}\tau_{\xi_3,m,n})\right)|\xi_3|d\xi_3$$

$$\geq\frac{1-C_1\varepsilon}{(2\pi)^2}|\Omega|\int_{-\infty}^{\infty}\sum_n\varphi\left(\mu_n(\xi_3)+\frac{C_2\varepsilon^{-2}}{1-c_1\varepsilon}\right)|\xi_3|d\xi_3$$

$$=\frac{1-C_1\varepsilon}{16}|\Omega|\int_0^{\infty}\left(\lambda-\frac{C_2\varepsilon^{-2}}{1-c_1\varepsilon}-s\right)_+^{\gamma}s\,ds$$

$$=(1-C_1\varepsilon)K_{\gamma}|\Omega|\left(\lambda-\frac{C_2\varepsilon^{-2}}{1-c_1\varepsilon}\right)^{\gamma+2}$$

similarly to the proof of Theorem 2.1. Hence (11.1) holds for cylinders. By the variational principle, the lower bound obtained when we approximate an arbitrary domain by decoupled cylinders subject to Dirichlet boundary conditions will not be less than the correct one. Since the constant actually coincides with that of the upper bound, we have proved the statement.

Acknowledgements

The authors are grateful for partial support from the ESF European Programme "SPECT" and have benefited from the excellent working environment at the Isaac Newton Institute in Cambridge. A. Laptev would also like to thank Prof. W. Hebisch for useful discussions.

Bibliography

[1] M. Aizenmann and E. H. Lieb, *On semi-classical bounds for eigenvalues of Schrödinger operators*. Phys. Lett. A **66** (1978) 427–429.

[2] F. A. Berezin, *Convex functions of operators*, Mat. sb. **88** (1972), 268–276.

[3] F. A. Berezin, *Covariant and contravariant symbols of operators*. Math. USSR Izv. **6** (1972), 1117–1151.

[4] M. Sh. Birman and M. Z. Solomyak, *The principal term of the spectral asymptotics formula for "non-smooth" elliptic problems*. Functional Anal. Appl., **4** (1970), 265–275.

[5] Z. Ciesielski, *On the spectrum of the Laplace operator*. Comment. Math. Prace Mat. **14** (1970), 41–50.

[6] L. Erdős, M. Loss and V. Vougalter, *Diamagnetic behavior of sums of Dirichlet eigenvalues*. Ann. Inst. Fourier (Grenoble) **50** (2000), 891–907.

[7] R. L. Frank, M. Loss and T. Weidl, *Eigenvalue estimates of the magnetic Laplacian in a domain*. In preparation.

[8] L. Hörmander, *The Analysis of Linear Partial Differential Operators III. Pseudo-Differential Operators.* Springer-Verlag, Berlin-Heidelberg-New York, 1985.

[9] T. Hupfer, H. Leschke and S. Warzel, *Upper bounds on the density of states of single Landau levels broadened by Gaussian random potentials.* J. Math. Phys. **42** (2001), 5626–5641.

[10] J. Karamata, *Neuer Beweis und Verallgemeinerung einiger Tauberian-Sätze.* Math. Z. **33** (1931), no. 1, 294–299.

[11] P. Kröger, *Estimates for sums of eigenvalues of the Laplacian.* J. Funct. Anal. **126** (1994), 217–227.

[12] A. Laptev, *Dirichlet and Neumann Eigenvalue Problems on Domains in Euclidean Spaces.* J. Func. Anal. **151** (1997), 531–545.

[13] A. Laptev and Yu. Safarov, *A generalization of the Berezin-Lieb inequality.* Amer. Math. Soc. Transl (2) **175** (1996), 69–79.

[14] A. Laptev and T. Weidl, *Recent results on Lieb-Thirring inequalities.* Journées "Équations aux Dérivées Partielles" (La Chapelle sur Erdre, 2000), Exp. No. XX, Univ. Nantes, Nantes (2000).

[15] E. H. Lieb, *The classical limit of quantum spin systems.* Comm. Math. Phys. **31** (1973), 327–340.

[16] E. H. Lieb, *The number of bound states of one-body Schrödinger operators and the Weyl problem.* Proc. Sym. Pure Math. **XXXVI**, 241–252. Amer. Math. Soc., Providence RI, 1980.

[17] E. H. Lieb and W. Thirring, *Inequalities for the moments of the eigenvalues of the Schrödinger Hamiltonian and their relation to Sobolev inequalities.* Studies in Mathematical Physics, Essays in Honor of Valentine Bargmann, 269–303. Princeton University Press, Princeton NJ, 1976.

[18] P. Li and S.-T. Yau, *On the Schrödinger equation and the eigenvalue problem.* Comm. Math. Phys. **88** (1983), 309–318.

[19] G. Métivier, *Fonction spectrale et valeurs propres d'une classe d'opérateurs non elliptiques.* Comm. Partial Differential Equations **1** (1976), 467–519.

[20] G. Métivier, *Valeurs propres de problèmes aux limites elliptiques irréguliers.* Bull. Soc. Math. France, Mem. **51–52** (1977), 125–229.

[21] G. Pólya, *On the eigenvalues of vibrating membranes.* Proc. London Math. Soc. **11** (1961), 419–433.

[22] G. D. Raikov and S. Warzel, *Quasi-classical versus non-classical spectral asymptotics for magnetic Schrödinger operators with decreasing electric potentials.* Rev. Math. Phys. **14** (2002), no. 10, 1051–1072.

[23] G. V. Rozenblyum, *On the eigenvalues of the first boundary problem in unbounded domains.* Mat. sb. **89** (1972), 234–247.

[24] R. S. Strichartz, *Estimates for Sums of Eigenvalues for Domains in Homogeneous Spaces.* J. Func. Anal. **137** (1996), 152–190.

[25] G. V. Rozenblyum and M. Z. Solomyak, *The Cwikel-Lieb-Rozenblyum estimator for generators of positive semigroups and semigroups dominated by positive semigroups.* St. Petersburg Math. J. **9** (1998), 1195–1211.

[26] M. E. Taylor, *Noncommutative harmonic analysis.* Mathematical Surveys and Monographs 22. American Mathematical Society, Providence RI, 1986.

6

Equidistribution for quadratic differentials

Ursula Hamenstädt

Mathematisches Institut
Rheinische Friedrich-Wilhelms-Universität Bonn
ursula@math.uni-bonn.de

1 Introduction

Hermann Weyl was one of the most influential mathematicians of the first half of the twentieth century. He was born in 1885 in Elmshorn. In 1933 he emigrated to the United States, and he died in 1955 in Zürich. The majority of his many fundamental contributions to mathematics belong to the area of analysis in the broadest possible sense. However, one of his earliest and most celebrated results can be viewed as the origin of the study of number theory with tools from dynamical systems.

His theorem, published in 1916 in the "Mathematische Annalen" [W16], is as follows.

Weyl's theorem: *Let $y_0 \in (0,1)$ be irrational. Then the sequence $(u_i)_{i \geq 1}$ defined by $u_i = iy_0 \mod 1$ is asymptotically equidistributed: For all $0 < a < b < 1$ we have*

$$\frac{|\{1 \leq i \leq n : a \leq u_i \leq b\}|}{n} \to b - a \quad (n \to \infty).$$

In this note we explain how the idea behind this theorem was used in the last quarter of the twentieth century to gain surprising insights into the interplay between number theory, geometry and dynamical systems.

2 Classical dynamical systems

In this section we discuss how Weyl's theorem can be reformulated in the language of dynamical systems, and we introduce some basic concepts which will be important in the later sections.

Let $S^1 = \{e^{it} \mid t \in [0, 2\pi)\} \subset \mathbb{C}$ be the standard unit circle in the complex plane. Then S^1 is an abelian group with multiplication $e^{it} \times$

$e^{is} = e^{i(t+s)}$. In particular, every angle $\alpha \in (0, 2\pi)$ defines a cyclic group T_α^n of rotations of S^1 ($n \in \mathbb{Z}$) via

$$T_\alpha^n(e^{it}) = e^{it+n\alpha}.$$

Thus the rotation T_α with angle α generates a *dynamical system* with phase space S^1. Global (or asymptotic) properties of a dynamical system can be investigated with the help of *invariant measures*.

Definition 2 *A* Radon measure μ *(i.e. a locally finite Borel measure) on a locally compact topological space X is* invariant *under a Borel map T if $\mu(T^{-1}(A)) = \mu(A)$ for every Borel set $A \subset X$.*

If X is a compact topological space then the space $\mathcal{P}(X)$ of *Borel probability measures* on X can be equipped with the *weak*-topology*. This weak*- topology is the weakest topology such that for every *continuous* function $f : X \to \mathbb{R}$ the function $\mu \in \mathcal{P}(X) \to \int f d\mu$ is continuous. In other words, a sequence $(\mu_i) \subset \mathcal{P}(X)$ converges to $\mu \in \mathcal{P}(X)$ if and only if for every open subset U of X we have $\liminf_{i\to\infty} \mu_i(U) \geq \mu(U)$. The space $\mathcal{P}(X)$ equipped with the weak*-topology is compact.

By compactness, every *continuous* transformation T of X admits an invariant Borel probability measure [Wa82]. Namely, for every point $x \in X$, any weak limit of the sequence of measures

$$\frac{1}{n} \sum_{i=0}^{n-1} \delta_{T^i x}$$

is T-invariant where δ_z is the *Dirac δ-measure* at z, defined by $\delta_z(\{z\}) = 1$ and $\delta_z(X - \{z\}) = 0$.

For our circle rotations, there are now two cases.

Case 1: α is a *rational multiple* of 2π, i.e $\alpha = 2p\pi/q$ for relatively prime $p, q \in \mathbb{N}$.

In this case we have $T_\alpha^q(e^{it}) = e^{it+2p\pi} = e^{it}$ for all t which means the following.

Every point in S^1 is periodic for T_α, with period independent of the point.

In particular, every point $y \in S^1$ is an *atom* of a T_α-invariant probability measure, namely the weighted counting measure on the orbit

$\{T_\alpha^i y \mid 0 \le i \le q-1\}$ of y. This measure is given by the formula

$$\mu = \frac{1}{q} \sum_{i=0}^{q-1} \delta_{T_\alpha^i y}.$$

Case 2: α is an *irrational multiple* of 2π, i.e. $\alpha = 2\pi\rho$ for an irrational number $\rho \in (0,1)$.

In this case, T_α does not have periodic points, and Weyl's theorem says precisely the following: For each $y \in S^1$,

$$\frac{1}{n} \sum_{i=0}^{n-1} \delta_{T_\alpha^i y} \to \lambda$$

weakly in the space of probability measures on S^1 where λ is the normalized standard Lebesgue measure on S^1 defined by $\lambda\{e^{is} \mid 0 \le \alpha < s < \beta \le 2\pi\} = (\beta - \alpha)/2\pi$.

For a continuous map T of a compact space X, the space $\mathcal{P}(X)_T$ of T-invariant Borel probability measures on X is convex: If μ_1, μ_2 are two such measures and if $s \in [0,1]$ then $s\mu_1 + (1-s)\mu_2 \in \mathcal{P}(X)_T$ as well. In other words, $\mathcal{P}(X)_T$ is a compact and convex subset of a topological vector space on which the dual separates points. Hence $\mathcal{P}(X)_T$ is the convex hull of the set of its *extreme* points.

An extreme point $\mu \in \mathcal{P}(X)_T$ is an *ergodic* invariant measure: If $A \subset X$ is a T-invariant Borel set then $\mu(A) = 0$ or $\mu(X - A) = 0$. By the *Birkhoff ergodic theorem* [Wa82], every extreme point $\mu \in \mathcal{P}(X)_T$ is a weak limit of measures of the form $\frac{1}{q} \sum_{i=0}^{q-1} \delta_{T^i y}$ for a suitable choice of $y \in X$. Note that the definition of ergodicity also makes sense for Radon measures on locally compact spaces which are invariant under a continuous transformation.

Definition 3 *A continuous transformation T of a compact space X is called* uniquely ergodic *if T admits a* unique *invariant Borel probability measure.*

An invariant Borel probability measure μ for a uniquely ergodic continuous transformation T of a compact space X is necessarily ergodic. Now Weyl's theorem can be rephrased as follows.

Irrational rotations of the circle are uniquely ergodic.

However, this also means the following.

If α is an irrational multiple of 2π then a measure which is invariant under T_α is invariant under the full *circle group of rotations.*

---

3 The modular group and hyperbolic geometry

About 1970, the significance of Weyl's theorem became apparent in a somewhat unexpected way and in a different context. This development began with the work of Hillel Furstenberg. Furstenberg was born in 1935 in Berlin and moved shortly later with his family to the United States. He now works at the Hebrew University in Jerusalem (Israel). Furstenberg was interested in lattices in semi-simple Lie groups G of non-compact type and their actions on homogeneous spaces associated to G. A large part of the structure theory for semi-simple Lie groups is due to Hermann Weyl, but it seems that he never attempted to draw a close connection between the structure of Lie groups, their actions on homogeneous spaces and his number theoretic result which we discussed in Section 1.

In this section we explain Furstenberg's work and its generalizations which are entirely in the spirit of Weyl's theorem.

Consider the *modular group*

$$SL(2, \mathbb{Z}) = \{ \begin{pmatrix} a & b \\ c & d \end{pmatrix} \mid a, b, c, d \in \mathbb{Z}, ad - bc = 1 \}$$

which acts as a group of linear transformations on $\mathbb{R}^2$ preserving the usual area form. This action is given by

$$\begin{pmatrix} a & b \\ c & d \end{pmatrix} \begin{pmatrix} x \\ y \end{pmatrix} \to \begin{pmatrix} ax + by \\ cx + dy \end{pmatrix}.$$

There is an obvious $SL(2, \mathbb{Z})$-invariant subset of $\mathbb{R}^2$, namely the countable set $\mathbb{R}\mathbb{Q}^2 \subset \mathbb{R}^2$ of points whose coordinates are *dependent over* $\mathbb{Q}$ (which means that their quotient is rational). Since $SL(2, \mathbb{Z})$ preserves the integral lattice $\mathbb{Z}^2 \subset \mathbb{R}^2$, each $SL(2, \mathbb{Z})$-orbit of a point whose coordinates are dependent over $\mathbb{Q}$ is a *discrete* subset of $\mathbb{R}^2$. Hence this orbit supports an $SL(2, \mathbb{Z})$-invariant purely atomic ergodic Radon measure. For example, the measure

$$\mu = \sum_{y \in \mathbb{Z}^2} \delta_y$$

is an $SL(2, \mathbb{Z})$-invariant Radon measure. However, it is not ergodic since $\mathbb{Z}^2$ contains countably many orbits for the action of $SL(2, \mathbb{Z})$. Namely, the $SL(2, \mathbb{Z})$-orbit of the point $(1, 0) \in \mathbb{Z}^2$ consists precisely of all points (p, q) such that $p, q \in \mathbb{Z}$ are relatively prime. As a consequence, there is an uncountable family of $SL(2, \mathbb{Z})$-invariant Radon measures on $\mathbb{R}^2$.

Each ergodic measure in this family is a sum of weighted Dirac masses on a single $SL(2,\mathbb{R})$-orbit in $\mathbb{R}\mathbb{Q}^2$.

In contrast, extending earlier work of Furstenberg [F72], Dani [D78] proved in 1978 the following *unique ergodicity result.*

Theorem 3.1 (Unique ergodicity for the standard linear action of $SL(2,\mathbb{Z})$):
An $SL(2,\mathbb{Z})$-invariant Radon measure on $\mathbb{R}^2$ which gives full mass to the set of points whose coordinates are independent over $\mathbb{Q}$ coincides with the Lebesgue measure up to scale.

As a consequence, we have.

A Radon measure on $\mathbb{R}^2$ which is invariant under $SL(2,\mathbb{Z})$ and which gives full measure to points whose coordinates are independent over $\mathbb{Q}$ is invariant under the full *group $SL(2,\mathbb{R})$.*

The proof of this result does not use directly the fact that $SL(2,\mathbb{Z})$ acts on $\mathbb{R}^2$ by linear transformations. Instead, the group $SL(2,\mathbb{Z})$ is viewed as a *lattice* in the simple Lie group $SL(2,\mathbb{R})$. By this we mean that $SL(2,\mathbb{Z})$ is a discrete subgroup of $SL(2,\mathbb{R})$ with the following property. The group $SL(2,\mathbb{R})$ admits a natural Radon measure which is invariant under the action of $SL(2,\mathbb{R})$ on itself by right or left translation. This measure is given by a biinvariant volume form. By biinvariance, this volume form projects to a volume form on the quotient orbifold $SL(2,\mathbb{Z})\backslash SL(2,\mathbb{R})$ of finite total volume.

The quotient group $PSL(2,\mathbb{R})$ under the center $\mathbb{Z}/2\mathbb{Z}$ of $SL(2,\mathbb{R})$ admits a natural simply transitive action on the unit tangent bundle $T^1\mathbf{H}^2$ of the *hyperbolic plane* $\mathbf{H}^2$ and hence this unit tangent bundle can be identified with $PSL(2,\mathbb{R})$. Namely, we have $\mathbf{H}^2 = \{z = x + iy \in \mathbb{C} \mid \mathrm{Im}(z) > 0\}$ with the Riemannian metric

$$Q = \frac{dx^2 + dy^2}{y^2}$$

which is invariant under the action of $SL(2,\mathbb{R})$ by *linear fractional transformations*

$$z \to \frac{az + b}{cz + d} \quad \text{where} \quad \begin{pmatrix} a & b \\ c & d \end{pmatrix} \in SL(2,\mathbb{R}).$$

The subgroup of $SL(2,\mathbb{R})$ acting trivially is just the center of $SL(2,\mathbb{R})$ and hence this action factors to an action of $PSL(2,\mathbb{R})$. The hyperbolic plane $\mathbf{H}^2$ admits a compactification by adding the circle $\partial\mathbf{H}^2 = \mathbb{R} \cup \infty$, and the action of $PSL(2, \mathbb{R})$ on $\mathbf{H}^2$ extends to a transitive action on this circle by homeomorphisms.

There are three characteristic one-parameter subgroups of $SL(2, \mathbb{R})$.

(i) The *diagonal subgroup*

$$A = \{ \begin{pmatrix} e^t & 0 \\ 0 & e^{-t} \end{pmatrix} \mid t \in \mathbb{R} \}$$

(ii) The *upper unipotent group*

$$N = \{ \begin{pmatrix} 1 & t \\ 0 & 1 \end{pmatrix} \mid t \in \mathbb{R} \}$$

(iii) The *lower unipotent group*

$$U = \{ \begin{pmatrix} 1 & 0 \\ t & 1 \end{pmatrix} \mid t \in \mathbb{R} \}$$

These groups project to one-parameter subgroups of $PSL(2, \mathbb{R})$ which we denote by the same symbols.

The right action of the diagonal subgroup A on $PSL(2, \mathbb{R})$ defines the *geodesic flow* on $T^1 \mathbf{H}^2$. The right action of the group N of upper triangular matrices of trace two is the *horocycle flow* on $T^1 \mathbf{H}^2$. The group $PSL(2, \mathbb{R})$ acts transitively from the left on the *homogeneous space* $PSL(2, \mathbb{R})/N$.

Recall that the linear action of $SL(2, \mathbb{R})$ on $\mathbb{R}^2$ naturally induces an action of $PSL(2, \mathbb{R})$ on the punctured cone $\mathbb{R}^2 - \{0\}/\pm 1$. We have.

Lemma 3.1 *There is a homeomorphism*

$$F : \mathbb{R}^2 - \{0\}/\pm 1 \to PSL(2, \mathbb{R})/N$$

which commutes with the action of $PSL(2, \mathbb{R})$. This means that we have $B(Fz) = F(Bz)$ for all $z \in \mathbb{R}^2 - \{0\}/\pm 1$ and for all $B \in PSL(2, \mathbb{R})$.

Proof A homeomorphism as required in the lemma can easily be determined explicitly (see e.g. the paper [LP03]). However, its existence can also be derived as follows. The group $PSL(2, \mathbb{R})$ acts transitively from the left on $\mathbb{R}^2 - \{0\}/\pm 1$ (this is immediate from transitivity of the left linear action of $SL(2, \mathbb{R})$ on $\mathbb{R}^2 - \{0\}$). Moreover, the stabilizer subgroup of the point $(1, 0)/\pm 1$ for this action is precisely the group N. $\square$

As a consequence, $PSL(2, \mathbb{Z})$-invariant Radon measures on $\mathbf{R}^2 - \{0\}/\pm 1$ correspond *precisely* to Radon measures on $PSL(2, \mathbb{R})/N$ which are invariant under the left action of $PSL(2, \mathbb{Z})$ or, equivalently, to finite Borel measures on the *unit tangent bundle* $T^1(\mathrm{Mod}) = PSL(2, \mathbb{Z}) \backslash PSL(2, \mathbb{R})$ of the *modular surface* $\mathrm{Mod} = PSL(2, \mathbb{Z}) \backslash \mathbf{H}^2$ which are invariant under the action of the *horocycle flow* h_t defined by the right action of the upper unipotent group N.

There is an obvious family of h_t-invariant Borel probability measures on the homogeneous space $T^1(\mathrm{Mod})$. Namely, a *fundamental domain* for the action of $PSL(2, \mathbb{Z})$ on the hyperbolic plane $\mathbf{H}^2$ by linear fractional transformations is the complement of the euclidean disc of radius one centered at the origin in the strip $\{z \in \mathbb{C} \mid \mathrm{Im}(z) > 0, \frac{1}{2} \leq \mathrm{Re}(z) \leq \frac{1}{2}\}$. The stabilizer of ∞ in the group $PSL(2, \mathbb{R})$ is the solvable subgroup G of all upper triangular matrices generated by A and N. This stabilizer is preserved by the action of N by right translation. The orbits of N in S project to the lines $\mathrm{Im} = \mathrm{const}$ in $\mathbf{H}^2$ and hence they project to *closed* orbits of the horocycle flow on $T^1(\mathrm{Mod})$. In particular, for every such orbit there is a unique h_t-invariant Borel probability measure supported on this orbit. Figure 1 shows a periodic orbit of the horocycle flow about the cusp in the standard fundamental domain of the action of the group $PSL(2, \mathbb{Z})$ on $\mathbf{H}^2$.

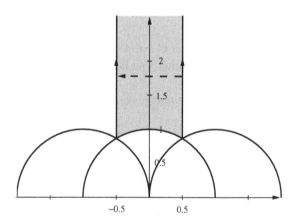

To describe the corresponding $PSL(2, \mathbb{Z})$-invariant Radon measure on the cone $\mathbb{R}^2 - \{0\}/\pm 1$, observe that the left action of $PSL(2, \mathbb{R})$ on the cone $\mathbb{R}^2 - \{0\}/\pm 1$ projects to the action of $PSL(2, \mathbb{R})$ on the *real projective line* $\mathbb{R}P^1 \sim S^1$ by projective transformations. This action is transitive, and the stabilizer of the real line $[1, 0]$ spanned by the point $(1,0) \in \mathbb{R}^2$ equals the subgroup G of $PSL(2,\mathbb{R})$. Thus the action of $PSL(2,\mathbb{R})$ on $\partial \mathbf{H}^2$ is just the action of $PSL(2,\mathbb{R})$ on $\mathbb{R}P^1$. In particular, the $PSL(2,\mathbb{Z})$-orbit of $[1,0]$ which consists of all points in $\mathbb{R}P^1$ spanned by vectors with integer coordinates (see the above discussion) naturally coincides with the $PSL(2,\mathbb{Z})$-orbit of the point $\infty \in \partial \mathbf{H}^2$. As a consequence, the h_t-invariant Borel probability measures on $PSL(2,\mathbb{Z})\backslash PSL(2,\mathbb{R})$ supported on the above described closed orbits of the horocycle flow correspond *precisely* to the invariant Radon measures on $\mathbb{R}^2 - \{0\}/\pm 1$ supported on points whose coordinates are dependent over $\mathbb{Q}$. Thus we have.

Borel probability measures supported on closed orbits of h_t on $T^1(\mathrm{Mod})$ correspond to $SL(2,\mathbb{Z})$-invariant Radon measures on $\mathbb{R}^2$ supported on orbits of points whose coordinates are dependent over $\mathbb{Q}$.

On the other hand, the Lebesgue Haar measure is invariant under the action of the whole group $PSL(2,\mathbb{R})$, and it is uniquely determined by this property up to scale. Thus Theorem 3.1 is an immediate consequence of the following result of Dani [D78].

Proposition 3.2 *Any h_t-invariant probability measure on $T^1(\mathrm{Mod})$ either is supported on a closed orbit for the horocycle flow or it is invariant under the whole group $PSL(2,\mathbb{R})$.*

In the early nineties, Ratner proved a far-reaching generalization of this result. We refer to the book [BM00] for an introduction to the subject and to [WM05] for a more detailed treatment of Ratner's celebrated work.

4 Uniquely ergodic unipotent flows on some homogeneous spaces of infinite volume

The results explained in Section 3 and their generalizations, in particular the work of Ratner, have many applications. However, they are only applicable in an algebraic setting and to invariant probability measures. The simplest extension of the questions discussed in Section 3 which can not be answered with these methods can be formulated as follows.

Let S be a closed oriented surface of genus $g \geq 2$. Choose a Riemannian metric g on S of constant sectional curvature -1. Then there is a discrete subgroup Γ of $PSL(2,\mathbb{R})$ such that our hyperbolic surface is just $\Gamma\backslash\mathbf{H}^2$, with unit tangent bundle $T^1 S = \Gamma\backslash PSL(2,\mathbb{R})$ as before. In particular, the horocycle flow h_t is defined on T^1S. For some $d \leq 2g$ choose a normal subgroup Λ of Γ with factor group Γ/Λ isomorphic to $\mathbb{Z}^d$. An example of such a group is the commutator subgroup of Γ. Consider the regular $\mathbb{Z}^d$-cover $\hat{S}$ of S with fundamental group Λ. Then the horocycle flow h_t on the unit tangent bundle $T^1\hat{S}$ of $\hat{S}$ is defined. By the work of Ratner, h_t-invariant Borel probability measures on $T^1\hat{S}$ can be classified. Namely, in the situation at hand, either they are supported on closed orbits of the horocycle flow or they are invariant under the full group $PSL(2,\mathbb{R})$ on $T^1\hat{S}$. In other words, since the volume of $T^1\hat{S}$ is infiniteand since there are no closed orbits for the horocycle flow, such invariant Borel probability measures do not exist. However, the Lebesgue measure (i.e. the measure induced by a Haar measure on $PSL(2,\mathbb{R})$) is a $PSL(2,\mathbb{R})$-invariant *Radon* measure on $T^1\hat{S}$. A natural problem is now to classify all invariant Radon measures for the horocycle flow on $T^1\hat{S}$.

Babillot and Ledrappier [BL98] constructed for every homomorphism $\varphi : \mathbb{Z}^d \to \mathbb{R}$ a Radon measures $\hat{\lambda}^\varphi$ on $T^1\hat{S}$ which is both invariant under the horocycle flow on $T^1\hat{S}$ and under the geodesic flow. The trivial homomorphism corresponds to the Lebesgue measure. Each of these measures is the lift of a Borel probability measure λ^φ on T^1S. If the homomorphism is nontrivial, then the measure λ^φ on T^1S is *not* invariant under the horocycle flow on T^1S. For $\varphi \neq \psi$ the measures $\hat{\lambda}^\varphi, \hat{\lambda}^\psi$ are singular. We call these measures *Babillot-Ledrappier measures*.

The Babillot-Ledrappier measures are all *absolutely continuous* with respect to the *stable foliation*, i.e. the foliation of $T^1\hat{S} = \Lambda\backslash PSL(2,\mathbb{R})$ into the orbits of the right action of the solvable subgroup G generated by the groups A, N. More precisely, the following holds true.

For every point $\xi \in S^1 = \partial\mathbf{H}^2$ there is a *Busemann function* θ_ξ : $\mathbf{H}^2 \to \mathbb{R}$ at ξ. Such a Busemann function is a one-Lipschitz function for the hyperbolic metric on $\mathbf{H}^2$ which is determined uniquely by ξ up to an additive constant. The function $\theta_\infty(z) = \log\mathrm{Im}(z)$ is a Busemann function at the point ∞. The Busemann functions are invariant under the action of $PSL(2,\mathbb{R})$ on $\partial\mathbf{H}^2 \times \mathbf{H}^2$ and therefore the images under the action of $PSL(2,\mathbb{R})$ of the function θ_∞ determines all Busemann functions on $\mathbf{H}^2$.

For a discrete subgroup Λ of $PSL(2,\mathbb{R})$ and a number $\alpha \geq 0$ define a *conformal density of dimension* $\alpha \geq 0$ for Λ to be an assignment which associates to every $x \in \mathbf{H}^2$ a finite measure μ^x on $\partial\mathbf{H}^2$ with the following properties.

(i) The measures μ^x $(x \in \mathbf{H}^2)$ are equivariant under the action of Λ on $\mathbf{H}^2 \times \partial \mathbf{H}^2$.

(ii) For $x, y \in \mathbf{H}^2$ the measures μ^x, μ^y are absolutely continuous, with Radon Nikodym derivative $\frac{d\mu^y}{d\mu^x}(\xi) = e^{\alpha(\theta_\xi(y) - \theta_\xi(x))}$ where θ_ξ is a Busemann function at ξ.

The Babillot-Ledrappier measures on $T^1 \hat{S}$ are related to conformal densities for the fundamental group Λ of $\hat{S}$ as follows.

Recall that there is a Λ-equivariant canonical projection $PSL(2, \mathbb{R}) \to \partial \mathbf{H}^2 = PSL(2, \mathbb{R})/G$. Let $\tilde{\mu}$ be the Λ-invariant measure on $PSL(2, \mathbb{R})$ which is the lift of a Babillot-Ledrappier measure $\hat{\lambda}^\varphi$ on

$$T^1 \hat{S} = \Lambda \backslash PSL(2, \mathbb{R}).$$

Then for every relatively compact open subset U of $PSL(2, \mathbb{R})$ the push-forward $\pi_*(\mu|U)$ is contained in the measure class of a conformal density for Λ. This conformal density $\{\mu^x\}$ has the additional property that

$$\mu^x \circ g = e^{\varphi(g)} \circ \mu^x \forall g \in \mathbb{Z}^d.$$

This additional condition determines the conformal density uniquely up to scale. The measure $\hat{\lambda}^\varphi$ in turn is uniquely determined by the conformal density up to scale.

A Radon measure μ on $T^1 \hat{S}$ is *quasi-invariant* under the geodesic flow Φ^t on $T^1 \hat{S}$ if for every $t \in \mathbb{R}$ the push-forward measure $\Phi^t_* \mu$ is absolutely continuous with respect to μ, i.e. the measures μ and $\Phi^t_* \mu$ have the same sets of measure zero. The following result is due to Babillot [B04] and Aaronson, Nakada, Sarig, Solomyak [ASS02].

Proposition 4.1 *Every h_t-invariant Radon measure on $T^1 \hat{S}$ which is quasi-invariant under the geodesic flow is a Babillot-Ledrappier measure.*

Now let λ by any h_t-invariant ergodic Radon measure on $T^1 \hat{S}$. Let Φ^t be the geodesic flow on $T^1 \hat{S}$ and let $H(\lambda) \subset \mathbb{R}$ be the set of all $t \in \mathbb{R}$ such that the measure $\Phi^t \lambda$ is contained in the measure class of λ. Then $H(\lambda)$ is a *closed* subgroup of $\mathbb{R}$ and hence if $H(\lambda) \neq \mathbb{R}$ then either $H(\lambda)$ is infinite cyclic or trivial. The measure λ is quasi-invariant under the flow Φ^t if and only if we have $H(\lambda) = \mathbb{R}$. Proposition 4.1 then shows that every h_t-invariant Radon measure λ with $H(\lambda) = \mathbb{R}$ is a Babillot-Ledrappier measure.

To classify the h_t-invariant measures with the property that $H(\lambda)$ is either infinite cyclic or trivial, Sarig [S04] proved a general structure theorem for cocycles. He uses this result to show that there is no h_t-invariant Radon measure λ on $T^1\hat{S}$ with $H_\lambda \neq \mathbb{R}$. Thus he obtains.

Theorem 4.1 *Every h_t-invariant ergodic Radon measure on $T^1\hat{S}$ is a Babillot-Ledrappier measure.*

In fact, the analog of Theorem 4.1 holds true for the horocycle flow on any $\mathbb{Z}^d$-cover of a closed surface S of higher genus equipped with a Riemannian metric of negative Gauß curvature. In other words, this result does not require any algebraic setting.

5 Moduli space

The first book written by Hermann Weyl is the monograph "Die Idee der Riemannschen Fläche" which appeared in 1913. In this section, a *Riemann surface* will be a closed oriented surface of genus $g \geq 1$ which is equipped with a complex structure. We are going to connect the *moduli space of Riemann surfaces* to the ideas discussed in Section 3 and Section 4.

Define a *marked Riemann surface* to be a Riemann surface M together with a homeomorphism $S \to M$ from a fixed closed oriented surface S of genus $g \geq 1$.

Definition 4

(i) *The* Teichmüller space $\mathcal{T}(S)$ *of S is the space of all* marked *Riemann surfaces which are homeomorphic to S up to biholomorphisms isotopic to the identity.*

(ii) *The* mapping class group $\mathcal{M}(S)$ *is the group of all isotopy classes of orientation preserving homeomorphisms of S.*

Every element of the mapping class group $\mathcal{M}(S)$ naturally induces a nontrivial automorphism of the fundamental group $\pi_1(S)$ of S. It is easy to see that this automorphism is not *inner*, i.e. it is not induced by a conjugation. Thus there is a natural homomorphism of $\mathcal{M}(S)$ into the group $\mathrm{Out}(\pi_1(S))$ of *outer* automorphisms of $\pi_1(S)$, i.e. the quotient of the group of all automorphisms of $\pi_1(S)$ by the normal subgroup of all inner automorphisms. By an old result of Nielsen, this map is in fact an isomorphism.

The mapping class group naturally acts on Teichmüller space by precomposition, i.e. by changing the markings. It is well known that $\mathcal{T}(S)$ can be equipped with a topology so that with respect to this topology,

$T(S)$ is homeomorphic to $\mathbb{R}^{6g-6}$ and that the mapping class group acts properly discontinuously on $T(S)$ by homeomorphisms. There is also an $\mathcal{M}(S)$-invariant complex structure on $T(S)$ which identifies $T(S)$ with a bounded domain Ω in $\mathbb{C}^{3g-3}$. The mapping class group is then just the group of all biholomorphic automorphisms of Ω.

By *uniformization*, if $g \geq 2$ then the *moduli space*

$$\text{Mod}(S) = \mathcal{M}(S)\backslash T(S)$$

of S can be identified with the space of all hyperbolic Riemannian metrics on S up to orientation preserving isometries. In the case $g = 1$, it is the space of all euclidean metrics of area one up to orientation preserving isometries.

Example:

In the case $g = 1$ (i.e. in the case of the 2-torus $S = T^2$), the fundamental group $\pi_1(S)$ of S is the lattice $\mathbb{Z}^2$ in $\mathbb{R}^2$. Then every automorphism of $\mathbb{Z}^2$ is induced by a linear isomorphism of $\mathbb{R}^2$ preserving the lattice $\mathbb{Z}^2$ and hence the mapping class group $\mathcal{M}(S)$ is just the group $SL(2, \mathbb{Z})$. Here the center $\mathbb{Z}/2\mathbb{Z}$ of $SL(2, \mathbb{Z})$ corresponds to the hyperelliptic involution which acts trivially on $T(T^2)$. More precisely, we have natural identifications as follows.

(i) $T(T^2) = \mathbf{H}^2 = \{z \in \mathbb{C} \mid \text{Im}(z) > 0\}$, the hyperbolic plane.
(ii) $\mathcal{M}(T^2) = SL(2, \mathbb{Z})$ acting on $\mathbf{H}^2$ by linear fractional transformations.
(iii) $\text{Mod}(T^2) = SL(2, \mathbb{Z})\backslash\mathbf{H}^2$, the modular surface.

A Riemann surface S is a one-dimensional complex manifold and hence it admits a natural holomorphic cotangent bundle $T'(S)$ whose fiber at a point x is the one-dimensional $\mathbb{C}$-vector space of all $\mathbb{C}$-linear maps $T_x S \to \mathbb{C}$.

Definition 5 *For a Riemann surface S, a* holomorphic quadratic differential q on S is a holomorphic section of the holomorphic line bundle $T'(S) \otimes T'(S)$.

In a holomorphic coordinate z on S, a holomorphic quadratic differential q can be written in the form $q(z) = f(z)dz^2$ with a local holomorphic function f. The bundle of all holomorphic quadratic differentials over all Riemann surfaces can be viewed as the *cotangent bundle* of Teichmüller space. It is a complex vector bundle of complex dimension $3g - 3$. The mapping class group $\mathcal{M}(S)$ acts properly discontinuously on this bundle as a group of bundle automorphisms.

A quadratic differential q defines a *singular euclidean metric* on S as follows. Near a *regular point* z, i.e. away from the zeros of the differential, there is a holomorphic coordinate z on S such that in this coordinate the differential is just dz^2. Such a chart is unique up to translation and multiplication with -1 and hence the euclidean metric defined by this chart is uniquely determined by q. We call such a chart *isometric*. The *area* of a quadratic differential is the area of the singular euclidean metric it defines. The mapping class group preserves the sphere bundle $\tilde{\mathcal{Q}}(S)$ over $\mathcal{T}(S)$ of all holomorphic quadratic differentials of area one and hence this bundle projects to the *moduli space* $\mathcal{Q}(S) = \mathcal{M}(S)\backslash\tilde{\mathcal{Q}}(S)$ of such differentials.

The real line bundles $q > 0, q < 0$ on the complement of the (finitely many) singular points of q define *transverse singular measured foliations* q_h, q_v on S called the *horizontal and vertical measured foliations* of q. By definition, a measured foliation of S is a foliation F with finitely many singularities together with a *transverse measure* which associates to every smooth compact arc which meets the leaves of the foliation F transversely a length which is invariant under a homotopy of the arc moving each endpoint of the arc within a single leaf of the foliation.

There is a natural action of the group $SL(2, \mathbb{R})$ on the space $\tilde{\mathcal{Q}}(S)$ of area one holomorphic quadratic differentials which is given as follows. For each quadratic differential $q \in \tilde{\mathcal{Q}}(S)$ choose a family of isometric charts near the regular points. For $B \in SL(2, \mathbb{R})$ define Bq to be the quadratic differential whose isometric charts are the compositions of the isometric charts for q with B. This collection of charts then defines a new holomorphic quadratic differential on a (different) Riemann surface. The action of $SL(2, \mathbb{R})$ commutes with the action of the mapping class group and hence it descends to an action of $SL(2, \mathbb{R})$ on $\mathcal{Q}(S)$. The diagonal subgroup of $SL(2, \mathbb{R})$ then defines a flow on $\mathcal{Q}(S)$ called the *Teichmüller flow* Φ^t, and the upper unipotent group defines the *horocycle flow* h_t. In the case of the Teichmüller space of surfaces of genus 1, these flows are precisely the geodesic flow and the horocycle flow on the unit tangent bundle of the modular surface. By construction, the horocycle flow preserves the horizontal measured foliation of the quadratic differentials since it preserves the lines $q > 0$ in our charts.

For the moduli space of surfaces of higher genus, the classification of h_t-invariant Borel probability measures is up to date impossible. There are lots of examples of such measures. For example, *Veech surfaces* in moduli space are holomorphically embedded (singular) Riemann surfaces of finite type. They correspond to *closed* $SL(2, \mathbb{R})$-orbits in $\mathcal{Q}(S)$. Any h_t-invariant Borel probability measure on such an orbit then defines a h_t-invariant Borel probability measure on $\mathcal{Q}(S)$.

However, we can ask for the easier question of a classification of $\mathcal{M}(S)$-invariant Radon measures on the space of *equivalence classes* of measured foliations. For this call two measured foliations on S are *equivalent* if they can be transformed into each other by so-called *collapses* of two singular points along a connecting compact singular arc and Whitehead moves. Figure 2 shows a modification of a singular foliation with such a Whitehead move.

The following fundamental fact was discovered by Hubbard and Masur in 1979 [HM79], see also [Hu06].

Theorem 5.1 *(Hubbard-Masur) Let $\mathcal{H}$ be the natural map which associates to a quadratic differential the equivalence class of its horizontal measured foliation. Then for every Riemann surface M the restriction of $\mathcal{H}$ to the truncated vector space of all nontrivial quadratic differentials on M is a bijection.*

Example: If $S = T^2$ then the space of equivalence classes of measured foliations is $\mathbb{R}^2$.

By the theorem of Hubbard and Masur (in a slightly stronger version than the one we described above), the natural topology on the bundle $\tilde{\mathcal{Q}}(S)$ induces a metrizable topology on the set $\mathcal{MF}$ of equivalence classes of measured foliations on S (which by construction is a purely topologically defined space). The mapping class group $\mathcal{M}(S)$ acts on the space of equivalence classes of measured foliations by homeomorphisms. If $S = T^2$ then this action can naturally be identified with the linear action of $SL(2, \mathbb{R})$ on $\mathbb{R}^2$.

Definition 6 *A measured foliation F on S fills up S if there is no simple closed curve on S whose F-length vanishes.*

Here the F-length of a simple closed curve c is the infimum of the transverse lengths of closed curves transverse to the foliation which are freely homotopic to c.

The following classification result which was independently and at the same time shown in [H07a] and in [LM07] extends unique ergodicity for the action of $SL(2,\mathbb{Z})$ on the set of irrational points in $\mathbb{R}^2$ to the framework of Teichmüller theory and moduli spaces. For the formulation of this result, define a *measured multi-cylinder* for a measured foliation of S to be a disjoint union of embedded annuli in S which are foliated by closed leaves of the foliation.

Theorem 5.2 *Let μ be an $\mathcal{M}(S)$-invariant ergodic Radon measure on $\mathcal{MF}$.*

(i) *If μ gives full mass to measured foliations which fill up S then μ coincides with the Lebesgue measure up to scale.*

(ii) *If μ is singular to the Lebesgue measure then there is a measured foliation F containing a nontrivial measured multi-cylinder c such that μ coincides with the translates of a $\text{Stab}(c)$-invariant measure on the space of measured foliations on $S - c$.*

Remark: The proof for genus $g \geq 2$ is not valid in the case $g = 1$, i.e. we do not obtain a new proof of the result of Dani. The argument for the first part of the theorem uses the structural result of Sarig discussed in Section 4 in an essential way. The proof of the second part relies on a result of Minsky and Weiss [MW02] which is motivated by an analogous classical result of Dani for the horocycle flow on non-compact hyperbolic surfaces of finite volume.

Finally we discuss some applications. We begin with two classical results of Margulis [M69] and Dani.

Theorem 5.3

(i) **(Margulis)** *For $\ell > 0$ let $n(\ell)$ be the number of closed geodesics on $M = SL(2,\mathbb{Z})\backslash\mathbf{H}^2$ of length at most ℓ. Then*

$$\lim_{\ell\to\infty} \frac{1}{\ell} \log n(\ell) = 1.$$

(ii) **(Dani)** *For a compact subset K of $SL(2,\mathbb{Z})\backslash\mathbf{H}^2$ and for $\ell > 0$ let*

$n_K(\ell)$ be the number of periodic orbits of length at most ℓ which are entirely contained in K. Then

$$1 = \sup\{\lim_{\ell\to\infty}\inf \frac{1}{\ell}\log n_K(\ell) \mid K \subset SL(2,\mathbb{Z})\backslash\mathbf{H}^2 \text{compact }\}.$$

Also, for a hyperbolic surface M a collar lemma holds: There is a compact subset K in M such that *every* geodesic in M intersects K.

For closed Teichmüller geodesics, Eskin and Mirzakhani [EM08] obtained recently the analog of Margulis' result.

Theorem 5.4 *Let $n(\ell)$ be the number of closed Teichmüller geodesics of length at most ℓ. Then*

$$\lim_{\ell\to\infty}\frac{1}{\ell}\log n(\ell) = 6g - 6.$$

We also have [H07b].

Theorem 5.5 *For a compact subset K of $\mathrm{Mod}(S)$ and for $\ell > 0$ let $n_K(\ell)$ be the number of closed geodesics in $\mathrm{Mod}(S)$ which are entirely contained in K; then*

$$6g - 6 = \sup\{\lim_{\ell\to\infty}\inf \frac{1}{\ell}\log n_K(\ell) \mid K \subset \mathrm{Mod}(S) \text{ compact }\}.$$

Bibliography

[ASS02] J. Aaronson, H. Nakada, O. Sarig, R. Solomyak, *Invariant measures and asymptotics for some skew products*, Isr. J. Math. 128 (2002), 93–134. Corrections: Isr. J. Math. 138 (2003), 377–379.

[B04] M. Babillot, *On the classification of invariant measures for horospherical foliations on nilpotent covers of negatively curved manifolds*, in "Random walks and geometry", Walter de Gruyter, Berlin 2004, 319–335.

[BL98] M. Babillot, F. Ledrappier, *Geodesic paths and horocycle flows on Abelian covers*, in "Lie groups and ergodic theory", Tata Inst. Fund. Res. Stud. Math 14, Bombay 1998, 1–32.

[BM00] B. Bekka, M. Mayer, *Ergodic theory and topological dynamics of group actions on homogeneous spaces*, London Math. Soc. Lec. Notes 269, Cambridge Univ. Press 2000.

[D78] S. G. Dani, *Invariant measures of horocycle flows on non compact homogeneous spaces*, Invent. Math. 47 (1978), 101–138.

[EM93] A. Eskin, C. McMullen, *Mixing, counting and equidistribution on Lie groups*, Duke Math. J. 71 (1993), 181–209.

[EM08] A. Eskin, M. Mirzakhani, to appear.

[FLP79] A. Fathi, F. Laudenbach, V. Poenaru, *Travaux de Thurston sur les surfaces*, Asterisque 66-67, 1979.

[F72] H. Furstenberg, *The unique ergodicity of the horocyclic flow*, Springer Lecture Notes in Math. 318 (1972), 95–115.

[H07a] U. Hamenstädt, *Invariant Radon measures on measured lamination space*, arXiv:math.DS/0703602.

[H07b] U. Hamenstädt, *Dynamics of the Teichmüller flow on compact invariant sets*, arXiv:0705.3812.

[Hu06] J. Hubbard, *Teichmüller theory*, Matrix edition, 2006.

[HM79] J. Hubbard, H. Masur, *Quadratic differentials and measured foliations*, Acta Math. 142 (1979), 221–274.

[LP03] F. Ledrappier, M. Pollicott, *Ergodic properties of linear actions of* (2×2)*-matrices*, Duke Math. J. 116 (2003), 353–388.

[LM07] E. Lindenstrauss, M. Mirzakhani, *Ergodic theory of the space of measured laminations*, preprint, April 2007.

[M69] G. Margulis, *On some application of ergodic theory to the study of manifolds of negative curvature*, Funct. Anal. Appl. 3 (1969), 335–336.

[MW02] Y. Minsky, B. Weiss, *Nondivergence of horocycle flows on moduli spaces*, J. reine angew. Math. 552 (2002), 131-177.

[S04] O. Sarig, *Invariant Radon measures for horocycle flows on Abelian covers*, Invent. Math. 157 (2004), 519–551.

[Wa82] P. Walters, *An introduction to ergodic theory*, Springer Graduate Texts in Math. 79 (1982).

[W16] H. Weyl, *Über die Gleichverteilung von Zahlen mod Eins*, Mathematische Annalen 77 (1916), 313–352.

[WM05] D. Witte Morris, *Ratner's theorem on unipotent flows*, Chicago University Press, Chicago 2005.

7

Weyl's law in the theory
of automorphic forms

Werner Müller

Universität Bonn
Mathematisches Institut
mueller@math.uni-bonn.de

1 Introduction

Let M be a smooth, compact Riemannian manifold of dimension n with smooth boundary ∂M (which may be empty). Let

$$\Delta = -\operatorname{div} \circ \operatorname{grad} = d^* d$$

be the Laplace-Beltrami operator associated with the metric g of M. We consider the Dirichlet eigenvalue problem

$$\Delta \varphi = \lambda \varphi, \quad \varphi\big|_{\partial M} = 0. \tag{1.1}$$

As is well known, (1.1) has a discrete set of solutions

$$0 \leq \lambda_0 \leq \lambda_1 \leq \cdots \to \infty$$

whose only accumulation point is at infinity and each eigenvalue occurs with finite multiplicity. The corresponding eigenfunctions φ_i can be chosen such that $\{\varphi_i\}_{i \in \mathbb{N}_0}$ is an orthonormal basis of $L^2(M)$. A fundamental problem in analysis on manifolds is to study the distribution of the eigenvalues of Δ and their relation to the geometric and topological structure of the underlying manifold. One of the first results in this context is Weyl's law for the asymptotic behavior of the eigenvalue counting function. For $\lambda \geq 0$ let

$$N(\lambda) = \#\{j \colon \sqrt{\lambda_j} \leq \lambda\}$$

be the counting function of the eigenvalues of $\sqrt{\Delta}$, where eigenvalues are counted with multiplicities. Denote by $\Gamma(s)$ the Gamma function. Then the Weyl law states

$$N(\lambda) = \frac{\operatorname{vol}(M)}{(4\pi)^{n/2} \Gamma\left(\frac{n}{2} + 1\right)} \lambda^n + o(\lambda^n), \quad \lambda \to \infty. \tag{1.2}$$

This was first proved by Weyl [We1] for a bounded domain $\Omega \subset \mathbb{R}^3$. Written in a slightly different form it is known in physics as the Rayleigh-Jeans law. Raleigh [Ra] derived it for a cube. Garding [Ga] proved Weyl's law for a general elliptic operator on a domain in $\mathbb{R}^n$. For a closed Riemannian manifold (1.2) was proved by Minakshisundaram and Pleijel [MP].

Formula (1.2) does not say very much about the finer structure of the eigenvalue distribution. The basic question is the estimation of the remainder term

$$R(\lambda) := N(\lambda) - \frac{\mathrm{vol}(M)}{(4\pi)^{n/2}\boldsymbol{\Gamma}\left(\frac{n}{2}+1\right)}\lambda^n.$$

That this is a deep problem shows the following example. Consider the flat 2-dimensional torus $T = \mathbb{R}^2/(2\pi\mathbb{Z})^2$. Then the eigenvalues of the flat Laplacian are $\lambda_{m,n} := m^2 + n^2$, $m, n \in \mathbb{Z}$ and the counting function equals

$$N(\lambda) = \#\{(m,n) \in \mathbb{Z}^2 : \sqrt{m^2 + n^2} \le \lambda\}.$$

Thus $N(\lambda)$ is the number of lattice points in the circle of radius λ. An elementary packing argument, attributed to Gauss, gives

$$N(\lambda) = \pi\lambda^2 + O(\lambda),$$

and the circle problem is to find the best exponent μ such that

$$N(\lambda) = \pi\lambda^2 + O_\varepsilon\left(\lambda^{\mu+\varepsilon}\right), \quad \forall \varepsilon > 0.$$

The conjecture of Hardy is $\mu = 1/2$. The first nontrivial result is due to Sierpinski who showed that one can take $\mu = 2/3$. Currently the best known result is $\mu = 131/208 \approx 0.629$ which is due to Huxley. Levitan [Le] has shown that for a domain in $\mathbb{R}^n$ the remainder term is of order $O(\lambda^{n-1})$.

For a closed Riemannian manifold, Avakumović [Av] proved the Weyl estimate with optimal remainder term:

$$N(\lambda) = \frac{\mathrm{vol}(M)}{(4\pi)^{n/2}\boldsymbol{\Gamma}\left(\frac{n}{2}+1\right)}\lambda^n + O(\lambda^{n-1}), \quad \lambda \to \infty. \qquad (1.3)$$

This result was extended to more general, and higher order operators by Hörmander [Ho]. As shown by Avakumović the bound $O(\lambda^{n-1})$ of the remainder term is optimal for the sphere. On the other hand, under certain assumption on the geodesic flow, the estimate can be slightly improved. Let S^*M be the unit cotangent bundle and let Φ_t be the geodesic flow. Suppose that the set of $(x, \xi) \in S^*M$ such that Φ_t has a contact of infinite order with the identity at (x, ξ) for some $t \neq 0$, has measure zero in S^*M. Then Duistermaat and Guillemin [DG] proved that the remainder term satisfies $R(\lambda) = o(\lambda^{n-1})$. This is a slight improvement over (1.3).

In [We3] Weyl formulated a conjecture which claims the existence of a second term in the asymptotic expansion for a bounded domain $\Omega \subset \mathbb{R}^3$, namely he predicted that

$$N(\lambda) = \frac{\mathrm{vol}(\Omega)}{6\pi^2}\lambda^3 - \frac{\mathrm{vol}(\partial\Omega)}{16\pi}\lambda^2 + o(\lambda^2).$$

This was proved for manifolds with boundary under a certain condition on the periodic billiard trajectories, by Ivrii [Iv] and Melrose [Me].

The purpose of this paper is to discuss Weyl's law in the context of locally symmetric spaces $\Gamma\backslash S$ of finite volume and non-compact type. Here $S = G/K$ is a Riemannian symmetric space, where G is a real semi-simple Lie group of non-compact type, and K a maximal compact subgroup of G. Moreover Γ is a lattice in G, i.e., a discrete subgroup of finite covolume. Of particular interest are arithmetic subgroups such as the principal congruence subgroup $\Gamma(N)$ of $\mathrm{SL}(2, \mathbb{Z})$ of level $N \in \mathbb{N}$. Spectral theory of the Laplacian on arithmetic quotients $\Gamma\backslash S$ is intimately related with the theory of automorphic forms. In fact, for a symmetric space S it is more natural and important to consider not only the Laplacian, but the whole algebra $\mathcal{D}(S)$ of G-invariant differential operators on S. It is known that $\mathcal{D}(S)$ is a finitely generated commutative algebra [He]. Therefore, it makes sense to study the joint spectral decomposition of $\mathcal{D}(S)$. Square integrable joint eigenfunctions of $\mathcal{D}(S)$ are examples of automorphic forms. Among them are the cusp forms which satisfy additional decay conditions. Cusps forms are the building blocks of the theory of automorphic forms and, according to deep and far-reaching conjectures of Langlands [La2], are expected to provide important relations between harmonic analysis and number theory.

Let $G = NAK$ be the Iwasawa decomposition of G and let $\mathfrak{a}$ be the Lie algebra of A. If $\Gamma\backslash S$ is compact, the spectrum of $\mathcal{D}(S)$ in $L^2(\Gamma\backslash S)$ is a discrete subset of the complexification $\mathfrak{a}_{\mathbb{C}}^*$ of $\mathfrak{a}^*$. It has been studied by Duistermaat, Kolk, and Varadarajan in [DKV]. The method is based on the Selberg trace formula. The results are more refined statements about the distribution of the spectrum than just the Weyl law. For example, one gets estimations for the distribution of the tempered and the complementary spectrum. We will review briefly these results in section 2.

If $\Gamma \backslash S$ is non-compact, which is the case for many important arithmetic groups, the Laplacian has a large continuous spectrum which can be described in terms of Eisenstein series [La1]. Therefore, it is not obvious that the Laplacian has any eigenvalue $\lambda > 0$, and an important problem in the theory of automorphic forms is the existence and construction of cusp forms for a given lattice Γ. This is were the Weyl law comes into play. Let $\mathbb{H}$ be the upper half-plane. Recall that $\mathrm{SL}(2, \mathbb{R})$ acts on $\mathbb{H}$ by fractional linear transformations. Using his trace formula [Se2], Selberg established the following version of Weyl's law for an arbitrary lattice Γ in $\mathrm{SL}(2, \mathbb{R})$

$$N_\Gamma(\lambda) + M_\Gamma(\lambda) \sim \frac{\mathrm{Area}(\Gamma \backslash \mathbb{H})}{4\pi} \lambda^2, \quad \lambda \to \infty \qquad (1.4)$$

[Se2, p. 668]. Here $N_\Gamma(\lambda)$ is the counting function of the eigenvalues and $M_\Gamma(\lambda)$ is the winding number of the determinant $\varphi(1/2 + ir)$ of the scattering matrix which is given by the constant Fourier coefficients of the Eisenstein series (see section 4). In general, the two functions on the left can not be estimated separately. However, for congruence groups like $\Gamma(N)$, the meromorphic function $\varphi(s)$ can be expressed in terms of well-known functions of analytic number theory. In this case, it is possible to show that the growth of $M_\Gamma(\lambda)$ is of lower order which implies Weyl's law for the counting function of the eigenvalues [Se2, p.668]. Especially it follows that Maass cusp forms exist in abundance for congruence groups. On the other hand, there are indications [PS1], [PS2] that the existence of many cusp forms may be restricted to arithmetic groups. This will be discussed in detail in section 4.

In section 5 we discuss the general case of a non-compact arithmetic quotient $\Gamma \backslash S$. There has been some recent progress with the spectral problems discussed above. Lindenstrauss and Venkatesh [LV] established Weyl's law without remainder term for congruence subgroups of a split adjoint semi-simple group $\mathbf{G}$. In [Mu3] this had been proved for congruence subgroups of $\mathrm{SL}(n)$ and for the Bochner-Laplace operator acting in sections of a locally homogeneous vector bundle over $S_n = \mathrm{SL}(n, \mathbb{R}) / \mathrm{SO}(n)$. For congruence subgroups of $\mathbf{G} = \mathrm{SL}(n)$, an estimation of the remainder term in Weyl's law has been established by E. Lapid and the author in [LM]. Using the approach of [DKV] combined with the Arthur trace formula, the results of [DKV] have been extended in [LM] to the cuspidal spectrum of $\mathcal{D}(S_n)$.

2 Compact locally symmetric spaces

In this section we review Hörmanders method of the derivation of Weyl's law with remainder term for the Laplacian Δ of a closed Riemannian manifold M of dimension n. Then we will discuss the results of [DKV] concerning spectral asymptotics for compact locally symmetric manifolds.

The method of Hörmander [Ho] to estimate the remainder term is based on the study of the kernel of $e^{-it\sqrt{\Delta}}$. The main point is the construction of a good approximate fundamental solution to the wave equation by means of the theory of Fourier integral operators and the analysis of the singularities of its trace

$$\operatorname{Tr} e^{-it\sqrt{\Delta}} = \sum_j e^{-it\sqrt{\lambda_j}},$$

which is well-defined as a distribution. The analysis of Hörmander of the "big" singularity of $\operatorname{Tr} e^{-it\sqrt{\Delta}}$ at $t = 0$ leads to the following key result [DG, (2.16)]. Let $\mu_j := \sqrt{\lambda_j}$, $j \in \mathbb{N}$. There exist $c_j \in \mathbb{R}$, $j = 0, ..., n-1$, and $\varepsilon > 0$ such that for every $h \in \mathcal{S}(\mathbb{R})$ with $\operatorname{supp} \hat{h} \subset [-\varepsilon, \varepsilon]$ and $\hat{h} \equiv 1$ in a neighborhood of 0 one has

$$\sum_j h(\mu - \mu_j) \sim (2\pi)^{-n} \sum_{k=0}^{n-1} c_k \mu^{n-1-k}, \quad \mu \to \infty, \qquad (2.1)$$

and rapidly decreasing as $\mu \to -\infty$. The constants c_k are of the form

$$c_k = \int_M \omega_k,$$

where the ω_k's are real valued smooth densities on M canonically associated to the Riemannian metric of M. Especially

$$c_0 = \operatorname{vol}(S^*M), \quad c_1 = (1-n) \int_{S^*M} \operatorname{sub} \Delta,$$

where S^*M is the unit co-tangent bundle, and $\operatorname{sub} \Delta$ denotes the sub-principal symbol of Δ. Consideration of the top term in (2.1) leads to the basic estimates for the eigenvalues.

If $M = \Gamma \backslash G/K$ is a locally symmetric manifold, the Selberg trace formula can be used to replace (2.1) by an exact formula [DKV]. Actually, if the rank of M is bigger than 1, the spectrum is multidimensional. Then the Selberg trace formula gives more refined information.

As example, we consider a compact hyperbolic surface $M = \Gamma \backslash \mathbb{H}$, where $\Gamma \subset \mathrm{PSL}(2,\mathbb{R})$ is a discrete, torsion-free, co-compact subgroup. Let Δ be the hyperbolic Laplace operator which is given by

$$\Delta = -y^2 \left(\frac{\partial^2}{\partial x^2} + \frac{\partial^2}{\partial y^2} \right), \quad z = x + iy. \tag{2.2}$$

Write the eigenvalues λ_j of Δ as

$$\lambda_j = \frac{1}{4} + r_j^2,$$

where $r_j \in \mathbb{C}$ and $\arg(r_j) \in \{0, \pi/2\}$. Let h be an analytic function in a strip $|\operatorname{Im}(z)| \leq \frac{1}{2} + \delta$, $\delta > 0$, such that

$$h(z) = h(-z), \quad |h(z)| \leq C(1 + |z|)^{-2-\delta}. \tag{2.3}$$

Let

$$g(u) = \frac{1}{2\pi} \int_{\mathbb{R}} h(r)e^{iru} \, dr.$$

Given $\gamma \in \Gamma$ denote by $\{\gamma\}_\Gamma$ its Γ-conjugacy class. Since Γ is co-compact, each $\gamma \neq e$ is hyperbolic. Each hyperbolic element γ is the power of a primitive hyperbolic element γ_0. A hyperbolic conjugacy class determines a closed geodesic τ_γ of $\Gamma \backslash \mathbb{H}$. Let $l(\gamma)$ denote the length of τ_γ. Then the Selberg trace formula [Sel] is the following identity:

$$\sum_{j=0}^{\infty} h(r_j)$$

$$= \frac{\operatorname{Area}(\Gamma \backslash \mathbb{H})}{4\pi} \int_{\mathbb{R}} h(r)r \tanh(\pi r) \, dr + \sum_{\{\gamma\}_\Gamma \neq e} \frac{l(\gamma_0)}{2 \sinh\left(\frac{l(\gamma)}{2}\right)} g(l(\gamma)). \tag{2.4}$$

Now let $g \in C_c^\infty(\mathbb{R})$ and $h(z) = \int_{\mathbb{R}} g(r)e^{-irz} \, dr$. Then h is entire and rapidly decreasing in each strip $|\operatorname{Im}(z)| \leq c$, $c > 0$. Let $t \in \mathbb{R}$ and set

$$h_t(z) = h(t - z) + h(t + z).$$

Then h_t is entire and satisfies (2.3). Note that $\hat{h}_t(r) = e^{-itr}g(r) + e^{itr}g(-r)$. We symmetrize the spectrum by $r_{-j} := -r_j$, $j \in \mathbb{N}$. Then by

(2.4) we get

$$
\sum_{j=-\infty}^{\infty} h(t - r_j)
$$

$$
= \frac{\text{Area}(\Gamma \backslash \mathbb{H})}{2\pi} \int_{\mathbb{R}} h(t - r) r \tanh(\pi r) \, dr \qquad (2.5)
$$

$$
+ \sum_{\{\gamma\}_\Gamma \neq e} \frac{l(\gamma_0)}{2 \sinh\left(\frac{l(\gamma)}{2}\right)} \left(e^{-itl(\gamma)} g(l(\gamma)) + e^{itl(\gamma)} g(-l(\gamma)) \right).
$$

Let $\varepsilon > 0$ be such that $l(\gamma) > \varepsilon$ for all hyperbolic conjugacy classes $\{\gamma\}_\Gamma$. The following lemma is an immediate consequence of (2.5).

Lemma 2.1 *Let $g \in C_c^\infty(\mathbb{R})$ such that $\operatorname{supp} g \subset (-\varepsilon, \varepsilon)$. Let $h(z) = \int_{\mathbb{R}} g(r) e^{-irz} \, dr$. Then for all $t \in \mathbb{R}$ we have*

$$
\sum_{j=-\infty}^{\infty} h(t - r_j) = \frac{\text{Area}(\Gamma \backslash \mathbb{H})}{2\pi} \int_{\mathbb{R}} h(t - r) r \tanh(\pi r) \, dr. \qquad (2.6)
$$

Changing variables in the integral on the right and using that

$$
\tanh(\pi(r + t)) = 1 - \frac{2e^{-2\pi(r+t)}}{1 + e^{-2\pi(r+t)}} = -1 + \frac{2e^{2\pi(r+t)}}{1 + e^{2\pi(r+t)}},
$$

we obtain the following asymptotic expansion

$$
\sum_{j=-\infty}^{\infty} h(t - r_j)
$$

$$
= \frac{\text{Area}(\Gamma \backslash \mathbb{H})}{2\pi} \left(|t| \int_{\mathbb{R}} h(r) \, dr - \operatorname{sign} t \int_{\mathbb{R}} h(r) r \, dr \right) + O\left(e^{-2\pi |t|} \right),
$$

$$
\tag{2.7}
$$

as $|t| \to \infty$. If h is even, the second term vanishes and the asymptotic expansion is related to (2.1). The asymptotic expansion (2.7) can be used to derive estimates for the number of eigenvalues near a given point $\mu \in \mathbb{R}$.

Lemma 2.2 *For every $a > 0$ there exists $C > 0$ such that*

$$
\#\{j \colon |r_j - \mu| \leq a\} \leq C(1 + |\mu|)
$$

for all $\mu \in \mathbb{R}$.

Proof We proceed as in the proof of Lemma 2.3 in [DG]. As shown in the proof, there exists $h \in \mathcal{S}(\mathbb{R})$ such that $h \geq 0$, $h > 0$ on $[-a,a]$, $\hat{h}(0) = 1$, and supp $\hat{h}$ is contained in any prescribed neighborhood of 0. Now observe that there are only finitely many eigenvalues $\lambda_j = 1/4 + r_j^2$ with $r_j \notin \mathbb{R}$. Therefore it suffices to consider $r_j \in \mathbb{R}$. Let $\mu \in \mathbb{R}$. By (2.7) we get

$$\#\{j \colon |r_j - \mu| \leq a,\ r_j \in \mathbb{R}\} \cdot \min\{h(u) \colon |u| \leq a\}$$
$$\leq \sum_{r_j \in \mathbb{R}} h(\mu - r_j) \leq C(1 + |\mu|).$$

$\square$

This lemma is the basis of the following auxiliary results.

Lemma 2.3 *For every h as above there exists $C > 0$ such that*

$$\sum_{|r_j| \leq \lambda} \left| \int_{\mathbb{R} - [-\lambda,\lambda]} h(t - r_j)\, dt \right| \leq C\lambda, \quad \sum_{|r_j| > \lambda} \left| \int_{-\lambda}^{\lambda} h(t - r_j)\, dt \right| \leq C\lambda,$$

(2.8)

for all $\lambda \geq 1$.

Proof Since h is rapidly decreasing, there exists $C > 0$ such that $|h(t)| \leq C(1 + |t|)^{-4}$, $t \in \mathbb{R}$. Let $[\lambda]$ be the largest integer $\leq \lambda$. Then we get

$$\sum_{|r_j| \leq \lambda} \left| \int_{\lambda}^{\infty} h(t - r_j)\, dt \right| \leq \sum_{|r_j| \leq \lambda} \int_{\lambda - r_j}^{\infty} |h(t)|\, dt \leq C \sum_{|r_j| \leq \lambda} \frac{1}{(1 + \lambda - r_j)^3}$$

$$= \sum_{k=-[\lambda]}^{[\lambda]-1} \sum_{k \leq r_j \leq k+1} \frac{1}{(1 + \lambda - r_j)^3} \leq \sum_{k=-[\lambda]}^{[\lambda]-1} \frac{\#\{j \colon |r_j - k| \leq 1\}}{(\lambda - k)^3},$$

and by Lemma 2.2 the right hand side is bounded by $C\lambda$ for $\lambda \geq 1$. Similarly we get

$$\sum_{|r_j| \leq \lambda} \left| \int_{-\infty}^{-\lambda} h(t - r_j)\, dt \right| \leq C_2 \lambda.$$

The second series can be treated in the same way. $\square$

Lemma 2.4 *Let h be as in Lemma 2.1 and such that $\hat{h}(0) = 1$. Then*

$$\int_{-\lambda}^{\lambda} \sum_{j=-\infty}^{\infty} h(t - r_j)\, dt = \frac{\mathrm{Area}(\Gamma \backslash \mathbb{H})}{2\pi} \lambda^2 + O(\lambda) \qquad (2.9)$$

as $\lambda \to \infty$.

Proof To prove the lemma, we integrate (2.6) and determine the asymptotic behavior of the integral on the right. Let $p(r)$ be a continuous function on $\mathbb{R}$ such that $|p(r)| \le C(1+|r|)$ and $p(r) = p(-r)$. Changing the order of integration and using that $\int_{\mathbb{R}} h(t-r)\, dt = \hat{h}(0) = 1$, we get

$$\int_{-\lambda}^{\lambda} \int_{\mathbb{R}} h(t-r)p(r)\, dr\, dt = \int_{-\lambda}^{\lambda} p(r)\, dr$$
$$- \int_{-\lambda}^{\lambda} \left(\int_{\mathbb{R}-[-\lambda,\lambda]} h(t-r)\, dt \right) p(r)\, dr$$
$$+ \int_{\mathbb{R}-[-\lambda,\lambda]} \left(\int_{-\lambda}^{\lambda} h(t-r)\, dt \right) p(r)\, dr.$$

Let $C_1 > 0$ be such that $|h(r)| \le C_1(1+|r|)^{-3}$, $r \in \mathbb{R}$. Then the second and the third integral can be estimated by $C(1+\lambda)$. Thus we get

$$\int_{-\lambda}^{\lambda} \int_{\mathbb{R}} h(t-r)p(r)\, dr\, dt = \int_{-\lambda}^{\lambda} p(r)\, dr + O(\lambda), \quad \lambda \to \infty. \qquad (2.10)$$

If we apply (2.10) to $p(r) = r\tanh(\pi r)$, we obtain

$$\int_{-\lambda}^{\lambda} \int_{\mathbb{R}} h(t-r)r\tanh(\pi r)\, dr\, dt = \lambda^2 + O(\lambda). \qquad (2.11)$$

This proves the lemma. $\qquad \square$

We are now ready to prove Weyl's law. We choose h such that $\hat{h}$ has sufficiently small support and $\hat{h}(0) = 1$. Then

$$\int_{-\lambda}^{\lambda} \sum_{j=-\infty}^{\infty} h(t-r_j)\, dt = \sum_{|r_j|\le\lambda} \int_{\mathbb{R}} h(t-r_j)\, dt$$
$$- \sum_{|r_j|\le\lambda} \int_{\mathbb{R}-[-\lambda,\lambda]} h(t-r_j)\, dt + \sum_{|r_j|>\lambda} \int_{-\lambda}^{\lambda} h(t-r_j)\, dt.$$

Using that $\int_{\mathbb{R}} h(t-r)\, dt = \hat{h}(0) = 1$, we get

$$2N_\Gamma(\lambda) = \int_{-\lambda}^{\lambda} \sum_j h(t-r_j)\, dt + \sum_{|r_j|\le\lambda} \int_{\mathbb{R}-[-\lambda,\lambda]} h(t-r_j)\, dt$$
$$- \sum_{|r_j|>\lambda} \int_{-\lambda}^{\lambda} h(t-r_j)\, dt.$$

By Lemmas 2.3 and 2.4 we obtain

$$N_\Gamma(\lambda) = \frac{\text{Area}(\Gamma\backslash\mathbb{H})}{4\pi}\lambda^2 + O(\lambda). \qquad (2.12)$$

We turn now to an arbitrary Riemannian symmetric space $S = G/K$ of non-compact type and we review the main results of [DKV]. The group of motions G of S is a semi-simple Lie group of non-compact type with finite center and K is a maximal compact subgroup of G. The Laplacian Δ of S is a G-invariant differential operator on S, i.e., Δ commutes with the left translations L_g, $g \in G$. Besides of Δ we need to consider the ring $\mathcal{D}(S)$ of all invariant differential operators on S. It is well-known that $\mathcal{D}(S)$ is commutative and finitely generated. Its structure can be described as follows. Let $G = NAK$ be the Iwasawa decomposition of G, W the Weyl group of (G, A) and $\mathfrak{a}$ be the Lie algebra of A. Let $S(\mathfrak{a}_\mathbb{C})$ be the symmetric algebra of the complexification $\mathfrak{a}_\mathbb{C} = \mathfrak{a}\otimes\mathbb{C}$ of $\mathfrak{a}$ and let $S(\mathfrak{a}_\mathbb{C})^W$ be the subspace of Weyl group invariants in $S(\mathfrak{a}_\mathbb{C})$. Then by a theorem of Harish-Chandra [He, Ch. X, Theorem 6.15] there is a canonical isomorphism

$$\mu\colon \mathcal{D}(S) \cong S(\mathfrak{a}_\mathbb{C})^W. \qquad (2.13)$$

This shows that $\mathcal{D}(S)$ is commutative. The minimal number of generators equals the rank of S which is $\dim\mathfrak{a}$ [He, Ch.X, §6.3]. Let $\lambda \in \mathfrak{a}_\mathbb{C}^*$. Then by (2.13), λ determines an character

$$\chi_\lambda\colon \mathcal{D}(S) \to \mathbb{C}$$

and $\chi_\lambda = \chi_{\lambda'}$ if and only if λ and λ' are in the same W-orbit. Since $S(\mathfrak{a}_\mathbb{C})$ is integral over $S(\mathfrak{a}_\mathbb{C})^W$ [He, Ch. X, Lemma 6.9], each character of $\mathcal{D}(S)$ is of the form χ_λ for some $\lambda \in \mathfrak{a}_\mathbb{C}^*$. Thus the characters of $\mathcal{D}(S)$ are parametrized by $\mathfrak{a}_\mathbb{C}^*/W$.

Let $\Gamma \subset G$ be a discrete, torsion-free, co-compact subgroup of G. Then Γ acts properly discontinuously on S without fixed points and the quotient $M = \Gamma\backslash S$ is a locally symmetric manifold which is equipped with the metric induced from the invariant metric of S. Then each $D \in \mathcal{D}(S)$ descends to a differential operator

$$D\colon C^\infty(\Gamma\backslash S) \to C^\infty(\Gamma\backslash S).$$

Let $\mathcal{E} \subset C^\infty(\Gamma\backslash S)$ be an eigenspace of the Laplace operator. Then $\mathcal{E}$ is a finite-dimensional vector space which is invariant under $D \in \mathcal{D}(S)$. For each $D \in \mathcal{D}(S)$, the formal adjoint D^* of D also belongs to $\mathcal{D}(S)$.

Thus we get a representation

$$\rho \colon \mathcal{D}(S) \to \operatorname{End}(\mathcal{E})$$

by commuting normal operators. Therefore, $\mathcal{E}$ decomposes into the direct sum of joint eigenspaces of $\mathcal{D}(S)$. Given $\lambda \in \mathfrak{a}_{\mathbb{C}}^*/W$, let

$$\mathcal{E}(\lambda) = \{\varphi \in C^\infty(\Gamma \backslash S) \colon D\varphi = \chi_\lambda(D)\varphi, \ D \in \mathcal{D}(S)\}.$$

Let $m(\lambda) = \dim \mathcal{E}(\lambda)$. Then the spectrum $\Lambda(\Gamma)$ of $\Gamma \backslash S$ is defined to be

$$\Lambda(\Gamma) = \{\lambda \in \mathfrak{a}_{\mathbb{C}}^*/W \colon m(\lambda) > 0\},$$

and we get an orthogonal direct sum decomposition

$$L^2(\Gamma \backslash S) = \bigoplus_{\lambda \in \Lambda(\Gamma)} \mathcal{E}(\lambda).$$

If we pick a fundamental domain for W, we may regard $\Lambda(\Gamma)$ as a subset of $\mathfrak{a}_{\mathbb{C}}^*$. If $\operatorname{rk}(S) > 1$, then $\Lambda(\Gamma)$ is multidimensional. Again the distribution of $\Lambda(\Gamma)$ is studied using the Selberg trace formula [Se1]. To describe it we need to introduce some notation. Let $C_c^\infty(G/\!/K)$ be the subspace of all $f \in C_c^\infty(G)$ which are K-bi-invariant. Let

$$\mathcal{A} \colon C_c^\infty(G/\!/K) \to C_c^\infty(A)^W$$

be the Abel transform which is defined by

$$\mathcal{A}(f)(a) = \delta(a)^{1/2} \int_N f(an) \, dn, \quad a \in A,$$

where δ is the modulus function of the minimal parabolic subgroup $P = NA$. Given $h \in C_c^\infty(A)^W$, let

$$\hat{h}(\lambda) = \int_A h(a)e^{\langle \lambda, H(a) \rangle} \, da.$$

Let $\beta(i\lambda)$, $\lambda \in \mathfrak{a}^*$, be the Plancherel density. Then the Selberg trace formula is the following identity

$$\sum_{\lambda \in \Lambda(\Gamma)} m(\lambda)\hat{h}(\lambda) = \frac{\operatorname{vol}(\Gamma \backslash G)}{|W|} \int_{\mathfrak{a}^*} \hat{h}(\lambda)\beta(i\lambda) \, d\lambda$$

$$+ \sum_{[\gamma]_\Gamma \neq e} \operatorname{vol}(\Gamma_\gamma \backslash G_\gamma) \int_{G_\gamma \backslash G} \mathcal{A}^{-1}(h)(x^{-1}\gamma x) \, d_\gamma \bar{x}.$$

$$(2.14)$$

This is still not the final form of the Selberg trace formula. The distributions

$$J_\gamma(f) = \text{vol}(\Gamma_\gamma \backslash G_\gamma) \int_{G_\gamma \backslash G} f(x^{-1}\gamma x)\, d_\gamma \bar{x}, \quad f \in C_c^\infty(G), \qquad (2.15)$$

are invariant distribution an G and can be computed using Harish-Chandra's Fourier inversion formula. This brings (2.14) into a form which is similar to (2.4). For the present purpose, however, it suffices to work with (2.14). Since for $\gamma \neq e$, the conjugacy class of γ in G is closed and does not intersect K, there exists an open neighborhood V of 1 in A satisfying $V = V^{-1}$, V is invariant under W, and $J_\gamma(\mathcal{A}^{-1}(h)) = 0$ for all $h \in C_c^\infty(V)$ [DKV, Proposition 3.8]. Thus we get

$$\sum_{\lambda \in \Lambda(\Gamma)} m(\lambda)\hat{h}(\lambda) = \frac{\text{vol}(\Gamma \backslash G)}{|W|} \int_{\mathfrak{a}^*} \hat{h}(\lambda)\beta(i\lambda)\, d\lambda \qquad (2.16)$$

for all $h \in C_c^\infty(V)$. One can now proceed as in the case of the upper half-plane. The basic step is again to estimate the number of $\lambda \in \Lambda(\Gamma)$ lying in a ball of radius r around a variable point $\mu \in i\mathfrak{a}^*$. This can be achieved by inserting appropriate test functions h into (2.16) [DKV, section 7]. Let

$$\Lambda_{\text{temp}}(\Gamma) = \Lambda(\Gamma) \cap i\mathfrak{a}^*, \quad \Lambda_{\text{comp}}(\Gamma) = \Lambda(\Gamma) \backslash \Lambda_{\text{temp}}(\Gamma)$$

be the tempered and complementary spectrum, respectively. Given an open bounded subset Ω of $\mathfrak{a}^*$ and $t > 0$, let

$$t\Omega := \{t\mu\colon \mu \in \Omega\}. \qquad (2.17)$$

One of the main results of [DKV] is the following asymptotic formula for the distribution of the tempered spectrum [DKV, Theorem 8.8]

$$\sum_{\lambda \in \Lambda_{\text{temp}}(\Gamma) \cap (it\Omega)} m(\lambda) = \frac{\text{vol}(\Gamma \backslash G)}{|W|} \int_{it\Omega} \beta(i\lambda)\, d\lambda + O(t^{n-1}), \quad t \to \infty,$$

$$(2.18)$$

Note that the leading term is of order $O(t^n)$. The growth of the complementary spectrum is of lower order. Let $B_t(0) \subset \mathfrak{a}_{\mathbb{C}}^*$ be the ball of radius $t > 0$ around 0. There exists $C > 0$ such that for all $t \geq 1$

$$\sum_{\lambda \in \Lambda_{\text{comp}}(\Gamma) \cap B_t(0)} m(\lambda) \leq Ct^{n-2} \qquad (2.19)$$

[DKV, Theorem 8.3]. The estimates (2.18) and (2.19) contain more information about the distribution of $\Lambda(\Gamma)$ than just the Weyl law. Indeed, the eigenvalue of Δ corresponding to $\lambda \in \Lambda_{\text{temp}}(\Gamma)$ equals $\|\lambda\|^2 + \|\rho\|^2$.

So if we choose Ω in (2.18) to be the unit ball, then (2.18) together with (2.19) reduces to Weyl's law for $\Gamma\backslash S$.

We note that (2.18) and (2.19) can also be rephrased in terms of representation theory. Let R be the right regular representation of G in $L^2(\Gamma\backslash G)$ defined by

$$(R(g_1)f)(g_2) = f(g_2 g_1), \quad f \in L^2(\Gamma\backslash G), \ g_1, g_2 \in G.$$

Let $\hat{G}$ be the unitary dual of G, i.e., the set of equivalence classes of irreducible unitary representations of G. Since $\Gamma\backslash G$ is compact, it is well known that R decomposes into direct sum of irreducible unitary representations of G. Given $\pi \in \hat{G}$, let $m(\pi)$ be the multiplicity with which π occurs in R. Let $\mathcal{H}_\pi$ denote the Hilbert space in which π acts. Then

$$L^2(\Gamma\backslash G) \cong \bigoplus_{\pi \in \hat{G}} m(\lambda)\mathcal{H}_\pi.$$

Now observe that $L^2(\Gamma\backslash S) = L^2(\Gamma\backslash G)^K$. Let $\mathcal{H}_\pi^K$ denote the subspace of K-fixed vectors in $\mathcal{H}_\pi$. Then

$$L^2(\Gamma\backslash S) \cong \bigoplus_{\pi \in \hat{G}} m(\lambda)\mathcal{H}_\pi^K.$$

Note that $\dim \mathcal{H}_\pi^K \leq 1$. Let $\hat{G}(1) \subset \hat{G}$ be the subset of all π with $\mathcal{H}_\pi^K \neq \{0\}$. This is the spherical dual. Given $\pi \in \hat{G}$, let λ_π be the infinitesimal character of π. If $\pi \in \hat{G}(1)$, then $\lambda_\pi \in \mathfrak{a}_\mathbb{C}^*/W$. Moreover $\pi \in \hat{G}(1)$ is tempered if π is unitarily induced from the minimal parabolic subgroup $P = NA$. In this case we have $\lambda_\pi \in i\mathfrak{a}^*/W$. So (2.18) can be rewritten as

$$\sum_{\substack{\pi \in \hat{G}(1) \\ \lambda_\pi \in it\Omega}} m(\pi) = \frac{\mathrm{vol}(\Gamma\backslash G)}{|W|} \int_{it\Omega} \beta(\lambda)\,d\lambda + O(t^{n-1}), \quad t \to \infty. \quad (2.20)$$

3 Automorphic forms

The theory of automorphic forms is concerned with harmonic analysis on locally symmetric spaces $\Gamma\backslash S$ of finite volume. Of particular interest are arithmetic groups Γ. This means that we consider a connected semi-simple algebraic group $\mathbf{G}$ defined over $\mathbb{Q}$ such that $G = \mathbf{G}(\mathbb{R})$ and Γ is a subgroup of $\mathbf{G}(\mathbb{Q})$ which is commensurable with $\mathbf{G}(\mathbb{Z})$, where $\mathbf{G}(\mathbb{Z})$ is defined with respect to some embedding $\mathbf{G} \subset \mathrm{GL}(m)$. The standard example is $\mathbf{G} = \mathrm{SL}(n)$ and $\Gamma(N) \subset \mathrm{SL}(n, \mathbb{Z})$ the principal congruence subgroup of level N. A basic feature of arithmetic groups is that the quotient $\Gamma\backslash S$ has finite volume [BH]. Moreover in many important cases it is non-compact. A typical example for that is $\Gamma(N)\backslash \mathrm{SL}(n,\mathbb{R})/\mathrm{SO}(n)$.

In this section we discuss only the case of the upper half-plane $\mathbb{H}$ and we consider congruence subgroups of $\mathrm{SL}(2,\mathbb{Z})$. For $N \geq 1$ the principal congruence subgroup $\Gamma(N)$ of level N is defined as

$$\Gamma(N) = \big\{\gamma \in \mathrm{SL}(2,\mathbb{Z})\colon \gamma \equiv \mathrm{Id} \bmod N\big\}.$$

A congruence subgroup Γ of $\mathrm{SL}(2,\mathbb{Z})$ is a subgroup for which there exists $N \in \mathbb{N}$ such that Γ contains $\Gamma(N)$. An example of a congruence subgroup is the Hecke group

$$\Gamma_0(N) = \left\{\begin{pmatrix} a & b \\ c & d \end{pmatrix} \in \mathrm{SL}(2,\mathbb{Z})\colon c \equiv 0 \bmod N\right\}.$$

If Γ is torsion free, the quotient $\Gamma\backslash\mathbb{H}$ is a finite area, non-compact, hyperbolic surface. It has a decomposition

$$\Gamma\backslash\mathbb{H} = M_0 \cup Y_1 \cup \cdots \cup Y_m, \tag{3.1}$$

into the union of a compact surface with boundary M_0 and a finite number of ends $Y_i \cong [a,\infty)\times S^1$ which are equipped with the Poincaré metric. In general, $\Gamma\backslash\mathbb{H}$ may have a finite number of quotient singularities. The quotient $\Gamma(N)\backslash\mathbb{H}$ is the modular surface $X(N)$.

Let Δ be the hyperbolic Laplacian (2.2). A Maass automorphic form is a smooth function $f\colon \mathbb{H} \to \mathbb{C}$ which satisfies

   a) $f(\gamma z) = f(z)$, $\gamma \in \Gamma$.
   b) There exists $\lambda \in \mathbb{C}$ such that $\Delta f = \lambda f$.
   c) f is slowly increasing.

Here the last condition means that there exist $C > 0$ and $N \in \mathbb{N}$ such that the restriction f_i of f to Y_i satisfies

$$|f_i(y,x)| \leq Cy^N, \quad y \geq a, \ i = 1,...,m.$$

Examples are the Eisenstein series. Let $a_1,...,a_m \in \mathbb{R}\cup\{\infty\}$ be representatives of the Γ-conjugacy classes of parabolic fixed points of Γ. The a_i's are called *cusps*. For each a_i let Γ_{a_i} be the stabilizer of a_i in Γ. Choose $\sigma_i \in \mathrm{SL}(2,\mathbb{R})$ such that

$$\sigma_i(\infty) = a_i, \quad \sigma_i^{-1}\Gamma_{a_i}\sigma_i = \left\{\begin{pmatrix} 1 & n \\ 0 & 1 \end{pmatrix} : n \in \mathbb{Z}\right\}.$$

Then the Eisenstein series $E_i(z, s)$ associated to the cusp a_i is defined as

$$E_i(z, s) = \sum_{\gamma \in \Gamma_{a_i} \backslash \Gamma} \operatorname{Im}(\sigma_i^{-1} \gamma z)^s, \quad \operatorname{Re}(s) > 1. \tag{3.2}$$

The series converges absolutely and uniformly on compact subsets of the half-plane $\operatorname{Re}(s) > 1$ and it satisfies the following properties.

1) $E_i(\gamma z, s) = E_i(z, s)$ for all $\gamma \in \Gamma$.
2) As a function of s, $E_i(z, s)$ admits a meromorphic continuation to $\mathbb{C}$ which is regular on the line $\operatorname{Re}(s) = 1/2$.
3) $E_i(z, s)$ is a smooth function of z and satisfies

$$\Delta_z E_i(z, s) = s(1 - s) E_i(z, s).$$

As an example consider the modular group $\Gamma(1)$ which has a single cusp ∞. The Eisenstein series attached to ∞ is the well-known series

$$E(z, s) = \sum_{\substack{(m,n) \in \mathbb{Z}^2 \\ (m,n)=1}} \frac{y^s}{|mz + n|^{2s}}.$$

In the general case, the Eisenstein series were first studied by Selberg [Sel]. The Eisenstein series are closely related with the study of the spectral resolution of Δ. Regarded as unbounded operator

$$\Delta \colon C_c^\infty(\Gamma \backslash \mathbb{H}) \to L^2(\Gamma \backslash \mathbb{H}),$$

Δ is essentially self-adjoint [Roe]. Let $\bar{\Delta}$ be the unique self-adjoint extension of Δ. The important new feature due to the non-compactness of $\Gamma \backslash \mathbb{H}$ is that $\bar{\Delta}$ has a large continuous spectrum which is governed by the Eisenstein series. The following basic result is due to Roelcke [Roe].

Proposition 3.1 *The spectrum of $\bar{\Delta}$ is the union of a pure point spectrum $\sigma_{pp}(\bar{\Delta})$ and an absolutely continuous spectrum $\sigma_{ac}(\bar{\Delta})$.*
1) The pure point spectrum consists of eigenvalues $0 = \lambda_0 < \lambda_1 \leq \cdots$ of finite multiplicities with no finite points of accumulation.
2) The absolutely continuous spectrum equals $[1/4, \infty)$ with multiplicity equal to the number of cusps of $\Gamma \backslash \mathbb{H}$.

Of particular interest are the eigenfunctions of $\bar{\Delta}$. They are Maass automorphic forms. This can be seen by studying the Fourier expansion of an eigenfunction in the cusps. As an example consider $f \in C^\infty(\Gamma_0(N)\backslash\mathbb{H})$ which satisfies

$$\Delta f = \lambda f, \quad f(z) = f(-\bar{z}), \quad \int_{\Gamma_0(N)\backslash\mathbb{H}} |f(z)|^2 \, dA(z) < \infty.$$

Assume that $\lambda = 1/4 + r^2$, $r \in \mathbb{R}$. Then $f(x + iy)$ admits a Fourier expansion w.r.t. x of the form

$$f(x + iy) = \sum_{n=1}^{\infty} a(n)\sqrt{y}K_{ir}(2\pi ny)\cos(2\pi nx), \qquad (3.3)$$

where $K_\nu(y)$ is the modified Bessel function which may be defined by

$$K_\nu(y) = \int_0^\infty e^{-y\cosh t}\cosh(\nu t)\, dt$$

and it satisfies

$$K_\nu''(y) + \frac{1}{y}K_\nu'(y) + \left(1 - \frac{\nu^2}{y^2}\right)K_\nu(y) = 0.$$

Now note that $K_\nu(y) = O(e^{-cy})$ as $y \to \infty$. This implies that f is rapidly decreasing in the cusp ∞. A similar Fourier expansion holds in the other cusps. This implies that f is rapidly decreasing in all cusps and therefore, it is a Maass automorphic form. In fact, since the zero Fourier coefficients vanish in all cusps, f is a Maass *cusp form*. In general, the space of cusp forms $L^2_{\text{cus}}(\Gamma\backslash\mathbb{H})$ is defined as the subspace of all $f \in L^2(\Gamma\backslash\mathbb{H})$ such that for almost all $y \in \mathbb{R}^+$:

$$\int_0^1 f(\sigma_k(x + iy))\, dx = 0, \quad k = 1, ..., m.$$

This is an invariant subspace of $\bar{\Delta}$ and the restriction of $\bar{\Delta}$ to $L^2_{\text{cus}}(\Gamma\backslash\mathbb{H})$ has pure point spectrum, i.e., $L^2_{\text{cus}}(\Gamma\backslash\mathbb{H})$ is the span of square integrable eigenfunctions of Δ. Let $L^2_{\text{res}}(\Gamma\backslash\mathbb{H})$ be the orthogonal complement of $L^2_{\text{cus}}(\Gamma\backslash\mathbb{H})$ in $L^2(\Gamma\backslash\mathbb{H})$. Thus

$$L^2_{pp}(\Gamma\backslash\mathbb{H}) = L^2_{\text{cus}}(\Gamma\backslash\mathbb{H}) \oplus L^2_{\text{res}}(\Gamma\backslash\mathbb{H}).$$

The subspace $L^2_{\text{res}}(\Gamma\backslash\mathbb{H})$ can be described as follows. The poles of the Eisenstein series $E_i(z, s)$ in the half-plane $\text{Re}(s) > 1/2$ are all simple and are contained in the interval $(1/2, 1]$. Let $s_0 \in (1/2, 1]$ be a pole of $E_i(z, s)$ and put

$$\psi = \text{Res}_{s=s_0} E_i(z, s).$$

Then ψ is a square integrable eigenfunction of Δ with eigenvalue $\lambda =$

$s_0(1 - s_0)$. The set of all such residues of the Eisenstein series $E_i(z, s)$, $i = 1, ..., m$, spans $L^2_{\mathrm{res}}(\Gamma \backslash \mathbb{H})$. This is a finite-dimensional space which is called the *residual subspace*. The corresponding eigenvalues form the *residual spectrum* of $\bar{\Delta}$. So we are left with the cuspidal eigenfunctions or *Maass cusp forms*. Cusp forms are the building blocks of the theory of automorphic forms. They play an important role in number theory. To illustrate this consider an even Maass cusp form f for $\Gamma(1)$ with eigenvalue $\lambda = 1/4 + r^2$, $r \in \mathbb{R}$. Let $a(n)$, $n \in \mathbb{N}$, be the Fourier coefficients of f given by (3.3). Put

$$L(s, f) = \sum_{n=1}^{\infty} \frac{a(n)}{n^s}, \quad \mathrm{Re}(s) > 1.$$

This Dirichlet series converges absolutely and uniformly in the half-plane $\mathrm{Re}(s) > 1$. Let

$$\Lambda(s, f) = \pi^{-s} \Gamma\left(\frac{s + ir}{2}\right) \Gamma\left(\frac{s - ir}{2}\right) L(s, f). \tag{3.4}$$

Then the modularity of f implies that $\Lambda(s, f)$ has an analytic continuation to the whole complex plane and satisfies the functional equation

$$\Lambda(s, f) = \Lambda(1 - s, f)$$

[Bu, Proposition 1.9.1]. Under additional assumptions on f, the Dirichlet series $L(s, f)$ is also an Euler product. This is related to the arithmetic nature of the groups $\Gamma(N)$. The surfaces $X(N)$ carry a family of algebraically defined operators T_n, the so called *Hecke operators*, which for $(n, N) = 1$ are defined by

$$T_n f(z) = \frac{1}{\sqrt{n}} \sum_{\substack{ad = n \\ b \bmod d}} f\left(\frac{az + b}{d}\right).$$

These are closely related to the cosets of the finite index subgroups

$$\begin{pmatrix} n & 0 \\ 0 & 1 \end{pmatrix} \Gamma(N) \begin{pmatrix} n & 0 \\ 0 & 1 \end{pmatrix}^{-1} \cap \Gamma(N)$$

of $\Gamma(N)$. Each T_n defines a linear transformation of $L^2(X(N))$. The T_n, $n \in \mathbb{N}$, are a commuting family of normal operators which also commute with Δ. Therefore, each T_n leaves the eigenspaces of Δ invariant. So we may assume that f is a common eigenfunction of Δ and T_n, $n \in \mathbb{N}$:

$$\Delta f = (1/4 + r^2)f, \quad T_n f = \lambda(n)f.$$

If $f \neq 0$, then $a(1) \neq 0$. So we can normalize f such that $a(1) = 1$. Then it follows that $a(n) = \lambda(n)$ and the Fourier coefficients satisfy the following multiplicative relations

$$a(m)a(n) = \sum_{d|(m,n)} a\left(\frac{mn}{d^2}\right).$$

This implies that $L(s, f)$ is an Euler product

$$L(s,f) = \sum_{n=1}^{\infty} a(n)n^{-s} = \prod_p \left(1 - a(p)p^{-s} + p^{-2s}\right)^{-1}, \qquad (3.5)$$

which converges for $\mathrm{Re}(s) > 1$. $L(s, f)$ is the basic example of an automorphic L-function. It is convenient to write this Euler product in a different way. Introduce roots α_p, β_p by

$$\alpha_p \beta_p = 1, \quad \alpha_p + \beta_p = a(p).$$

Let

$$A_p = \begin{pmatrix} \alpha_p & 0 \\ 0 & \beta_p \end{pmatrix}.$$

Then

$$L(s,f) = \prod_p \det\left(\mathrm{Id} - A_p p^{-s}\right)^{-1}.$$

Now let $\rho \colon \mathrm{GL}(2, \mathbb{C}) \to \mathrm{GL}(N, \mathbb{C})$ be a representation. Then we can form a new Euler product by

$$L(s, f, \rho) = \prod_p \det\left(\mathrm{Id} - \rho(A_p)p^{-s}\right)^{-1},$$

which converges in some half-plane. It is part of the general conjectures of Langlands [La2] that each of these Euler products admits a meromorphic extension to $\mathbb{C}$ and satisfies a functional equation. The construction of Euler products for Maass cusp forms can be extended to other groups $\mathbf{G}$, in particular to cusp forms on $\mathrm{GL}(n)$. It is also conjectured that $L(s, f, \rho)$ is an automorphic L-function of an automorphic form on some $\mathrm{GL}(n)$. This is part of the functoriality principle of Langlands. Furthermore, all L-functions that occur in number theory and algebraic geometry are expected to be automorphic L-functions. This is one of the main reasons for the interest in the study of cusp forms. Other applications are discussed in [Sa1].

4 The Weyl law and existence of cusp forms

Since $\Gamma(N)\backslash\mathbb{H}$ is not compact, it is not clear that there exist any eigenvalues $\lambda > 0$. By Proposition 3.1 the continuous spectrum of $\tilde{\Delta}$ equals $[1/4, \infty)$. Thus all eigenvalues $\lambda \geq 1/4$ are embedded in the continuous spectrum. It is well-known in mathematical physics, that embedded eigenvalues are unstable under perturbations and therefore, are difficult to study.

One of the basic tools to study the cuspidal spectrum is the Selberg trace formula [Se2]. The new terms in the trace formula, which are due to the non-compactness of $\Gamma\backslash\mathbb{H}$ arise from the parabolic conjugacy classes in Γ and the Eisenstein series. The contribution of the Eisenstein series is given by their zeroth Fourier coefficients of the Fourier expansion in the cusps. The zeroth Fourier coefficient of the Eisenstein series $E_k(z, s)$ in the cusp a_l is given by

$$\int_0^1 E_k(\sigma_l(x + iy), s)\, dx = y^s + C_{kl}(s)y^{1-s},$$

where $C_{kl}(s)$ is a meromorphic function of $s \in \mathbb{C}$. Put

$$C(s) := (C_{kl}(s))_{k,l=1}^m.$$

This is the so called scattering matrix. Let

$$\varphi(s) := \det C(s).$$

Let the notation be as in (2.4) and assume that Γ has no torsion. Then the trace formula is the following identity.

$$\sum_j h(r_j) = \frac{\text{Area}(\Gamma\backslash\mathbb{H})}{4\pi} \int_{\mathbb{R}} h(r)r\tanh(\pi r)\, dr + \sum_{\{\gamma\}_\Gamma} \frac{l(\gamma_0)}{2\sinh\left(\frac{l(\gamma)}{2}\right)} g(l(\gamma))$$

$$+ \frac{1}{4\pi} \int_{-\infty}^{\infty} h(r)\frac{\varphi'}{\varphi}(1/2 + ir)\, dr - \frac{1}{4}\varphi(1/2)h(0)$$

$$- \frac{m}{2\pi} \int_{-\infty}^{\infty} h(r)\frac{\Gamma'}{\Gamma}(1 + ir)dr + \frac{m}{4}h(0) - m\ln 2\, g(0).$$

$$(4.1)$$

The trace formula holds for every discrete subgroup $\Gamma \subset SL(2,\mathbb{R})$ with finite coarea. In analogy to the counting function of the eigenvalues we introduce the winding number

$$M_\Gamma(\lambda) = -\frac{1}{4\pi} \int_{-\lambda}^{\lambda} \frac{\varphi'}{\varphi}(1/2 + ir)\, dr$$

which measures the continuous spectrum. Using the cut-off Laplacian of Lax-Phillips [CV] one can deduce the following elementary bounds

$$N_\Gamma(\lambda) \ll \lambda^2, \quad M_\Gamma(\lambda) \ll \lambda^2, \quad \lambda \geq 1. \tag{4.2}$$

These bounds imply that the trace formula (4.1) holds for a larger class of functions. In particular, it can be applied to the heat kernel $k_t(u)$. Its spherical Fourier transform equals $h_t(r) = e^{-t(1/4+r^2)}$, $t > 0$. If we insert h_t into the trace formula we get the following asymptotic expansion as $t \to 0$.

$$\sum_j e^{-t\lambda_j} - \frac{1}{4\pi} \int_{\mathbb{R}} e^{-t(1/4+r^2)} \frac{\varphi'}{\varphi}(1/2+ir)\, dr$$
$$= \frac{\text{Area}(\Gamma\backslash\mathbb{H})}{4\pi t} + \frac{a \log t}{\sqrt{t}} + \frac{b}{\sqrt{t}} + O(1) \tag{4.3}$$

for certain constants $a, b \in \mathbb{R}$. Using [Se2, (8.8), (8.9)] it follows that the winding number $M_\Gamma(\lambda)$ is monotonic increasing for $r \gg 0$. Therefore we can apply a Tauberian theorem to (4.3) and we get the Weyl law (1.4).

In general, we cannot estimate separately the counting function and the winding number. For congruence subgroups, however, the entries of the scattering matrix can be expressed in terms of well-known analytic functions. For $\Gamma(N)$ the determinant of the scattering matrix $\varphi(s)$ has been computed by Huxley [Hu]. It has the form

$$\varphi(s) = (-1)^l A^{1-2s} \left(\frac{\Gamma(1-s)}{\Gamma(s)} \right)^k \prod_\chi \frac{L(2-2s, \bar\chi)}{L(2s, \chi)}, \tag{4.4}$$

where $k, l \in \mathbb{Z}$, $A > 0$, the product runs over Dirichlet characters χ to some modulus dividing N and $L(s, \chi)$ is the Dirichlet L-function with character χ. Especially for $\Gamma(1)$ we have

$$\varphi(s) = \sqrt{\pi} \frac{\Gamma(s-1/2)\zeta(2s-1)}{\Gamma(s)\zeta(2s)}, \tag{4.5}$$

where $\zeta(s)$ denotes the Riemann zeta function.

Using Stirling's approximation formula to estimate the logarithmic derivative of the Gamma function and standard estimations for the logarithmic derivative of Dirichlet L-functions on the line $\text{Re}(s) = 1$ [Pr, Theormem 7.1], we get

$$\frac{\varphi'}{\varphi}(1/2+ir) = O(\log(4+|r|)), \quad |r| \to \infty. \tag{4.6}$$

This implies that

$$M_{\Gamma(N)}(\lambda) \ll \lambda \log \lambda. \tag{4.7}$$

Together with (1.4) we obtain Weyl's law for the point spectrum

$$N_{\Gamma(N)}(\lambda) \sim \frac{\text{Area}(X(N))}{4\pi}\lambda^2, \quad \lambda \to \infty, \tag{4.8}$$

which is due to Selberg [Se2, p.668]. A similar formula holds for other congruence groups such as $\Gamma_0(N)$. In particular, (4.8) implies that for congruence groups Γ there exist infinitely many linearly independent Maass cusp forms.

A proof of the Weyl law (4.8) which avoids the use of the constant terms of the Eisenstein series has recently been given by Lindenstrauss and Venkatesh [LV]. Their approach is based on the construction of convolution operators with purely cuspidal image.

Neither of these methods give good estimates of the remainder term. One approach to obtain estimates of the remainder term is based on the Selberg zeta function

$$Z_{\Gamma}(s) = \prod_{\{\gamma\}_{\Gamma}} \prod_{k=0}^{\infty} \left(1 - e^{-(s+k)\ell(\gamma)}\right), \quad \text{Re}(s) > 1,$$

where the outer product runs over the primitive hyperbolic conjugacy classes in Γ and $\ell(\gamma)$ is the length of the closed geodesic associated to $\{\gamma\}_{\Gamma}$. The infinite product converges absolutely in the indicated half-plane and admits an analytic continuation to the whole complex plane. If $\lambda = 1/4 + r^2$, $r \in \mathbb{R} \cup i(1/2, 1]$, is an eigenvalue of Δ, then $s_0 = 1/2 + ir$ is a zero of $Z_{\Gamma}(s)$. Using this fact and standard methods of analytic number theory one can derive the following strong form of the Weyl law [Hj, Theorem 2.28], [Ve, Theorem 7.3].

Theorem 4.1 *Let m be the number of cusps of $\Gamma\backslash\mathbb{H}$. There exists $c > 0$ such that*

$$N_{\Gamma}(\lambda) + M_{\Gamma}(\lambda) = \frac{\text{Area}(\Gamma\backslash\mathbb{H})}{4\pi}\lambda^2 - \frac{m}{\pi}\lambda \log \lambda + c\lambda + O\left(\lambda(\log\lambda)^{-1}\right)$$

as $\lambda \to \infty$.

Together with (4.7) we obtain Weyl's law with remainder term.

Theorem 4.2 *For every $N \in \mathbb{N}$ we have*

$$N_{\Gamma(N)}(\lambda) = \frac{\text{Area}(X(N))}{4\pi}\lambda^2 + O(\lambda \log \lambda) \quad as \quad \lambda \to \infty.$$

The use of the Selberg zeta function to estimate the remainder term is limited to rank one cases. However, the remainder term can also be estimated by Hörmander's method using the trace formula as in the compact case. We describe the main steps. Let $\varepsilon > 0$ such that $\ell(\gamma) > \varepsilon$ for all hyperbolic conjugacy classes $\{\gamma\}_{\Gamma(N)}$. Choose $g \in C_c^\infty(\mathbb{R})$ to be even and such that $\operatorname{supp} g \subset (-\varepsilon, \varepsilon)$. Let $h(z) = \int_{\mathbb{R}} g(r)e^{-irz}\, dr$. Then the hyperbolic contribution in the trace formula (4.1) drops out. We symmetrize the spectrum by $r_{-j} = -r_j$, $j \in \mathbb{N}$. Then for each $t \in \mathbb{R}$ we have

$$\sum_j h(t - r_j) = \frac{\operatorname{Area}(X(N))}{2\pi} \int_{\mathbb{R}} h(t - r)r \tanh(\pi r)\, dr$$

$$+ \frac{1}{2\pi} \int_{-\infty}^{\infty} h(t - r)\frac{\varphi'}{\varphi}(1/2 + ir)\, dr - \frac{1}{2}\varphi(1/2)h(t)$$

$$- \frac{m}{\pi} \int_{-\infty}^{\infty} h(t - r)\frac{\Gamma'}{\Gamma}(1 + ir)dr + \frac{m}{2}h(t) - 2m \ln 2\, g(0).$$

$$(4.9)$$

Now we need to estimate the behavior of the terms on the right hand side as $|t| \to \infty$. The first integral has been already considered in (2.7). It is of order $O(|t|)$. To deal with the second integral we use (4.6). This implies

$$\int_{\mathbb{R}} h(t - r)\frac{\varphi'}{\varphi}(1/2 + ir)\, dr = O(\log(|t|)), \quad |t| \to \infty. \qquad (4.10)$$

Using Stirling's formula we get

$$\int_{\mathbb{R}} h(t - r)\frac{\Gamma'}{\Gamma}(1 + ir)\, dr = O(\log(|t|)), \quad |t| \to \infty.$$

The remaining terms are bounded. Combining these estimations, we get

$$\sum_j h(t - r_j) = O(|t|), \quad |t| \to \infty.$$

Therefore, Lemma 2.2 holds also in the present case. It remains to establish the analog of Lemma 2.4. Using (2.10) and (4.6) we get

$$\int_{-\lambda}^{\lambda} \int_{\mathbb{R}} h(t - r)\frac{\varphi'}{\varphi}(1/2 + it)\, dr\, dt = O(\lambda \log \lambda). \qquad (4.11)$$

Similarly, using Stirling's formula and (2.10), we obtain

$$\int_{-\lambda}^{\lambda} \int_{\mathbb{R}} h(t - r)\frac{\Gamma'}{\Gamma}(1 + it)\, dr\, dt = O(\lambda \log \lambda). \qquad (4.12)$$

The integral of the remaining terms is of order $O(\lambda)$. Thus we obtain

$$\int_{-\lambda}^{\lambda} \sum_{j=-\infty}^{\infty} h(t - r_j)\, dt = \frac{\text{Area}(X(N))}{2\pi} \lambda^2 + O(\lambda \log \lambda) \qquad (4.13)$$

as $\lambda \to \infty$. Now we proceed in exactly the same way as in the compact case. Using Lemma 2.3 and (4.13), Theorem 4.2 follows. $\qquad\square$

The Weyl law shows that for congruence groups Maass cusp forms exist in abundance. In general very little is known. Let Γ be any discrete, co-finite subgroup of $\text{SL}(2, \mathbb{R})$. Then by Donnelly [Do] the following general bound is known

$$\limsup_{\lambda \to \infty} \frac{N_{\Gamma}^{\text{cus}}(\lambda)}{\lambda} \leq \frac{\text{Area}(\Gamma \backslash \mathbb{H})}{4\pi}.$$

A group Γ for which the equality is attained is called *essentially cuspidal* by Sarnak [Sa2]. By (4.8), $\Gamma(N)$ is essentially cuspidal. The study of the behavior of eigenvalues under deformations of Γ, initiated Phillips and Sarnak [PS1], [PS2], supports the conjecture that essential cuspidality may be limited to special arithmetic groups.

The consideration of the behavior of cuspidal eigenvalues under deformations was started by Colin de Verdiere [CV] in the more general context of metric perturbations. One of his main results [CV, Théorème 7] states that under a generic compactly supported conformal perturbation of the hyperbolic metric of $\Gamma \backslash \mathbb{H}$ all Maass cusp forms are dissolved. This means that each point $s_j = 1/2 + i r_j$, $r_j \in \mathbb{R}$, such that $\lambda_j = s_j(1 - s_j)$ is an eigenvalue moves under the perturbation into the half-plane $\text{Re}(s) < 1/2$ and becomes a pole of the scattering matrix $C(s)$.

In the present context we are only interested in deformations such that the curvature stays constant. Such deformations are given by curves in the Teichmüller space $\mathcal{T}(\Gamma)$ of Γ. The space $\mathcal{T}(\Gamma)$ is known to be a finite-dimensional and therefore, it is by no means clear that the results of [CV] will continue to hold for perturbations of this restricted type. For $\Gamma(N)$ this problem has been studied in [PS1], [PS2]. One of the main results is an analog of Fermi's golden rule which gives a sufficient condition for a cusp form of $\Gamma(N)$ to be dissolved under a deformation in $\mathcal{T}(\Gamma(N))$. Based on these results, Sarnak made the following conjecture [Sa2]:

Conjecture.

(a) *The generic Γ in a given Teichmüller space of finite area hyperbolic surfaces is not essentially cuspidal.*

(b) *Except for the Teichmüller space of the once punctured torus, the generic Γ has only finitely many eigenvalues.*

5 Higher rank

In this section we consider an arbitrary locally symmetric space $\Gamma\backslash S$ defined by an arithmetic subgroup $\Gamma \subset \mathbf{G}(\mathbb{Q})$, where $\mathbf{G}$ is a semi-simple algebraic group over $\mathbb{Q}$ with finite center, $G = \mathbf{G}(\mathbb{R})$ and $S = G/K$. The basic example will be $\mathbf{G} = \mathrm{SL}(n)$ and $\Gamma = \Gamma(N)$, the principal congruence subgroup of $\mathrm{SL}(n,\mathbb{Z})$ of level N which consists of all $\gamma \in \mathrm{SL}(n,\mathbb{Z})$ such that $\gamma \equiv \mathrm{Id} \bmod N$.

Let Δ be the Laplacian of $\Gamma\backslash S$, and let $\bar{\Delta}$ be the closure of Δ in L^2. Then $\bar{\Delta}$ is a non-negative self-adjoint operator in $L^2(\Gamma\backslash S)$. The properties of its spectral resolution can be derived from the known structure of the spectral resolution of the regular representation R_Γ of G on $L^2(\Gamma\backslash G)$ [La1], [BG]. In this way we get the following generalization of Proposition 3.1.

Proposition 5.1 *The spectrum of $\bar{\Delta}$ is the union of a point spectrum $\sigma_{pp}(\bar{\Delta})$ and an absolutely continuous spectrum $\sigma_{ac}(\bar{\Delta})$.*
1) The point spectrum consists of eigenvalues $0 = \lambda_0 < \lambda_1 \leq \cdots$ of finite multiplicities with no finite point of accumulation.
2) The absolutely continuous spectrum equals $[b,\infty)$ for some $b > 0$.

The theory of Eisenstein series [La1] provides a complete set of generalized eigenfunctions for Δ. The corresponding wave packets span the absolutely continuous subspace $L^2_c(\Gamma\backslash S)$. This allows us to determine the constant b explicitly in terms of the root structure. The statement about the point spectrum was proved in [BG, Theorem 5.5].

Let $L^2_{\mathrm{dis}}(\Gamma\backslash S)$ be the closure of the span of all eigenfunctions. It contains the subspace of cusp forms $L^2_{\mathrm{cus}}(\Gamma\backslash S)$. We recall its definition. Let $\mathbf{P} \subset \mathbf{G}$ be a parabolic subgroup defined over $\mathbb{Q}$ [Bo]. Let $P = \mathbf{P}(\mathbb{R})$. This is a cuspidal parabolic subgroup of G and all cuspidal parabolic subgroups arise in this way. Let $\mathbf{N}_P$ be the unipotent radical of $\mathbf{P}$ and let $N_P = \mathbf{N}_P(\mathbb{R})$. Then $N_P \cap \Gamma\backslash N_P$ is compact. A cusp form is a smooth function φ on $\Gamma\backslash S$ which is a joint eigenfunction of the ring

$\mathcal{D}(S)$ of invariant differential operators on S, and which satisfies

$$\int_{N_P \cap \Gamma \backslash N_P} \varphi(nx)\, dn = 0 \tag{5.1}$$

for all cuspidal parabolic subgroups $P \neq G$. Each cusp form is rapidly decreasing and hence square integrable. Let $L^2_{\text{cus}}(\Gamma \backslash S)$ be the closure in $L^2(\Gamma \backslash S)$ of the linear span of all cusp forms. Then $L^2_{\text{cus}}(\Gamma \backslash S)$ is an invariant subspace of $\bar{\Delta}$ which is contained in $L^2_{\text{dis}}(\Gamma \backslash S)$.

Let $L^2_{\text{res}}(\Gamma \backslash S)$ be the orthogonal complement of $L^2_{\text{cus}}(\Gamma \backslash S)$ in $L^2_{\text{dis}}(\Gamma \backslash S)$, i.e., we have an orthogonal decomposition

$$L^2_{\text{dis}}(\Gamma \backslash S) = L^2_{\text{cus}}(\Gamma \backslash S) \oplus L^2_{\text{res}}(\Gamma \backslash S).$$

It follows from Langlands's theory of Eisenstein systems that $L^2_{\text{res}}(\Gamma \backslash S)$ is spanned by *iterated residues* of cuspidal Eisenstein series [La1, Chapter 7]. Therefore $L^2_{\text{res}}(\Gamma \backslash S)$ is called the *residual subspace*.

Let $N^{\text{dis}}_\Gamma(\lambda)$, $N^{\text{cus}}_\Gamma(\lambda)$, and $N^{\text{res}}_\Gamma(\lambda)$ be the counting function of the eigenvalues with eigenfunctions belonging to the corresponding subspaces. The following general results about the growth of the counting functions are known for any lattice Γ in a real semi-simple Lie group. Let $n = \dim S$. Donnelly [Do] has proved the following bound for the cuspidal spectrum

$$\limsup_{\lambda \to \infty} \frac{N^{\text{cus}}_\Gamma(\lambda)}{\lambda^n} \leq \frac{\text{vol}(\Gamma \backslash S)}{(4\pi)^{n/2} \Gamma\left(\frac{n}{2} + 1\right)}, \tag{5.2}$$

where $\Gamma(s)$ denotes the Gamma function. For the full discrete spectrum, we have at least an upper bound for the growth of the counting function. The main result of [Mu2] states that

$$N^{\text{dis}}_\Gamma(\lambda) \ll (1 + \lambda^{4n}). \tag{5.3}$$

This result implies that invariant integral operators are trace class on the discrete subspace which is the starting point for the trace formula. The proof of (5.3) relies on the description of the residual subspace in terms of iterated residues of Eisenstein series. One actually expects that the growth of the residual spectrum is of lower order than the cuspidal spectrum. For $\text{SL}(n)$ the residual spectrum has been determined by Moeglin and Waldspurger [MW]. Combined with (5.2) it follows that for $G = \text{SL}(n)$ we have

$$N^{\text{res}}_{\Gamma(N)}(\lambda) \ll \lambda^{d-1}, \tag{5.4}$$

where $d = \dim \text{SL}(n, \mathbb{R}) / \text{SO}(n)$.

In [Sa2] Sarnak conjectured that if $\operatorname{rk}(G/K) > 1$, each irreducible lattice Γ in G is essentially cuspidal in the sense that Weyl's law holds for $N_\Gamma^{\mathrm{cus}}(\lambda)$, i.e., equality holds in (5.2). This conjecture has now been established in quite generality. A. Reznikov proved it for congruence groups in a group G of real rank one, S. Miller [Mi] proved it for $\mathbf{G} = \mathrm{SL}(3)$ and $\Gamma = \mathrm{SL}(3,\mathbb{Z})$, the author [Mu3] established it for $\mathbf{G} = \mathrm{SL}(n)$ and a congruence group Γ. The method of [Mu3] is an extension of the heat equation method described in the previous section for the case of the upper half-plane. More recently, Lindenstrauss and Venkatesh [LV] proved the following result.

Theorem 5.1 *Let* $\mathbf{G}$ *be a split adjoint semi-simple group over* $\mathbb{Q}$ *and let* $\Gamma \subset \mathbf{G}(\mathbb{Q})$ *be a congruence subgroup. Let* $n = \dim S$. *Then*

$$N_\Gamma^{\mathrm{cus}}(\lambda) \sim \frac{\mathrm{vol}(\Gamma \backslash S)}{(4\pi)^{n/2}\,\mathbf{\Gamma}\left(\frac{n}{2}+1\right)}\lambda^n, \quad \lambda \to \infty.$$

The method is based on the construction of convolution operators with pure cuspidal image. It avoids the delicate estimates of the contributions of the Eisenstein series to the trace formula. This proves existence of many cusp forms for these groups.

The next problem is to estimate the remainder term. For $\mathbf{G} = \mathrm{SL}(n)$, this problem has been studied by E. Lapid and the author in [LM]. Actually, we consider not only the cuspidal spectrum of the Laplacian, but the cuspidal spectrum of the whole algebra of invariant differential operators.

As $\mathcal{D}(S)$ preserves the space of cusp forms, we can proceed as in the compact case and decompose $L_{\mathrm{cus}}^2(\Gamma \backslash S)$ into joint eigenspaces of $\mathcal{D}(S)$. Given $\lambda \in \mathfrak{a}_{\mathbb{C}}^*/W$, let

$$\mathcal{E}_{\mathrm{cus}}(\lambda) = \left\{\varphi \in L_{\mathrm{cus}}^2(\Gamma \backslash S) \colon D\varphi = \chi_\lambda(D)\varphi\right\}$$

be the associated eigenspace. Each eigenspace is finite-dimensional. Let $m(\lambda) = \mathcal{E}_{\mathrm{cus}}(\lambda)$. Define the cuspidal spectrum $\Lambda_{\mathrm{cus}}(\Gamma)$ to be

$$\Lambda_{\mathrm{cus}}(\Gamma) = \{\lambda \in \mathfrak{a}_{\mathbb{C}}^*/W : m(\lambda) > 0\}.$$

Then we have an orthogonal direct sum decomposition

$$L_{\mathrm{cus}}^2(\Gamma \backslash S) = \bigoplus_{\lambda \in \Lambda_{\mathrm{cus}}(\Gamma)} \mathcal{E}_{\mathrm{cus}}(\lambda).$$

Let the notation be as in (2.18) and (2.19). Then in [LM] we established the following extension of main results of [DKV] to congruence quotients of $S = \mathrm{SL}(n, \mathbb{R})/\mathrm{SO}(n)$.

Theorem 5.2 *Let $d = \dim S$. Let $\Omega \subset \mathfrak{a}^*$ be a bounded domain with piecewise smooth boundary. Then for $N \geq 3$ we have*

$$\sum_{\substack{\lambda \in \Lambda_{\mathrm{cus}}(\Gamma(N)) \\ \lambda \in it\Omega}} m(\lambda) = \frac{\mathrm{vol}(\Gamma(N)\backslash S)}{|W|} \int_{t\Omega} \beta(i\lambda)\, d\lambda + O\left(t^{d-1}(\log t)^{\max(n,3)}\right),$$

(5.5)

as $t \to \infty$, and

$$\sum_{\substack{\lambda \in \Lambda_{\mathrm{cus}}(\Gamma(N)) \\ \lambda \in B_t(0)\backslash i\mathfrak{a}^*}} m(\lambda) = O\left(t^{d-2}\right), \quad t \to \infty.$$

(5.6)

If we apply (5.5) and (5.6) to the unit ball in $\mathfrak{a}^*$, we get the following corollary.

Corollary 5.3 *Let $\mathbf{G} = \mathrm{SL}(n)$ and let $\Gamma(N)$ be the principal congruence subgroup of $\mathrm{SL}(n, \mathbb{Z})$ of level N. Let $S = \mathrm{SL}(n, \mathbb{R})/\mathrm{SO}(n)$ and $d = \dim S$. Then for $N \geq 3$ we have*

$$N_{\Gamma(N)}^{\mathrm{cus}}(\lambda) = \frac{\mathrm{vol}(\Gamma(N)\backslash S)}{(4\pi)^{d/2}\Gamma\left(\frac{d}{2}+1\right)}\lambda^d + O\left(\lambda^{d-1}(\log \lambda)^{\max(n,3)}\right), \quad \lambda \to \infty.$$

The condition $N \geq 3$ is imposed for technical reasons. It guarantees that the principal congruence subgroup $\Gamma(N)$ is neat in the sense of Borel, and in particular, has no torsion. This simplifies the analysis by eliminating the contributions of the non-unipotent conjugacy classes in the trace formula.

Note that $\Lambda_{\mathrm{cus}}(\Gamma(N)) \cap i\mathfrak{a}^*$ is the cuspidal tempered spherical spectrum. The Ramanujan conjecture [Sa3] for $\mathrm{GL}(n)$ at the Archimedean place states that

$$\Lambda_{\mathrm{cus}}(\Gamma(N)) \subset i\mathfrak{a}^*$$

so that (5.6) is empty, if the Ramanujan conjecture is true. However, the Ramanujan conjecture is far from being proved. Moreover, it is known to be false for other groups $\mathbf{G}$ and (5.6) is what one can expect in general.

The method to prove Theorem 5.2 is an extension of the method of [DKV]. The Selberg trace formula, which is one of the basic tools in [DKV], needs to be replaced by the Arthur trace formula [A1], [A2]. This requires to change the framework and to work with the adelic setting. It is also convenient to replace $\mathrm{SL}(n)$ by $\mathrm{GL}(n)$.

Again, one of the main issues is to estimate the terms in the trace formula which are associated to Eisenstein series. Roughly speaking, these terms are a sum of integrals which generalize the integral

$$\int_{-\infty}^{\infty} h(r) \frac{\varphi'}{\varphi}(1/2 + ir)\, dr$$

in (4.1). The sum is running over Levi components of parabolic subgroups and square integrable automorphic forms on a given Levi component. The functions which generalize $\varphi(s)$ are obtained from the constant terms of Eisenstein series. In general, they are difficult to describe. The main ingredients are logarithmic derivatives of automorphic L-functions associated to automorphic forms on the Levi components. As example consider $\mathbf{G} = \mathrm{SL}(3)$, $\Gamma = \mathrm{SL}(3, \mathbb{Z})$, and a standard maximal parabolic subgroup P which has the form

$$P = \left\{ \begin{pmatrix} m_1 & X \\ 0 & m_2 \end{pmatrix} \ \middle|\ m_i \in \mathrm{GL}(n_i, \mathbb{R}),\ \det m_1 \cdot \det m_2 = 1 \right\},$$

with $n_1 + n_2 = 3$. Thus there are exactly two standard maximal parabolic subgroups. The standard Levi component of P is

$$L = \left\{ \begin{pmatrix} m_1 & 0 \\ 0 & m_2 \end{pmatrix} \ \middle|\ m_i \in \mathrm{GL}(n_i, \mathbb{R}),\ \det m_1 \cdot \det m_2 = 1 \right\}.$$

So L is isomorphic to $\mathrm{GL}(2, \mathbb{R})$. The Eisenstein series are associated to Maass cusp forms on $\Gamma(1) \backslash \mathbb{H}$. The constant terms of the Eisenstein series are described in [Go, Proposition 10.11.2]. Let f be a Maass cusp form for $\Gamma(1)$ and let $\Lambda(s, f \times \tilde{f})$ be the completed Rankin-Selberg L-function associated to f (cf. [Bu]). Then the relevant constant term of the Eisenstein series associated to f is given by

$$\frac{\Lambda(s, f \times \tilde{f})}{\Lambda(1 + s, f \times \tilde{f})}.$$

To proceed one needs a bound similar to (4.6). Assume that $\Delta f = (1/4 + r^2)f$. Using the analytic properties of $\Lambda(s, f \times \tilde{f})$ one can show that for $T \geq 1$

$$\int_{-T}^{T} \frac{\Lambda'}{\Lambda}(1 + it, f \times \tilde{f})\, dt \ll T \log(T + |r|). \tag{5.7}$$

This is the key result that is needed to deal with the contribution of the Eisenstein series to the trace formula.

The example demonstrates a general feature of spectral theory on locally symmetric spaces. Harmonic analysis on higher rank spaces requires the knowledge of the analytic properties of automorphic L-functions attached to cusp forms on lower rank groups. For $\mathrm{GL}(n)$, the corresponding L-functions are Rankin-Selberg convolutions $L(s, \varphi_1 \times \varphi_2)$ of automorphic cusp forms on $\mathrm{GL}(n_i)$, $i = 1, 2$, where $n_1 + n_2 = n$ (cf. [Bu], [Go] for their definition). The analytic properties of these L-functions are well understood so that estimates similar to (5.7) can be established. For other groups $\mathbf{G}$ (except for some low dimensional cases) our current knowledge of the analytic properties of the corresponding L-functions is not sufficient to prove estimates like (5.7). Only partial results exist [CPS]. This is one of the main obstacles to extend Theorem 5.2 to other groups.

Bibliography

[A1] J. Arthur, *A trace formula for reductive groups I: terms associated to classes in $G(\mathbb{Q})$*, Duke. Math. J. **45** (1978), 911–952.

[A2] J. Arthur, *A trace formula for reductive groups II: applications of a truncation operator*, Comp. Math. **40** (1980), 87–121.

[Av] V. G. Avakumović, *Über die Eigenfunktionen auf geschlossenen Riemannschen Mannigfaltigkeiten.* Math. Z. **65** (1956), 327–344.

[Bo] A. Borel, *Linear algebraic groups*, Second edition. Graduate Texts in Mathematics, 126. Springer-Verlag, New York, 1991.

[BG] A. Borel, H. Garland, *Laplacian and the discrete spectrum of an arithmetic group.* Amer. J. Math. **105** (1983), no. 2, 309–335.

[BH] A. Borel, Harish-Chandra, *Arithmetic subgroups of algebraic groups.* Ann. of Math. (2) **75** (1962), 485–535.

[Bu] D. Bump, *Automorphic forms and representations*, Cambridge Studies in Advanced Mathematics, 55. Cambridge University Press, Cambridge, 1997.

[CPS] J.W. Cogdell, H.H. Kim, I.I. Piatetski-Shapiro, F. Shahidi, *Functoriality for the classical groups.* Publ. Math. Inst. Hautes Études Sci. **99** (2004), 163–233.

[CV] Y. Colin de Verdiere, *Pseudo-Laplacians* II, Ann. Inst. Fourier, Grenoble, **33** (1983), 87–113.

[Do] H. Donnelly, *On the cuspidal spectrum for finite volume symmetric spaces.* J. Differential Geom. **17** (1982), no. 2, 239–253.

[DG] J.J. Duistermaat, V. Guillemin, *The spectrum of positive elliptic operators and periodic bicharacteristics.* Invent. Math. **29** (1975), no. 1, 39–79.

[DKV] Duistermaat, J.J., Kolk, J.A.C., Varadarajan, V.S.: *Spectra of compact locally symmetric manifolds of negative curvature*, Inventiones math. **52** (1979), 27–93.

[Ga] L. Garding, *Dirichlet's problem for linear elliptic partial differential equations.* Math. Scand. **1**, (1953). 55–72.

[Go] D. Goldfeld, *Automorphic forms and L-functions for the group* $\mathrm{GL}(n, \mathbb{R})$.

Cambridge Studies in Advanced Mathematics, 99. Cambridge University Press, Cambridge, 2006.

[Hj] D. A. Hejhal, *The Selberg trace formula for* PSL(2, ℝ). Lecture Notes in Math. **548** (1796) and **1001** (1983).

[He] S. Helgason, *Differential geometry and symmetric spaces*, New York, Academic Press, 1962.

[Ho] L. Hörmander, *The spectral function of an elliptic operator.* Acta Math. **121** (1968), 193–218.

[Hu] Huxley, M.: *Scattering matrices for congruence subgroups.* In: Modular forms, R. Rankin ed., Horwood, Chichester, 1984, pp. 141–156.

[Iv] V. Ivrii, *The second term of the spectral asymptotics for a Laplace-Beltrami operator on manifolds with boundary.* Funktsional. Anal. i Prilozhen. **14** (1980), no. 2, 25–34.

[La1] R.P. Langlands, *On the functional equations satisfied by Eisenstein series*, LNM **544**, Springer, Berlin-Heidelberg-New York, 1976.

[La2] R.P, Langlands, *Problems in the theory of automorphic forms.* Lectures in modern analysis and applications, III, pp. 18–61. Lecture Notes in Math., Vol. **170**, Springer, Berlin, 1970.

[LM] E. Lapid, W. Müller, *Spectral asymptotics for arithmetic quotients of the symmetric space of positive-definite matrices*, Preprint, 2007.

[Le] B.M. Levitan, *On expansion in characteristic functions of the Laplace operator.* Doklady Akad. Nauk SSSR (N.S.) **90** (1953). 133–135.

[LV] E. Lindenstrauss, A. Venkatesh, *Existence and Weyl's law for spherical cusps forms,* Geom. and Funct. Analysis, to appear.

[Me] R.B. Melrose, *Weyl's conjecture for manifolds with concave boundary.* Geometry of the Laplace operator pp. 257–274, Proc. Sympos. Pure Math., 36, Amer. Math. Soc., Providence, R.I., 1980.

[Mi] S.D. Miller, *On the existence and temperedness of cusp forms for* SL$_3$(ℤ). J. Reine Angew. Math. **533** (2001), 127–169.

[MP] S. Minakshisundaram, A. Pleijel, *Some properties of the eigenfunctions of the Laplace-operator on Riemannian manifolds.* Canadian J. Math. **1** (1949), 242–256.

[MW] C. Moeglin et J.-L. Waldspurger, *Le spectre résiduel de GL(n)*, Ann. scient. Éc. Norm. Sup., 4e série, t. **22** (1989), 605–674.

[Mu1] W. Müller, *Spectral theory for Riemannian manifolds with cusps and a related trace formula.* Math. Nachr. **111** (1983), 197–288.

[Mu2] W. Müller, *The trace class conjecture in the theory of automorphic forms.* Ann. of Math. (2) **130** (1989), no. 3, 473–529.

[Mu3] W. Müller, *Weyl's law for the cuspidal spectrum of* SL$_n$. Ann. of Math. (2) **165** (2007), no. 1, 275–333.

[MS] W. Müller, B. Speh, *Absolute convergence of the spectral side of the Arthur trace formula for* GL$_n$. With an appendix by E. M. Lapid. Geom. Funct. Anal. **14** (2004), no. 1, 58–93.

[PS1] R. Phillips, P. Sarnak, *On cusp forms for co-finite subgroups of* PSL(2, R). Invent. Math. **80** (1985), 339–364.

[PS2] R. Phillips, P. Sarnak, *Perturbation theory for the Laplacian on automorphic functions.* J. Amer. Math. Soc. **5** (1992), 1–32.

[Pr] K. Prachar, *Primzahlverteilung*, Grundlehren der math. Wissenschaften, vol. 91, Springer, 1957.

[Ra] Lord Raleigh, *The dynamical theory of gases and of radiation,* Nature **72** (1905), 54–55, 243–244.

[Rez] A. Reznikov,*Eisenstein matrix and existence of cusp forms in rank one symmetric spaces.* Geom. Funct. Anal. **3** (1993), no. 1, 79–105.

[Roe] W. Roelcke, *Das Eigenwertproblem der automorphen Formen in der hyperbolischen Ebene.* I, II. Math. Ann. **167** (1966), 292–337; ibid. **168** (1966), 261–324.

[Sa1] P. Sarnak, *Spectra of hyperbolic surfaces*, Bull. Amer. Math. Soc. **40** (2003), no. 4, 441–478.

[Sa2] P. Sarnak, *On cusp forms*, In: The Selberg trace formula and related topics (Brunswick, Maine, 1984), 393–407, Contemp. Math., **53**, Amer. Math. Soc., Providence, RI, 1986.

[Sa3] P. Sarnak, *Notes on the generalized Ramanujan conjectures.* Harmonic analysis, the trace formula, and Shimura varieties, 659–685, Clay Math. Proc., 4, Amer. Math. Soc., Providence, RI, 2005.

[Se1] A. Selberg, *Harmonic analysis and discontinuous groups in weakly symmetric Riemannian spaces with applications to Dirichlet series*, in "Collected Papers", Vol. I, Springer-Verlag, Berlin-Heidelberg-New York (1989), 423–492.

[Se2] A. Selberg, *Harmonic analysis*, in "Collected Papers", Vol. I, Springer-Verlag, Berlin-Heidelberg-New York (1989), 626–674.

[Ve] A.B. Venkov, *Spectral theory of automorphic functions and its applications.* Math. and its Applications (Soviet Series), **51**, Kluwer, Dordrecht, 1990.

[We1] H. Weyl, *Das asymptotische Verteilungsgesetz der Eigenwerte linearer partieller Differentialgleichungen (mit einer Anwendung auf die Theorie der Hohlraumstrahlung).* Math. Ann. **71** (1912), 441–479.

[We2] H. Weyl, *Über die Abhängigkeit der Eigenschwingungen einer Membran von der Begrenzung*, J. Reine Angew. Math. **141** (1912), 1–11.

[We3] H. Weyl, *Über die Randwertaufgabe der Strahlungstheorie und asymptotische Spektralgesetze*, J. Reine. Angew. Math. **143** (1913), 177–202.

[We4] H. Weyl, *Das asymptotische Verteilungsgesetz der Eigenschwingungen eines beliebig gestalteten elastischen Körpers*, Rend. Circ. Math. Palermo **39** (1915), 1–50.

[We5] H. Weyl, *Ramifications, old and new, of the eigenvalue problem*, Bull. Amer. Math. Soc. **56** (1950), 115–139.

8

Weyl's Lemma, One of Many

M.I.T.
2-272, Cambridge, MA 02140, U.S.A.
dws@math.mit.edu

ABSTRACT This note is a brief, and somewhat biased, account of the evolution of what people working in P.D.E.'s call Weyl's Lemma about the regularity of solutions to second order elliptic equations. As distinguished from most modern treatments, which are based on pseudod-ifferential operator technology, the approach adopted here is potential theoretic, like Weyl's own.

1 Where It All Started

Given a bounded, connected open region $\Omega \subseteq \mathbb{R}^N$ with smooth boundary $\partial\Omega$ and a smooth $\Phi : \Omega \longrightarrow \mathbb{R}^N$, consider the problem of smoothly decomposing Φ into a divergence free part Φ_0 and an exact part Φ_1. That is, the problem of writing $\Phi = \Phi_0 + \Phi_1$, where Φ_0 and Φ_1 are smooth, $\text{div}\Phi_0$ vanishes, and $\Phi_1 = \nabla\varphi$ for some φ which vanishes at $\partial\Omega$.

To solve this problem, one should begin by observing that, if one ignores questions of smoothness, then it is reasonably clear how to proceed. Namely, because Φ_0 is divergence free, and Φ_1 is exact, $\text{div}\Phi_0 = 0$ and $\Phi_1 = \nabla\varphi$. Hence, if φ vanishes at $\partial\Omega$, the divergence theorem says that $\nabla\varphi$ is perpendicular to Φ_0 in $L^2(\Omega; \mathbb{R}^N)$. With this in mind, let Φ_1 denote the orthognal projection in $L^2(\Omega; \mathbb{R}^N)$ of Φ onto the closure in $L^2(\Omega; \mathbb{R}^N)$ of $\{\nabla\psi : \psi \in C_c^\infty(\Omega; \mathbb{R}^N)\}$. Next, choose $\{\varphi_n : n \geq 0\} \subseteq C_c^\infty(\Omega; \mathbb{R})$ so that $\nabla\varphi_n \longrightarrow \Phi_1$ in $L^2(\Omega; \mathbb{R}^N)$. Because the φ_n's vanish at $\partial\Omega$, $\lambda_0\|\varphi_n - \varphi_m\|^2_{L^2(\Omega; \mathbb{R})} \leq \|\nabla\varphi_n - \varphi_m\|^2_{L^2(\Omega; \mathbb{R}^N)}$, where $-\lambda_0 < 0$ is the largest Dirichlet eigenvalue of Laplacian Δ on $L^2(\Omega; \mathbb{R})$. Hence, there is a $\varphi \in L^2(\Omega; \mathbb{R})$ to which the φ_n's converge. Morever, if $\psi \in C_c^\infty(\Omega; \mathbb{R})$,

1 The author acknowledges support from NSF grant DMS 0244991.

then

$$\int_{\Omega} \Delta\psi(x)\varphi(x)\,dx = \lim_{n\to\infty} \int_{\Omega} \Delta\psi(x)\varphi_n(x)\,dx$$

$$= -\lim_{n\to\infty} \int_{\Omega} \nabla\psi(x)\cdot\nabla\varphi_n(x)\,dx$$

$$= -\int_{\Omega} \nabla\psi(x)\cdot\Phi_1(x)\,dx$$

$$= -\int_{\Omega} \nabla\psi(x)\cdot\Phi(x)\,dx = \int_{\Omega} \psi(x)\mathrm{div}\Phi(x)\,dx$$

That is, $\Delta\varphi = \mathrm{div}\Phi$ in the sense of (Schwartz) distributions.

In view of the preceding, we will be done as soon as we show that φ is smooth. Indeed, if φ is smooth, then $\Delta\varphi = \mathrm{div}(\Phi)$ in the classical sense, and so, when $\Phi_1 = \nabla\varphi$, $\Phi_0 \equiv \Phi - \nabla\varphi$ satisfies $\mathrm{div}(\Phi_0) = \mathrm{div}(\Phi) - \Delta\varphi = 0$.

2 Weyl's Lemma

As we saw in §1, the problem posed there will be solved as soon as we show that φ is smooth, and it is at this point that Weyl made a crucial contribution. Namely, he proved (cf. [6]) the following statement.

Lemma 2.1 *(Weyl's Lemma) Let $\Omega \subseteq \mathbb{R}^N$ be open. If $u \in \mathfrak{D}'(\Omega;\mathbb{R})$ (the space of Schwartz distributions on Ω) satisfies $\Delta u = f \in C^{\infty}(\Omega;\mathbb{R})$ in the sense that*

$$\langle \Delta\psi, u\rangle = \langle\psi, f\rangle, \quad \psi \in C_c^{\infty}(\Omega;\mathbb{R}),$$

then $u \in C^{\infty}(\Omega;\mathbb{R})$.

Proof: Set

$$\gamma_t(x) = (4\pi t)^{-\frac{N}{2}} \exp\left[-\frac{|x|^2}{4t}\right].$$

Given $x_0 \in \Omega$, choose $r > 0$ so that $\bar{B}(x_0, 3r) \subset\subset \Omega$ and

$$\eta \in C_c^{\infty}\big(B(x_0, 3r); [0,1]\big)$$

so that $\eta = 1$ on $\bar{B}(x_0, 2r)$.

Set $v = \eta u$ and $w = \Delta v - \eta f$. Then w is supported in $B(x_0, 3r) \setminus \bar{B}(x_0, 2r)$. Now take

$$v_t(x) = \gamma_t \star v(x) \equiv \langle\gamma_t(\cdot - x), v\rangle \text{ and } w_t(x) = \gamma_t \star w \equiv \langle\gamma_t(\cdot - x), w\rangle.$$

For each $t > 0$, v_t is smooth. Moreover,

$$\dot{v}_t(x) = \langle \gamma_t(\cdot - x), \eta f \rangle + \langle \gamma_t(\cdot - x), w \rangle,$$

and so

$$v_t(x) = v_1(x) - \int_t^1 \gamma_\tau \star (\eta f)(x) \, d\tau - \int_t^1 w_\tau(x) \, d\tau.$$

The first term on the right causes no problems, since

$$\eta f \in C_c^\infty \big(B(x_0, 3); \mathbb{R} \big).$$

Finally,

$$\sup_{(\tau, x) \in (0,1) \times B(x_0, r)} \big| \partial^\alpha w_\tau(x) \big| < \infty$$

for all $\alpha \in \mathbb{N}^N$. Hence, we have now shown that every derivative of $v_t \restriction B(x_0, 1)$ is uniformly bounded by a bound which is independent of $t \in (0, 1]$. Since, as $t \searrow 0$, $v_t \longrightarrow v$ on $B(x_0, 1)$ in the sense of distributions, this means that $v \restriction B(x_0, 1)$ is smooth. □

So far as I know, Weyl's Lemma is the first definitive statement of what are now known as *elliptic regularity* results. More precisely, it is the statement that Δ is *hypoelliptic* in the sense that the singular support of a distribution u is contained in that of Δu.

The spirit of Weyl's own proof is very much like that of the one just given. Namely, it is based on an analysis of the singularity in the Green's function. The only difference is that he dealt with the Green's function directly, whereas we have used the mollification of the Green's function provided by the heat flow. Most modern proofs of hypoellipticity prove a more quantitative statement. Namely, they prove hypoellipticity as a consequence of a *elliptic* estimate which says that the s-order Sobolev norm $(I - \Delta)u$ can be used to dominate the $(s + 2)$-order Sobolev norm of u.

3 Weyl's Lemma for Heat Equation

As we will see, there are various directions in which Weyl's Lemma has been extended. The following is an important example of such an extension.

Theorem 3.1 *(Weyl's Lemma for the Heat Equation) Let $\Omega \subseteq \mathbb{R}^1 \times \mathbb{R}^N$ be open. If $u \in \mathfrak{D}'(\Omega; \mathbb{R})$ satisfies $(\partial_\xi + \Delta)u = f \in C^\infty(\Omega; \mathbb{R})$ in the* *sense that*

$$\langle(-\partial_\xi + \Delta)\psi, u\rangle = \langle\psi, f\rangle, \quad \psi \in C_c^\infty(\Omega; \mathbb{R}),$$

then $u \in C^\infty(\Omega; \mathbb{R})$.

Proof Given $(\xi_0, x_0) \in \Omega$, set $P(r) = (\xi_0 - r, \xi_0 + r) \times B(x_0, r)$, and choose $r > 0$ so that $\bar{P}(3r) \subset\subset \Omega$. Next, choose $\eta \in C_c^\infty(P(3r); [0, 1])$ so that $\eta = 1$ on $P(2r)$, and set $v = \eta u$ and $w = (\partial_\xi + \Delta)v - \eta f$. Also, choose $\rho \in C_c^\infty((2, 3); [0, \infty))$ with total integral 1, and set $\rho_t(x) = t^{-1}\rho(t^{-1}x)$. Finally, for $t \in (0, 1]$, set

$$v_t(\xi, x) = \langle\rho_t(\cdot - \xi)\gamma_{\cdot - \xi}(* - x), v\rangle \quad \text{and}$$
$$w_t(\xi, x) = \langle\rho_t(\cdot - \xi)\gamma_{\cdot - \xi}(* - x), w\rangle.$$

Because $\frac{d}{dt}\rho_t(\xi) = -\tilde{\rho}_t'(\xi)$, where $\tilde{\rho}(\xi) = \xi\rho(\xi)$ and $\tilde{\rho}_t(\xi) = t^{-1}\tilde{\rho}(t^{-1}\xi)$,

$$\begin{aligned}
\frac{d}{dt}v_t(\xi, x) &= -\langle\tilde{\rho}_t'(\cdot - \xi)\gamma_{\cdot - \xi}(* - x), v\rangle \\
&= \langle\tilde{\rho}_t(\cdot - \xi)\gamma_{\cdot - \xi}(* - x), \eta f\rangle + w_t(\xi, x).
\end{aligned}$$

The first term causes no problem as $t \searrow 0$ because $\eta f \in C_c^\infty(\Omega; \mathbb{R})$ and $\tilde{\rho} \in L^1(\mathbb{R})$. As for the second term, so long as $(\xi, x) \in P(r)$, its derivatives are controlled independent of $t \in (0, 1]$ because

(i) $\text{supp}(w) \subseteq P(3r) \setminus P(2r)$.
(ii) Derivatives of $\tilde{\rho}_t$ are bounded by powers of t^{-1}.
(iii) For $(\xi, x) \in P(r)$ and $(\xi', x') \notin P(2r)$ with $0 < \xi' - \xi < 3t$, all derivatives of $\gamma_{\xi'-\xi}(x' - x)$ are bounded uniformly by any power of t.

Thus, just as before, we can conclude that $v \in C^\infty(P(r); \mathbb{R})$. $\qquad\square$

4 A General Result

If one examines the proofs given in §§ 2 & 3, one sees that they turn on two properties of the classic heat flow. The first of these is that the heat flow does "no damage" to initial data. That is, if one starts with smooth data, then it evolves smoothly. The second property is that so long as one stays away from the diagonal, the heat kernel $\gamma_t(y - x)$ remains smooth as $t \searrow 0$. When dealing with Δ, these two suffice. When dealing with $\Delta + \partial_\xi$, one needs a more quantitative statement of the latter property. Namely, one needs to know that away from the diagonal, the heat kernel goes to 0 faster than any power of t. Based on this discussion, we formulate the following general principle.

Suppose that L is a linear partial differential operator from $C^\infty(\mathbb{R}^N; \mathbb{R})$ to itself, and assume that associated with L there is a kernel

$$(t, x, y) \in (0, 2) \times \mathbb{R}^N \times \mathbb{R}^N \longmapsto q(t, x, y) \in \mathbb{R}$$

and the operators $t \rightsquigarrow Q_t$ given by

$$Q_t \varphi(x) = \int \varphi(y) q(t, x, y) \, dy, \quad \varphi \in C_c^\infty(\mathbb{R}^N; \mathbb{R})$$

with the properties that

(i) For each $x \in \mathbb{R}^N$, $(t, y) \rightsquigarrow q(t, x, y)$ satisfies the adjoint equation with initial value δ_x. That is, $\partial_t q(t, x, y) = [L^* q(t, x, \cdot)](y)$ and $q(t, x, \cdot) \longrightarrow \delta_x$ as $t \searrow 0$.

(ii) For each $n \geq 0$, there exists a $C_n < \infty$ such that

$$\sup_{t \in (0, 2]} \|Q_t \varphi\|_{C_b^n} \vee \|Q_t^* \varphi\|_{C_b^n} \leq C_n \|\varphi\|_{C_b^n}.$$

Using the same ideas on which we based the proofs in §§ 2 & 3, one can prove the following.

Theorem 4.1 *If, in addition to (i) and (ii), for each $n \geq 0$ and $\epsilon > 0$,*

$$\max_{\|\alpha\| + \|\beta\| \leq n} \sup_{\substack{t \in (0, 1] \\ |y - x| \geq \epsilon}} \left| \partial_x^\alpha \partial_y^\beta q(t, x, y) \right| < \infty,$$

then L is hypoelliptic. If, for each $n \geq 0$, $\epsilon > 0$, and $\nu > 0$,

$$\max_{m + \|\alpha\| + \|\beta\| \leq n} \sup_{\substack{t \in (0, 1] \\ |y - x| \geq \epsilon}} t^{-\nu} \left| \partial_t^m \partial_x^\alpha \partial_y^\beta q(t, x, y) \right| < \infty,$$

then $L + \partial_\xi$ is hypoelliptic.

5 Elliptic Operators

The original generization of Weyl's Lemma was to replace the Laplace operator by a variable coefficient, second order, elliptic partial differential operator. That is, let $a : \mathbb{R}^N \longrightarrow \mathbb{R}^N \otimes \mathbb{R}^N$, $b : \mathbb{R}^N \longrightarrow \mathbb{R}^N$, and $c : \mathbb{R}^N \longrightarrow \mathbb{R}$ be smooth, bounded functions with bounded derivatives of all orders, and set

$$L = \sum_{i,j=1}^N a(x)_{ij} \partial_{x_i} \partial_{x_j} + \sum_{i=1}^N b(x)_i \partial_{x_i} + c(x).$$

Without loss in generality, we assume that $a(x)_{ij} = a(x)_{ji}$. The operator L is said to be *uniformly elliptic* if $a(x) \geq \delta I$ for some $\delta > 0$.

Theorem 5.1 *If L is uniformly elliptic, then there is a $q \in C^\infty\big((0, \infty) \times \mathbb{R}^N \times \mathbb{R}^N; \mathbb{R}\big)$ such that $\partial_t q(t, x, y) = [L^* q(t, x, \cdot)](y)$, $q(t, x, \cdot) \longrightarrow \delta_x$ as $t \searrow 0$ for each $x \in \mathbb{R}^N$, and, for each $m \geq 0$ and $(\alpha, \beta) \in (\mathbb{N}^N)^2$, there is a $K_{m,\alpha,\beta} < \infty$ such that*

$$\left| \partial_t^m \partial_x^\alpha \partial_y^\beta q(t, x, y) \right| \leq K_{m,\alpha,\beta} t^{-\frac{N + 2m + \|\alpha\| + \|\beta\|}{2}} \exp\left[-\frac{|y - x|^2}{K_{m,\alpha,\beta} t} \right].$$

In particular, $L + \partial_\xi$ is hypoelliptic.

To see how such a result gets applied, consider the following contruction. Take a and b as in the preceding and $c \equiv 0$. Let $\Gamma(t, x)$ denote the Gaussian probability measure on $\mathbb{R}^N$ with mean $x + tb(x)$ and covariance $2ta(x)$. That is, $\Gamma(t, x)$ has density given by

$$\big[(4\pi t)^N \det(a(x))\big]^{-\frac{1}{2}} \exp\left[-\frac{(y - x - tb(x)) \cdot a(x)^{-1}(y - x - tb(x))}{4t} \right].$$

Then

$$\frac{d}{dt} \int \varphi(y)\, \Gamma(t, x, dy) = \int L_x \varphi(y)\, \Gamma(t, x, dy),$$

where L_x is the constant coefficient operator obtained by freezing the coefficients of L at x.

For $n \geq 0$, define $P_n(t, x) = \delta_x$ and

$$P_n(t, x) = \int \Gamma\big(t - [t]_n, x'\big)\, P_n\big([t]_n, x, dx'\big),$$

where $[t]_n = 2^{-n}[2^n t]$ is the largest dyadic $m2^{-n}$ dominated by t. Then,

$$\langle \varphi, P_n(t, x) \rangle - \varphi(x) = \int_0^t \left(\int \langle L_y \varphi, \Gamma(\{\tau\}_n, y) \rangle\, P_n\big([\tau]_n, x, dy\big) \right) d\tau,$$

where $\{\tau\}_n = \tau - [\tau]_n$. Using elementary facts about weak convergence of probability measures, one can show that there is a continuous map $(t, x) \rightsquigarrow P(t, x)$ such that $P_n(t, x) \longrightarrow P(t, x)$ uniformly on compacts. Moreover, because, uniformly on compacts,

$$\langle L_y \varphi, \Gamma(\{\tau\}_n, y) \rangle \longrightarrow L\varphi(y) \quad \text{as } n \to \infty,$$

$$\langle \varphi, P(t, x) \rangle = \varphi(x) + \int_0^t \langle L\varphi, P(\tau, x) \rangle\, d\tau.$$

Thus, $\partial_t P(t, x) = L^* P(t, x)$ and $P(t, x) \longrightarrow \delta_x$. Finally, given $T > 0$, define the distribution u on $(0, T) \times \mathbb{R}^N$ by

$$\langle \varphi, u \rangle = \int_0^T \left(\int \varphi(y) \, P(T - t, x, dy) \right) dt.$$

Then $(L + \partial_\xi)u = 0$, and so u is smooth on $\{(t, y) : a(y) > 0\}$. That is, $P(t, x, dy) = p(t, x, y)dy$ where $(t, y) \rightsquigarrow p(t, x, y)$ is a smooth function there. Similar reasoning shows that $(t, x, y) \rightsquigarrow p(t, x, y)$ is smooth on $\{(t, x, y) : a(x) > 0 \ \& \ a(y) > 0\}$, and more delicate considerations show that it is smooth on $\{(t, x, y) : a(x) > 0 \text{ or } a(y) > 0\}$.

6 Kolmogorov's Example

Although ellipticity guarantees hypoellipticity, hypoellipticity holds in for many operators which are not elliptic. The following example due to Kolmogorov is seminal.

Take $N = 2$, and consider $L = \partial_{x_1}^2 + x_1 \partial_{x_2}$, which is severely non-elliptic. As Kolmogorov realized, the corresponding diffusion has coordinates

$$X_1(t) = x_1 + \sqrt{2}\, B(t) \ \& \ X_2(t) = x_2 + \int_0^t X_1(\tau)\, d\tau,$$

where $t \rightsquigarrow B(t)$ is a standard, 1-dimensional Brownian motion. In particular, this means that the distribution of $(X_1(t), X_2(t))$ is the Gaussian measure on $\mathbb{R}^2$ with mean

$$\mathbf{m}(t, x) = \begin{pmatrix} x_1 \\ x_2 + tx_2 \end{pmatrix}$$

and covariance

$$C(t) = \begin{pmatrix} 2t & t^2 \\ t^2 & \frac{2t^3}{3} \end{pmatrix}.$$

Thus, the fundamental solution to the heat equation $\partial_t u = Lu$ is

$$q(t, x, y) = \frac{1}{2\pi\sqrt{\det C(t)}} \times \exp\left[-\frac{\big(y - \mathbf{m}(t, x)\big) \cdot C(t)^{-1}\big(y - \mathbf{m}(t, x)\big)}{2} \right].$$

In particular, $L + \partial_\xi$ is hypoelliptic. One can use the same reasoning to draw the same conclusion about when

$$L = \partial_{x_1}^2 + \sum_{i=2}^N x_{i-1} \partial_{x_i}.$$

7 Results of Hörmander Type

Kolmogorov's example was put into context by a remarkable result proved by Hörmander. To state his result, let $\{X_0, \ldots, X_r\}$ be a set of smooth vector fields on $\mathbb{R}^N$ and set

$$L = X_0 + \sum_{i=1}^{r} X_i^2,$$

where the X_k's are interpreted as directional derivative operators and X_k^2 is the composition of X_k with itself. Equivalently, if

$$X_i(x) = \sum_{j=1}^{N} \sigma(x)_{ij} \partial_{x_j}, \quad 1 \le i \le r, \quad \text{and} \quad X_0(x) = \sum_{j=1}^{N} \beta_j(x) \partial_{x_j},$$

then the matrix a of second order coefficients equals $\sigma \sigma^\top$ and the vector b of first order coeficient part equals

$$\begin{pmatrix} \beta_1(x) \\ \vdots \\ \beta_N(x) \end{pmatrix} + \sum_{i=1}^{r} \begin{pmatrix} X_i \sigma_{i1}(x) \\ \vdots \\ X_i \sigma_{iN}(x) \end{pmatrix}.$$

Let $\mathcal{L}$ and $\mathcal{L}'$ denote the Lie algebras generated by, respectively,

$$\{X_0, \ldots, X_r\}$$

and

$$\{[X_0, X_1], \ldots, [X_0, X_r], X_1, \ldots, X_r\}.$$

Hörmander proved the following result in [1].

Theorem 7.1 *(Hörmander's Theorem) If $\mathcal{L}(x)$ has dimension N at each $x \in \Omega$, then L is hypoelliptic on Ω. If $\mathcal{L}'(x)$ has dimension N at each $x \in \Omega$, then $\partial_\xi + L$ is hypoelliptic in $\mathbb{R}^1 \times \Omega$.*

First Oleinik and Radekevich [5] and later (see [2]) Fefferman, Phong, and others extended and sharpened this theorem to cover cases when L cannot be represented in terms of vector fields. That is, when there is no smooth square root of a. There are situations in which more detailed information is available. For instance, suppose that $X_0 = \sum_1^r c_k X_k$ for some smooth $c_1, \ldots, c_k$ with bounded derivatives of all orders and that the vector fields X_k have bounded derivatives of all orders. Further, assume that there is an $n \in \mathbb{N}$ and $\epsilon > 0$ such that

$$\sum_{m=1}^{n} \sum_{\alpha \in \{1,\ldots,r\}^m} \left(V_\alpha(x), \xi\right)_{\mathbb{R}^N}^2 \geq \epsilon |\xi|^2, \quad (x,\xi) \in \mathbb{R}^N \times \mathbb{R}^N,$$

where, for $\alpha \in \{1,\ldots,r\}^m$, $V_\alpha(x) = X_\alpha x$ and, for $m \geq 2$,

$$X_\alpha = \left[X_{\alpha_m}, X_{\alpha'}\right] \quad \text{when } \alpha' = (\alpha_1,\ldots,\alpha_{m-1}).$$

Then Rothschield and Stein showed that the operator L can be interpreted in terms of a degenerate Riemannian geometry in which the model space is a nilpotent Lie group instead of Euclidean space. Variations on and extensions of their ideas can be found in [2] and [4].

8 Concluding Remarks

One can show that if there is smooth differentiable manifold M for which $\mathcal{L}(x)$ is the tangent space at each $x \in M$, then a diffusion process generated by L which starts on at an $x \in M$ will stay on M for a positive length of time. As a consequence, one can see that when such a manifold exists, L cannot be hypoellictic in a neighborhood of M. Thus, when one combines (cf. § 7 in part II of [4]) this with Nagumo's Theorem about integral manifolds for real analytic vector fields, one realizes that the criterion in Hörmander's Theorem is necessary and sufficient when the X_k's are real analytic.

When the X_k's are not real analytic, Hörmander's criterion is necessary and sufficient for subellipticiy but not for hypoellipticity, For example, take $N = 3$ and consider

$$L = \partial_{x_1}^2 + \left(\alpha(x_1)\partial_{x_2}\right)^2 + \partial_{x_3}^2,$$

where α is a smooth function which vanishes only at 0 but vanishes to all orders there. Further, assume that α^2 is an even function on $\mathbb{R}$ which is non-decreasing on $[0,\infty)$. Then (cf. the last part of § 8 in part II of [4]) L is hypoelliptic in a neighborhood of 0 if and only if

$$\lim_{\xi \searrow 0} \xi \log\left(|\alpha(\xi)|\right) = 0.$$

When dealing with elliptic operators, hypoellipticity extends easily to systems. However, the validity of a Hörmander type theorem for systems remains an open question. Indeed, there is considerable doubt about what criterion replaces Hörmander's for systems. Recently, J.J. Kohn [3] has made some progress in this direction. Namely, he has found examples of complex vector fields for which hypoellipticity holds in the absence of ellipticity. Perhaps the most intriguing aspect of Kohn's example is that, from a subelliptic standpoint, his operators "lose derivatives."

Bibliography

[1] Hörmander, L., Hypoelliptic second order differential operators, Acta. Math., vol. 119 (1967), 147–171

[2] Jerison, D. and Sánchez-Calle A., Subelliptic second-order differential operators, Lecture Notes in Math. #1277 (1986), 46–77

[3] Kohn, J.J., Hypoellipticity with loss of derivatives, Ann. of Math.vol. 162 # 2 (2005), 943-986

[4] Kusuoka, S. and Stroock, D., Applicatons of the Malliavin calculus, II & III, J. Fac. Sci. of Tokyo Univ., Sec IA vol. 32 & 34 #1 & #2 (1985 & 1987), 1–76 & 391–442

[5] Oleinik, O.A. and Radekevich, E.V., Second Order Equations with Non-negative Characteristic Form, Plenum, NY (1973)

[6] Weyl, H., The method of orthogonal projection in potential theory, Duke Math. J., vol. 7 (1940), 414–444

9

Analogies between analysis on foliated spaces and arithmetic geometry

Christopher Deninger

Mathematisches Institut
Einsteinstr. 62
48149 Münster, Germany
c.deninger@math.uni-muenster.de

1 Introduction

For the arithmetic study of varieties over finite fields powerful cohomo-logical methods are available which in particular shed much light on the nature of the corresponding zeta functions. For algebraic schemes over spec $\mathbb{Z}$ and in particular for the Riemann zeta function no cohomology theory has yet been developed that could serve similar purposes. For a long time it had even been a mystery how the formalism of such a theory could look like. This was clarified in [D1]. However until now the conjectured cohomology has not been constructed.

There is a simple class of dynamical systems on foliated manifolds whose reduced leafwise cohomology has several of the expected structural prop-erties of the desired cohomology for algebraic schemes. In this analogy, the case where the foliation has a dense leaf corresponds to the case where the algebraic scheme is flat over spec $\mathbb{Z}$ e.g. to spec $\mathbb{Z}$ itself. In this situation the foliation cohomology which in general is infinite di-mensional is not of a topological but instead of a very analytic nature. This can also be seen from its description in terms of global differential forms which are harmonic along the leaves. An optimistic guess would be that for arithmetic schemes $\mathcal{X}$ there exist foliated dynamical systems X whose reduced leafwise cohomology gives the desired cohomology of $\mathcal{X}$. If $\mathcal{X}$ is an elliptic curve over a finite field this is indeed the case with X a generalized solenoid, not a manifold, [D3].

We illustrate this philosophy by comparing the "explicit formulas" in analytic number theory to a transversal index theorem. We also give a dynamical analogue of a recent conjecture of Lichtenbaum on special values of Hasse–Weil zeta functions by using the Cheeger-Müller the-orem equating analytic and Reidemeister torsion. As a new example

174

we point out an analogy between properties of Cramer's function made from zeroes of the Riemann zeta function and properties of the trace of a wave operator. Incidentally, in all cases analytic number theory suggests questions in the theory of partial differential operators on foliated manifolds.

Since the entire approach is not yet in a definitive state our style is deliberately a bit brief in places. For more details we refer to the references provided.

Hermann Weyl was very interested in number theory and one of the pioneers in the spectral theory of differential operators. In particular the theory of zeta-regularized determinants and their application to analytic torsion evolved from his work. In my opinion the analogies reviewed in the present note and in particular the analytic nature of foliation cohomology show that a deeper understanding of number theoretical zeta functions will ultimately require much more analysis of partial differential operators and of Weyl's work than presently conceived.

I would like to thank the referee very much for thoughtful comments.

2 Gradient flows

We introduce zeta functions and give some arithmetic motivation for the class of dynamical systems considered in the sequel.

Consider the Riemann zeta function

$$\zeta(s) = \prod_p (1 - p^{-s})^{-1} = \sum_{n=1}^{\infty} n^{-s} \quad \text{for } \mathrm{Re}\, s > 1 \ .$$

It has a holomorphic continuation to $\mathbb{C} \setminus \{1\}$ with a simple pole at $s = 1$. The famous Riemann hypothesis asserts that all zeroes of $\zeta(s)$ in the critical strip $0 < \mathrm{Re}\, s < 1$ should lie on the line $\mathrm{Re}\, s = 1/2$. A natural generalization of $\zeta(s)$ to the context of arithmetic geometry is the Hasse–Weil zeta function $\zeta_{\mathcal{X}}(s)$ of an algebraic scheme $\mathcal{X}/\mathbb{Z}$

$$\zeta_{\mathcal{X}}(s) = \prod_{x \in |\mathcal{X}|} (1 - Nx^{-s})^{-1} \quad \text{for } \mathrm{Re}\, s > \dim \mathcal{X} \ .$$

Here $|\mathcal{X}|$ is the set of closed points x of $\mathcal{X}$ and Nx is the number of elements in the residue field $\kappa(x)$ of x. For the one-dimensional scheme $\mathcal{X} = \mathrm{spec}\,\mathbb{Z}$ we recover $\zeta(s)$ and for $\mathcal{X} = \mathrm{spec}\,\mathfrak{o}_K$ where $\mathfrak{o}_K$ is the ring of integers in a number field we obtain the Dedekind zeta function $\zeta_K(s)$. It is expected that $\zeta_{\mathcal{X}}(s)$ has a meromorphic continuation to all of $\mathbb{C}$. This is known in many interesting cases but by no means in general.

If $\mathcal{X}$ has characteristic p then the datum of $\mathcal{X}$ over $\mathbb{F}_p$ is equivalent to the pair $(\mathcal{X} \otimes \overline{\mathbb{F}}_p, \varphi)$ where $\varphi = F \otimes \mathrm{id}_{\overline{\mathbb{F}}_p}$ is the $\overline{\mathbb{F}}_p$-linear base extension of the absolute Frobenius morphism F of $\mathcal{X}$. For example, the set of closed points $|\mathcal{X}|$ of $\mathcal{X}$ corresponds to the set of finite φ-orbits $\mathfrak{o}$ on the $\overline{\mathbb{F}}_p$-valued points $(\mathcal{X} \otimes \overline{\mathbb{F}}_p)(\overline{\mathbb{F}}_p) = \mathcal{X}(\overline{\mathbb{F}}_p)$ of $\mathcal{X} \otimes \overline{\mathbb{F}}_p$. We have $\log Nx = |\mathfrak{o}| \log p$ under this correspondence. Pairs $(\mathcal{X} \otimes \overline{\mathbb{F}}_p, \varphi)$ are roughly analogous to pairs (M, φ) where M is a smooth manifold of dimension $2 \dim \mathcal{X}$ and φ is a diffeomorphism of M. A better analogy would be with Kähler manifolds and covering self maps of degree greater than one. See remark 2.1 below.

The $\mathbb{Z}$-action on M via the powers of φ can be suspended as follows to an $\mathbb{R}$-action on a new space. Consider the quotient

$$X = M \times_{p^{\mathbb{Z}}} \mathbb{R}_+^* \qquad (2.1)$$

where the subgroup $p^{\mathbb{Z}}$ of $\mathbb{R}_+^*$ acts on $M \times \mathbb{R}_+^*$ as follows:

$$p^{\nu} \cdot (m, u) = (\varphi^{-\nu}(m), p^{\nu} u) \quad \text{for } \nu \in \mathbb{Z}, m \in M, u \in \mathbb{R}_+^* .$$

The group $\mathbb{R}$ acts smoothly on X by the formula

$$\varphi^t [m, u] = [m, e^t u] .$$

Here φ^t is the diffeomorphism of X corresponding to $t \in \mathbb{R}$. Note that the closed orbits γ of the $\mathbb{R}$-action on X are in bijection with the finite φ-orbits $\mathfrak{o}$ on M in such a way that the length $l(\gamma)$ of γ satisfies the relation $l(\gamma) = |\mathfrak{o}| \log p$.

Thus in an analogy of $\mathcal{X}/\mathbb{F}_p$ with X, the closed points x of $\mathcal{X}$ correspond to the closed orbits of the $\mathbb{R}$-action φ on X and $Nx = p^{|\mathfrak{o}|}$ corresponds to $e^{l(\gamma)} = p^{|\mathfrak{o}|}$. Moreover if $d = \dim \mathcal{X}$ then $\dim X = 2d + 1$.

The system (2.1) has more structure. The fibres of the natural projection of X to $\mathbb{R}_+^*/p^{\mathbb{Z}}$ form a 1-codimensional foliation $\mathcal{F}$ (in fact a fibration). The leaves (fibres) of $\mathcal{F}$ are the images of M under the immersions for every $u \in \mathbb{R}_+^*$ sending m to $[m, u]$. In particular, the leaves are transversal to the flow lines of the $\mathbb{R}$-action and φ^t maps leaves to leaves for every $t \in \mathbb{R}$.

Now the basic idea is this: If $\mathcal{X}/\mathbb{Z}$ is flat, there is no Frobenius and hence no analogy with a discrete time dynamical system i.e. an action of $\mathbb{Z}$. However one obtains a reasonable analogy with a continuous time dynamical system by the following correspondence

Dictionary, part 1

$\mathcal{X}$ d-dimensional regular algebraic scheme over spec $\mathbb{Z}$	triple $(X, \mathcal{F}, \varphi^t)$, where X is a $2d+1$-dimensional smooth manifold with a smooth $\mathbb{R}$-action $\varphi : \mathbb{R} \times X \to X$ and a one-codimensional foliation $\mathcal{F}$. The leaves of $\mathcal{F}$ should be transversal to the $\mathbb{R}$-orbits and every diffeomorphism φ^t should map leaves to leaves.
closed point x of $\mathcal{X}$	closed orbit γ of the $\mathbb{R}$ action
Norm Nx of closed point x	$N\gamma = \exp l(\gamma)$ for closed orbit γ
Hasse–Weil zeta function $\zeta_{\mathcal{X}}(s) = \prod_{x \in |\mathcal{X}|}(1 - Nx^{-s})^{-1}$	Ruelle zeta function (4.3) $\zeta_X(s) = \prod_{\gamma}(1 - N\gamma^{-s})^{\pm 1}$ (if the product makes sense)
$\mathcal{X} \to \operatorname{spec} \mathbb{F}_p$	triples $(X, \mathcal{F}, \varphi^t)$ where $\mathcal{F}$ is given by the fibres of an $\mathbb{R}$-equivariant fibration $X \to \mathbb{R}_+^*/p^{\mathbb{Z}}$

Remark 2.1 A more accurate analogy can be motivated as follows: If (M, φ) is a pair consisting of a manifold with a self covering φ of degree $\deg \varphi \geq 2$ one can form the generalized solenoid $\hat{M} = \varprojlim(\ldots \xrightarrow{\varphi} M \xrightarrow{\varphi} M \to \ldots)$, cf. [CC] 11.3. On $\hat{M}$ the map φ becomes the shift isomorphism and as before we may consider a suspension $X = \hat{M} \times_{p^{\mathbb{Z}}} \mathbb{R}_+^*$. This example suggests that more precisely schemes $\mathcal{X}/\mathbb{Z}$ should correspond to triples $(X, \mathcal{F}, \varphi^t)$ where X is a $2d+1$ dimensional generalized solenoid which is also a foliated space with a foliation $\mathcal{F}$ by Kähler manifolds of complex dimension d. It is possible to do analysis on such generalized spaces cf. [CC]. Analogies with arithmetic are studied in more detail in [D2] and [Le].

Construction 2.2 Let us consider a triple $(X, \mathcal{F}, \varphi^t)$ as in the dictionary above, let Y_φ be the vector field generated by the flow φ^t and let $T\mathcal{F}$ be the tangent bundle to the foliation. Let ω_φ be the one-form on X defined by

$$\omega_\varphi|_{T\mathcal{F}} = 0 \quad \text{and} \quad \langle \omega_\varphi, Y_\varphi \rangle = 1 .$$

One checks that $d\omega_\varphi = 0$ and that ω_φ is φ^t-invariant i.e. $\varphi^{t*}\omega_\varphi = \omega_\varphi$ for all $t \in \mathbb{R}$. We may view the cohomology class $\psi = [\omega_\varphi]$ in $H^1(X, \mathbb{R})$ defined by ω_φ as a homomorphism

$$\psi : \pi_1^{\mathrm{ab}}(X) \longrightarrow \mathbb{R} .$$

Its image $\Lambda \subset \mathbb{R}$ is called the group of periods of $(X, \mathcal{F}, \varphi^t)$.

It is known that $\mathcal{F}$ is a fibration if and only if rank $\Lambda = 1$. In this case there is an $\mathbb{R}$-equivariant fibration $X \longrightarrow \mathbb{R}/\Lambda$ whose fibres are the leaves of $\mathcal{F}$. Incidentally, a good reference for the dynamical systems we are considering is [Fa].

Consider another motivation for our foliation setting: For a number field K and any $f \in K^*$ we have the product formula where $\mathfrak{p}$ runs over the places of K

$$\prod_{\mathfrak{p}} \|f\|_{\mathfrak{p}} = 1 \; .$$

Here for the finite places i.e. the prime ideals $\mathfrak{p}$ we have

$$\|f\|_{\mathfrak{p}} = N\mathfrak{p}^{-\mathrm{ord}_{\mathfrak{p}} f} \; .$$

Now look at triples $(X, \mathcal{F}, \varphi^t)$ where the leaves of $\mathcal{F}$ are Riemann surfaces varying smoothly in the transversal direction. In particular we have $\dim X = 3 = 2d + 1$ where $d = 1 = \dim(\mathrm{spec}\, \mathfrak{o}_K)$. Consider smooth functions f on X which are meromorphic on leaves and have their divisors supported on closed orbits of the flow. For compact X it follows from a formula of Ghys that

$$\prod_{\gamma} \|f\|_{\gamma} = 1 \; .$$

In the product γ runs over the closed orbits and

$$\|f\|_{\gamma} = (N\gamma)^{-\mathrm{ord}_{\gamma} f}$$

where $N\gamma = e^{l(\gamma)}$ and $\mathrm{ord}_{\gamma} f = \mathrm{ord}_z (f|_{F_z})$. Here z is any point on γ and F_z is the leaf through z.

So the foliation setting allows for a product formula where the $N\gamma$ are not all powers of the same number. If one wants an infinitely generated Λ, one must allow the flow to have fixed points ($\hat{=}$ infinite places). A product formula in this more general setting is given in [Ko].

We end this motivational section listing these and more analogies some of which will be explained later.

Dictionary, part 2

Number of residue characteristics $|\text{char}(\mathcal{X})|$ of $\mathcal{X}$	rank of period group Λ of $(X, \mathcal{F}, \varphi^t)$
Weil étale cohomology of $\mathcal{X}$	Sheaf cohomology of X
Arakelov compactification $\overline{\mathcal{X}} = \mathcal{X} \cup \mathcal{X}_\infty$ where $\mathcal{X}_\infty = \mathcal{X} \otimes \mathbb{R}$	Triples $(\overline{X}, \mathcal{F}, \varphi^t)$ where $\overline{X}$ is a $2d+1$-dimensional compactification of X with a flow φ and a 1-codimensional foliation. The flow maps leaves to leaves, however φ may have fixed points on $\overline{X}$
$\mathcal{X}(\mathbb{C})/F_\infty$, where F_∞ denotes complex conjugation	set of fixed points of φ^t. Note that the leaf of $\mathcal{F}$ containing a fixed point is φ-invariant.
$\text{spec}\,\kappa(x) \hookrightarrow \mathcal{X}$ for $x \in |\mathcal{X}|$	Embedded circle i.e. knot $\mathbb{R}/l(\gamma)\mathbb{Z} \hookrightarrow X$ corresponding to a periodic orbit γ (map $t + l(\gamma)\mathbb{Z}$ to $\varphi^t(x)$ for a chosen point x of γ).
Product formula for number fields $\prod_{\mathfrak{p}} \|f\|_{\mathfrak{p}} = 1$	Kopei's product formula [Ko]
Explicit formulas of analytic number theory	transversal index theorem for $\mathbb{R}$-action on X and Laplacian along the leaves of $\mathcal{F}$, cf. [D3]
$-\log|d_{K/\mathbb{Q}}|$	Connes' Euler characteristic $\chi_{\mathcal{F}}(X, \mu)$ for Riemann surface laminations with respect to transversal measure μ defined by φ^t, cf. [D2] section 4

3 Explicit formulas and transversal index theory

A simple version of the explicit formulas in number theory asserts the following equality of distributions in $\mathcal{D}'(\mathbb{R}^{>0})$:

$$1 - \sum_\rho e^{t\rho} + e^t = \sum_p \log p \sum_{k \geq 1} \delta_{k \log p} + (1 - e^{-2t})^{-1}. \qquad (3.1)$$

Here ρ runs over the zeroes of $\zeta(s)$ in $0 < \text{Re}\,s < 1$, the Dirac distribution in $a \in \mathbb{R}$ is denoted by δ_a and the functions $e^{t\rho}$ are viewed as distributions. Note that in the space of distributions the sum $\sum_\rho e^{t\rho}$ converges as one sees by partial integration since $\sum_\rho \rho^{-2}$ converges.

We want to compare this formula with a transversal index theorem in geometry.

With our dictionary in mind consider triples $(X, \mathcal{F}, \varphi^t)$ where X is a smooth compact manifold of odd dimension $2d+1$, equipped with a one-codimensional foliation $\mathcal{F}$ and φ^t is a flow mapping leaves of $\mathcal{F}$ to leaves. Moreover we assume that the flow lines meet the leaves transversally in every point so that φ has no fixed points. Consider the (reduced) foliation cohomology:

$$\bar{H}_{\mathcal{F}}^{\bullet}(X) := \operatorname{Ker} d_{\mathcal{F}} / \overline{\operatorname{Im} d_{\mathcal{F}}} .$$

Here $(\mathcal{A}_{\mathcal{F}}^{\bullet}(X), d_{\mathcal{F}})$ with $\mathcal{A}_{\mathcal{F}}^i(X) = \Gamma(X, \Lambda^i T^* \mathcal{F})$ is the "de Rham complex along the leaves", (differentials only in the leaf direction). Moreover $\overline{\operatorname{Im} d_{\mathcal{F}}}$ is the closure in the smooth topology of $\mathcal{A}_{\mathcal{F}}^{\bullet}(X)$, cf. [AK1].

The groups $\bar{H}_{\mathcal{F}}^{\bullet}(X)$ have a smooth linear $\mathbb{R}$-action φ^* induced by the flow φ^t. The infinitesimal generator $\theta = \lim_{t \to 0} \frac{1}{t}(\varphi^{t*} - \operatorname{id})$ exists on $\bar{H}_{\mathcal{F}}^{\bullet}(X)$. It plays a similar role as the Frobenius morphism on étale or crystalline cohomology.

In general, the cohomologies $\bar{H}_{\mathcal{F}}^{\bullet}(X)$ will be infinite dimensional Fréchet spaces. They are related to harmonic forms. Assume for simplicity that X and $\mathcal{F}$ are oriented in a compatible way and choose a metric $g_{\mathcal{F}}$ on $T\mathcal{F}$. This gives a Hodge scalar product on $\mathcal{A}_{\mathcal{F}}^{\bullet}(X)$. We define the leafwise Laplace operator by

$$\Delta_{\mathcal{F}} = d_{\mathcal{F}} d_{\mathcal{F}}^* + d_{\mathcal{F}}^* d_{\mathcal{F}} \quad \text{on } \mathcal{A}_{\mathcal{F}}^{\bullet}(X) .$$

Then we have by a deep theorem by Álvarez López and Kordyukov [AK1]:

$$\bar{H}_{\mathcal{F}}^{\bullet}(X) = \operatorname{Ker} \Delta_{\mathcal{F}} . \tag{3.2}$$

Note that $\Delta_{\mathcal{F}}$ is not elliptic but only elliptic along the leaves of $\mathcal{F}$. Hence the standard regularity theory of elliptic operators does not suffice for (3.2).

The isomorphism (3.2) is a consequence of the Hodge decomposition proved in [AK1]

$$\mathcal{A}_{\mathcal{F}}^{\bullet}(X) = \operatorname{Ker} \Delta_{\mathcal{F}} \oplus \overline{\operatorname{Im} d_{\mathcal{F}} d_{\mathcal{F}}^*} \oplus \overline{\operatorname{Im} d_{\mathcal{F}}^* d_{\mathcal{F}}} . \tag{3.3}$$

Consider $\Delta_{\mathcal{F}}$ as an unbounded operator on the space $\mathcal{A}_{\mathcal{F},(2)}^{\bullet}(X)$ of leafwise forms which are L^2 on X. Then $\Delta_{\mathcal{F}}$ is symmetric and its closure $\overline{\Delta}_{\mathcal{F}}$ is selfadjoint. The orthogonal projection P of $\mathcal{A}_{\mathcal{F},(2)}^{\bullet}(X)$ to $\operatorname{Ker} \overline{\Delta}_{\mathcal{F}}$ restricts to the projection P of $\mathcal{A}_{\mathcal{F}}^{\bullet}(X)$ to $\operatorname{Ker} \Delta_{\mathcal{F}}$ in (3.3). There is also an L^2-version $\bar{H}_{\mathcal{F},(2)}^{\bullet}(X)$ of the reduced leafwise cohomology and it follows from a result of von Neumann that we have

$$\bar{H}_{\mathcal{F},(2)}^{\bullet}(X) = \operatorname{Ker} \overline{\Delta}_{\mathcal{F}} .$$

In order to get a formula analogous to (3.1) we have to take $\dim X = 3$. Then $\mathcal{F}$ is a foliation by surfaces so that $\bar{H}^i_{\mathcal{F}}(X) = 0$ for $i > 2$. The following transversal index theorem is contained in [AK2] and [DS]:

Theorem 3.1 *Let $(X, \mathcal{F}, \varphi^t)$ be as above with $\dim X = 3$ such that $\mathcal{F}$ has a dense leaf. Assume that there is a metric $g_{\mathcal{F}}$ on $T\mathcal{F}$ such that φ^t acts with conformal factor $e^{\alpha t}$. (This condition is very strong and can be replaced by weaker ones.) Then the spectrum of θ on $\bar{H}^1_{\mathcal{F},(2)}(X)$ consists of eigenvalues. If all periodic orbits of φ^t are non-degenerate we have an equality of distributions in $\mathcal{D}'(\mathbb{R}^{>0})$ for suitable (known) signs:*

$$1 - \sum_\rho e^{t\rho} + e^{\alpha t} = \sum_\gamma l(\gamma) \sum_{k \geq 1} \pm \delta_{kl(\gamma)} . \qquad (3.4)$$

In the sum ρ runs over the spectrum of θ on $\bar{H}^1_{\mathcal{F},(2)}(X)$ and all ρ satisfy $\operatorname{Re} \rho = \frac{\alpha}{2}$.

Comparing (3.1) and (3.4) we see again that the primes p and the closed orbits γ should correspond with $\log p \,\hat{=}\, l(\gamma)$. Since we assumed that the flow φ^t has no fixed points there is no contribution in (3.4) corresponding to the term $(1 - e^{-2t})^{-1}$ in (3.1). Ongoing work of Álvarez López and Korduykov leads to optimism that there is an extension of the formula in the theorem to the case where φ^t may have fixed points and where the right hand side looks similar to (3.1). However reduced leafwise L^2-cohomology may have to be replaced by "adiabatic cohomology".

The conditions in the theorem actually force $\alpha = 0$ so that φ^t is isometric. This is related to remark 2. In fact in [D3] using elliptic curves over finite fields we constructed examples of systems $(X, \mathcal{F}, \varphi^t)$ where X is a solenoidal space and we have $\alpha = 1$.

For simplicity we have only considered the explicit formula on $\mathbb{R}^{>0}$. The comparison works also on all of $\mathbb{R}$ and among other features one gets a beautiful analogy between Connes' Euler characteristic $\chi_{\mathcal{F}}(X, \mu)$ and $-\log |d_{K/\mathbb{Q}}|$ cf. [D2] section 4.

4 A conjecture of Lichtenbaum and a dynamical analogue

Consider a regular scheme $\mathcal{X}$ which is separated and of finite type over $\operatorname{spec} \mathbb{Z}$ and assume that $\zeta_{\mathcal{X}}(s)$ has an analytic continuation to $s = 0$.

Lichtenbaum conjectures the existence of a certain "Weil-étale" cohomology theory with and without compact supports, $H^i(\mathcal{X}, A)$ and

$H^i(\mathcal{X}, A)$ for topological rings A. See [Li1], [Li2]. It should be related to the zeta function of $\mathcal{X}$ as follows.

Conjecture 4.1 (Lichtenbaum) *Let $\mathcal{X}/\mathbb{Z}$ be as above. Then the groups $H^i_c(\mathcal{X}, \mathbb{Z})$ are finitely generated and vanish for $i > 2 \dim \mathcal{X} + 1$. Giving $\mathbb{R}$ the usual topology we have*

$$H^i_c(\mathcal{X}, \mathbb{Z}) \otimes_{\mathbb{Z}} \mathbb{R} = H^i_c(\mathcal{X}, \mathbb{R}) \ .$$

Moreover, there is a canonical element ψ in $H^1(\mathcal{X}, \mathbb{R})$ which is functorial in $\mathcal{X}$ and such that we have:
a *The complex*

$$\ldots \xrightarrow{D} H^i_c(\mathcal{X}, \mathbb{R}) \xrightarrow{D} H^{i+1}_c(\mathcal{X}, \mathbb{R}) \to \ldots$$

where $Dh = \psi \cup h$, is acyclic. Note that $D^2 = 0$ because $\deg \psi = 1$.

b $\operatorname{ord}_{s=0} \zeta_{\mathcal{X}}(s) = \sum_i (-1)^i i \operatorname{rk} H^i_c(\mathcal{X}, \mathbb{Z})$.

c *For the leading coefficient $\zeta^*_{\mathcal{X}}(0)$ of $\zeta_{\mathcal{X}}(s)$ in the Taylor development at $s = 0$ we have the formula:*

$$\zeta^*_{\mathcal{X}}(0) = \pm \prod_i |H^i_c(\mathcal{X}, \mathbb{Z})_{\mathrm{tors}}|^{(-1)^i} / \det(H^{\bullet}_c(\mathcal{X}, \mathbb{R}), D, \mathfrak{f}^{\bullet}) \ .$$

Here, $\mathfrak{f}^i$ is a basis of $H^i_c(\mathcal{X}, \mathbb{Z})/\mathrm{tors}$.

Explanation For an acyclic complex of finite dimensional $\mathbb{R}$-vector spaces

$$0 \to V^0 \xrightarrow{D} V^1 \xrightarrow{D} \ldots \xrightarrow{D} V^r \to 0 \qquad (4.1)$$

and bases $\mathfrak{b}^i$ of V^i a determinant $\det(V^{\bullet}, D, \mathfrak{b}^{\bullet})$ in $\mathbb{R}^*_+$ is defined as follows: For bases $\mathfrak{a} = (w_1, \ldots, w_n)$ and $\mathfrak{b} = (v_1, \ldots, v_n)$ of a finite dimensional vector space V set $[\mathfrak{b}/\mathfrak{a}] = \det M$ where $v_i = \sum_j m_{ij} w_j$ and $M = (m_{ij})$. Thus we have $[\mathfrak{c}/\mathfrak{a}] = [\mathfrak{c}/\mathfrak{b}][\mathfrak{b}/\mathfrak{a}]$.

For all i choose bases $\mathfrak{c}^i$ of $D(V^{i-1})$ in V^i and a linearly independent set $\tilde{\mathfrak{c}}^{i-1}$ of vectors in V^{i-1} with $D(\tilde{\mathfrak{c}}^{i-1}) = \mathfrak{c}^i$. Then $(\mathfrak{c}^i, \tilde{\mathfrak{c}}^i)$ is a basis of V^i since (4.1) is acyclic and one defines:

$$\det(V^{\bullet}, D, \mathfrak{b}^{\bullet}) = \prod_i |[\mathfrak{b}^i/(\mathfrak{c}^i, \tilde{\mathfrak{c}}^i)]|^{(-1)^i} \ . \qquad (4.2)$$

Note that $\det(V^{\bullet}, D, \mathfrak{b}^{\bullet})$ is unchanged if we replace the bases $\mathfrak{b}^i$ with bases $\mathfrak{a}^i$ such that $|[\mathfrak{b}^i/\mathfrak{a}^i]| = 1$ for all i. Thus the $\mathfrak{b}^i$ could be replaced by unimodularly or orthogonally equivalent bases. In particular the determinant

$$\det(H_c^\bullet(\mathcal{X}, \mathbb{R}), D, \mathfrak{f}^\bullet)$$

does not depend on the choice of bases $\mathfrak{f}^i$ of $H_c^i(\mathcal{X}, \mathbb{Z})/\text{tors}$.

The following formula is obvious:

Proposition 4.2 *Let $\mathfrak{a}^i$ and $\mathfrak{b}^i$ be bases of the V^i in (4.1). Then we have:*

$$\det(V^\bullet, D, \mathfrak{b}^\bullet) = \det(V^\bullet, D, \mathfrak{a}^\bullet) \prod_i |[\mathfrak{b}^i/\mathfrak{a}^i]|^{(-1)^i} .$$

For smooth projective varieties over finite fields using the Weil-étale topology Lichtenbaum has proved his conjecture in [Li1]. See also [Ge] for generalizations to the singular case. The formalism also works nicely in the study of $\zeta_{\mathcal{X}}(s)$ at $s = 1/2$, cf. [Ra]. If $\mathcal{X}$ is the spectrum of a number field, Lichtenbaum gave a definition of "Weil-étale" cohomology groups using the Weil group of the number field. Using the formula

$$\zeta_K^*(0) = -\frac{hR}{w}$$

he was able to verify his conjecture except for the vanishing of cohomology in degrees greater than three, [Li2]. In fact, his cohomology does not vanish in higher degrees as was recently shown by Flach and Geisser so that some modification will be necessary.

For a dynamical analogue let us look at a triple $(X, \mathcal{F}, \varphi^t)$ as in section 2 with X a closed manifold of dimension $2d + 1$ and φ^t everywhere transversal to $\mathcal{F}$. We then have a decomposition $TX = T\mathcal{F} \oplus T_0 X$ where $T_0 X$ is the rank one bundle of tangents to the flow lines.

In this situation the role of Lichtenbaum's Weil-étale cohomology is played by the ordinary singular cohomology with $\mathbb{Z}$ or $\mathbb{R}$-coefficients of X. Note that because X is compact we do not have to worry about compact supports. From the arithmetic point of view we are dealing with a very simple analogue only!

Lichtenbaum's complex is replaced by

$$(H^\bullet(X, \mathbb{R}), D) \quad \text{where } Dh = \psi \cup h \quad \text{and } \psi = [\omega_\varphi] .$$

Here ω_φ is the 1-form introduced in construction 3 above.

Now assume that the closed orbits of the flow are non-degenerate.

Then at least formally we have:

$$
\begin{aligned}
\zeta_R(s) &:= \prod_\gamma (1 - e^{-sl(\gamma)})^{-\varepsilon_\gamma} \\
&= \prod_i \det{}_\infty(s \cdot \mathrm{id} - \theta \mid \bar{H}^i_{\mathcal{F}}(X))^{(-1)^{i+1}} .
\end{aligned}
\tag{4.3}
$$

Here γ runs over the closed orbits, $\varepsilon_\gamma = \mathrm{sgn}\, \det(1 - T_x \varphi^{l(\gamma)} \mid T_x \mathcal{F})$ for any $x \in \gamma$, the Euler product converges in some right half plane and $\det_\infty$ is the zeta-regularized determinant. The functions $\det_\infty(s \cdot \mathrm{id} - \theta \mid \bar{H}^i_{\mathcal{F}}(X))$ should be entire.

If the closed orbits of φ are degenerate one can define a Ruelle zeta function via Fuller indices [Fu] and relation (4.3) should still hold. In the present note we assume for simplicity that the action of the flow φ on $T\mathcal{F}$ is isometric with respect to $g_{\mathcal{F}}$ and we do not insist on the condition that the closed orbits should be non-degenerate. Then θ has pure eigenvalue spectrum with finite multiplicities on $\bar{H}^\bullet_{\mathcal{F}}(X) = \mathrm{Ker}\,\Delta^\bullet_{\mathcal{F}}$ by [DS], Theorem 2.6. We define $\zeta_R(s)$ by the formula

$$
\zeta_R(s) = \prod_i \det{}_\infty(s \cdot \mathrm{id} - \theta \mid \bar{H}^i_{\mathcal{F}}(X))^{(-1)^{i+1}} ,
$$

if the individual regularized determinants exist and define entire functions. The following result is proved in [D4].

Theorem 4.3 *Consider a triple $(X, \mathcal{F}, \varphi^t)$ as above with $\dim X = 2d+1$ and compatible orientations of X and $\mathcal{F}$ such that the flow acts isometrically with respect to a metric $g_{\mathcal{F}}$ on $T\mathcal{F}$. For $\mathbf{b}$ and $\mathbf{c}$ assume that the above zeta-regularized determinants exist. Then the following assertions hold:*

a *The complex*

$$
\ldots \xrightarrow{D} H^i(X, \mathbb{R}) \xrightarrow{D} H^{i+1}(X, \mathbb{R}) \longrightarrow \ldots
$$

where $Dh = \psi \cup h$ with $\psi = [\omega_\varphi]$ is acyclic.

b $\mathrm{ord}_{s=0}\zeta_R(s) = \sum_i (-1)^i i\, \mathrm{rk}\, H^i(X, \mathbb{Z}).$

c *For the leading coefficient in the Taylor development at $s = 0$ we have the formula:*

$$
\zeta_R^*(0) = \prod_i |H^i(X, \mathbb{Z})_{\mathrm{tors}}|^{(-1)^i} / \det(H^\bullet(X, \mathbb{R}), D, \mathfrak{f}^\bullet) .
$$

Here, $\mathfrak{f}^i$ is a basis of $H^i(X, \mathbb{Z})/\mathrm{tors}$.

Idea of a proof of c We define a metric g on X by $g = g_{\mathcal{F}} + g_0$ (orthogonal sum) on $TX = T\mathcal{F} \oplus T_0 X$ with g_0 defined by $\|Y_{\varphi,x}\| = 1$ for all $x \in X$. Using techniques from the heat equation proof of the index theorem and assuming the existence of zeta-regularized determinants one can prove the following identity:

$$\zeta_R^*(0) = T(X,g)^{-1} . \tag{4.4}$$

Here $T(X,g)$ is the analytic torsion introduced by Ray and Singer using the spectral zeta functions of the Laplace operators Δ^i on i-forms on X

$$T(X,g) = \exp \sum_i (-1)^i \frac{i}{2} \zeta'_{\Delta^i}(0) . \tag{4.5}$$

Equation (4.4) is the special case of Fried's conjecture in [Fr3], see also [Fr1], [Fr2] where the flow has an *integrable* complementary distribution. Next, we use the famous Cheeger–Müller theorem:

$$T(X,g) = \tau(X,g) \qquad \text{cf. [Ch], [M]} \tag{4.6}$$

where $\tau(X,g) = \tau(X,\mathfrak{h}_\bullet)$ is the Reidemeister torsion with respect to the volume forms on homology given by the following bases $\mathfrak{h}_i$ of $H_i(X,\mathbb{R})$. Choose orthonormal bases $\mathfrak{h}^i$ of $\operatorname{Ker} \Delta^i$ with respect to the Hodge scalar product and view them as bases of $H^i(X,\mathbb{R})$ via $\operatorname{Ker} \Delta^i \xrightarrow{\sim} H^i(X,\mathbb{R})$. Let $\mathfrak{h}_i$ be the dual base to $\mathfrak{h}^i$.

If we are given two bases $\mathfrak{a}$ and $\mathfrak{b}$ of a real vector space we have the formula

$$\tau(X,\mathfrak{b}_\bullet) = \tau(X,\mathfrak{a}_\bullet) \prod_i |[\mathfrak{b}_i/\mathfrak{a}_i]|^{(-1)^i}$$

for any choices of bases $\mathfrak{a}_i, \mathfrak{b}_i$ of $H_i(X,\mathbb{R})$. Let $\mathfrak{f}_\bullet, \mathfrak{f}^\bullet$ be dual bases of $H_\bullet(X,\mathbb{Z})/\text{tors}$ resp. $H^\bullet(X,\mathbb{Z})/\text{tors}$. Then we have in particular:

$$\tau(X,\mathfrak{h}_\bullet) = \tau(X,\mathfrak{f}_\bullet) \prod_i |[\mathfrak{h}_i/\mathfrak{f}_i]|^{(-1)^i} . \tag{4.7}$$

Now it follows from the definition of Reidemeister torsion that we have:

$$\tau(X,\mathfrak{f}_\bullet) = \prod_i |H_i(X,\mathbb{Z})_{\text{tors}}|^{(-1)^i} .$$

The cap-isomorphism:

$$\_ \cap [X] : H^j(X,\mathbb{Z}) \xrightarrow{\sim} H_{2d+1-j}(X,\mathbb{Z})$$

now implies that

$$\tau(X, \mathfrak{f}_\bullet) = \prod_i |H^i(X, \mathbb{Z})_{\text{tors}}|^{(-1)^{i+1}} . \tag{4.8}$$

Next let us look at the acyclic complex

$$(H^\bullet(X, \mathbb{R}), D = \psi \cup \_) \cong (\text{Ker}\, \Delta^\bullet, \omega_\varphi \wedge \_) .$$

Using the canonical decomposition into bidegrees

$$\text{Ker}\, \Delta^n = \omega_\varphi \wedge (\text{Ker}\, \Delta_\mathcal{F}^{n-1})^{\theta=0} \oplus (\text{Ker}\, \Delta_\mathcal{F}^n)^{\theta=0}$$

we see that it is isometrically isomorphic to

$$(M^{\bullet-1} \oplus M^\bullet, D) \tag{4.9}$$

where $M^i = (\text{Ker}\, \Delta_\mathcal{F}^i)^{\theta=0}$ and $D(m', m) = (m, 0)$. We may choose the orthonormal basis $\mathfrak{h}^i$ above to be of the form $\mathfrak{h}^i = (\omega_\varphi \wedge \tilde{\mathfrak{h}}^{i-1}, \tilde{\mathfrak{h}}^i)$ where $\tilde{\mathfrak{h}}^i$ is an orthonormal basis of $(\text{Ker}\, \Delta_\mathcal{F}^i)^{\theta=0}$. For this basis, i.e. $(\tilde{\mathfrak{h}}^{i-1}, \tilde{\mathfrak{h}}^i)$ in the version (4.9) it is trivial from the definition (4.2) that

$$|\det(H^\bullet(X, \mathbb{R}), D, \mathfrak{h}^\bullet)| = 1 .$$

Using proposition 4.2 for $\mathfrak{b}^\bullet = \mathfrak{h}^\bullet$ and $\mathfrak{a}^\bullet = \mathfrak{f}^\bullet$ we find

$$1 = |\det(H^\bullet(X, \mathbb{R}), D, \mathfrak{f}^\bullet)| \prod_i |[\mathfrak{h}^i/\mathfrak{f}^i]|^{(-1)^i} . \tag{4.10}$$

Hence we get

$$\prod_i |H^i(X, \mathbb{Z})_{\text{tors}}|^{(-1)^i} / \det(H^\bullet(X, \mathbb{R}), D, \mathfrak{f}^\bullet)$$

$$\overset{(4.10)}{=} \prod_i |H^i(X, \mathbb{Z})_{\text{tors}}|^{(-1)^i} \prod_i |[\mathfrak{h}^i/\mathfrak{f}^i]|^{(-1)^i}$$

$$\overset{(4.8)}{=} \tau(X, \mathfrak{f}_\bullet)^{-1} \prod_i |[\mathfrak{h}_i/\mathfrak{f}_i]|^{(-1)^{i+1}} \overset{(4.7)}{=} \tau(X, \mathfrak{h}_\bullet)^{-1}$$

$$\overset{(4.6)}{=} T(X, g)^{-1} \overset{(4.4)}{=} \zeta_R^*(0) .$$

$\square$

The simplest example For $q > 1$ let X be the circle $\mathbb{R}/(\log q)\mathbb{Z}$ foliated by points and with $\mathbb{R}$ acting by translation. The Ruelle zeta function is given by

$$\zeta_R(s) = (1 - e^{-s \log q})^{-1} = (1 - q^{-s})^{-1} .$$

The operator $\theta = d/dx$ on $\bar{H}^0_{\mathcal{F}}(X) = C^\infty(X)$ defines an unbounded operator on $\bar{H}^0_{\mathcal{F},(2)}(X) = L^2(X)$ with pure eigenvalue spectrum $2\pi i \nu / \log q$ for $\nu \in \mathbb{Z}$. It follows from a formula of Lerch e.g. [D1] §2 that we have

$$\mathrm{det}_\infty (s \cdot \mathrm{id} - \theta \,|\, \bar{H}^0_{\mathcal{F}}(X)) = 1 - q^{-s}$$

and hence formula (4.3) holds in this case.

Next, we have $\omega_\varphi = dt$ and hence $\psi = [dt]$ is a generator of $H^1(X,\mathbb{R})$. The complex

$$\ldots \xrightarrow{D} H^i(X,\mathbb{R}) \xrightarrow{D} H^{i+1}(X,\mathbb{R}) \to \ldots$$

is therefore acyclic. Moreover we have

$$\mathrm{ord}_{s=0} \zeta_R(s) = -1 = \sum_i (-1)^i i \,\mathrm{rk}\, H^i(X,\mathbb{Z}) \;.$$

The leading coefficient $\zeta_R^*(0)$ is given by $\zeta_R^*(0) = (\log q)^{-1}$. We have $H^i(X,\mathbb{Z})_{\mathrm{tors}} = 0$ for all i. As $\psi / \log q$ is a basis of $H^1(X,\mathbb{Z})$ the formula $D(1) = \psi = (\log q)(\psi / \log q)$ for $1 \in H^0(X,\mathbb{R})$ shows that we have

$$\mathrm{det}(H^\bullet(X,\mathbb{R}), D, \mathfrak{f}^\bullet) = \log q \;.$$

This illustrates part **c** of Theorem 4.3.

5 Cramérs function and the transversal wave equation

In this section we observe a new analogy and mention some further directions of research in analysis suggested by this analogy. Let us write the non-trivial zeroes of $\zeta(s)$ as $\rho = \frac{1}{2} + \gamma$. In [C] Cramér studied the function $W(z)$ which is defined for $\mathrm{Im}\, z > 0$ by the absolutely and locally uniformly convergent series

$$W(z) = \sum_{\mathrm{Im}\, \gamma > 0} e^{\gamma z} \;.$$

According to Cramér the function W has a meromorphic continuation to $\mathbb{C}_- = \mathbb{C} \setminus \{xi \,|\, x \leq 0\}$ with poles only for $z = m \log p$ with $m \in \mathbb{Z}, m \neq 0$ and p a prime number. The poles are of first order. If we view the locally integrable function $e^{\gamma t}$ of t as a distribution on $\mathbb{R}$ the series

$$W_{\mathrm{dis}} = \sum_{\mathrm{Im}\, \gamma > 0} e^{\gamma t}$$

converges in $\mathcal{D}'(\mathbb{R})$. Using the mentioned results of Cramér on W one sees that the singular support of W_{dis} consists of $t = 0$ and the numbers $t = m \log p$ for p a prime and $m \in \mathbb{Z}, m \neq 0$, cf. [DSch] section 1. One

my ask if there is an analytic counterpart to this result in the spirit of
the dictionary in section 2.

Presently this is not the case although the work [K] is related to the
problem. According to our dictionary we consider a system $(X, \mathcal{F}, \varphi^t)$
with a 3-manifold X. Let us first assume that X is compact and that φ^t
is isometric and has no fixed points. Then we have $-\theta^2 = \Delta_0$ on $\mathcal{A}_{\mathcal{F}}^\bullet(X)$
where $\theta = \lim_{t \to 0} \frac{1}{t}(\varphi^{t*} - \mathrm{id})$ is the infinitesimal generator of the induced
group of operators φ^{t*} on the Fréchet space $\mathcal{A}_{\mathcal{F}}^\bullet(X)$. Moreover Δ_0 is the
Laplacian along the flow lines with coefficients in $\Lambda^\bullet T^* \mathcal{F}$. Note that it is
transversally elliptic with respect to $\mathcal{F}$. Since $\Delta_0 \mid_{\mathrm{Ker}\,\Delta_{\mathcal{F}}} = \Delta \mid_{\mathrm{Ker}\,\Delta_{\mathcal{F}}}$ by
isometry, it follows from the corresponding result for the spectrum of the
Laplacian that the spectrum $\{\gamma\}$ of θ on $\bar{H}_{\mathcal{F},(2)}^1(X) = \mathrm{Ker}\,\overline{\Delta}_{\mathcal{F}}^1$ consists
of eigenvalues and that all γ are purely imaginary. (This proves a part
of theorem 5.) For φ in $\mathcal{D}(\mathbb{R})$ the operator $\int_\mathbb{R} \varphi(t) e^{it|\theta|}\, dt$ on $\mathrm{Ker}\,\overline{\Delta}_{\mathcal{F}}^1$
is trace class and the map $\varphi \mapsto \mathrm{tr} \int_\mathbb{R} \varphi(t) e^{it|\theta|}$ defines a distribution
denoted by $\mathrm{tr}(e^{it|\theta|} \mid \mathrm{Ker}\,\overline{\Delta}_{\mathcal{F}}^1)$. Then we have:

$$2 \sum_{\mathrm{Im}\,\gamma > 0} e^{t\gamma} = \mathrm{tr}(e^{it|\theta|} \mid \mathrm{Ker}\,\overline{\Delta}_{\mathcal{F}}^1)$$

$$= \mathrm{tr}(e^{it\sqrt{\Delta}} \mid \mathrm{Ker}\,\overline{\Delta}_{\mathcal{F}}^1) \quad \text{in } \mathcal{D}'(\mathbb{R}) \, .$$

Here we have assumed that zero is not in the spectrum of θ on $\mathrm{Ker}\,\overline{\Delta}_{\mathcal{F}}^1$,
corresponding to the fact that $\zeta(1/2) \neq 0$. But this is not important.

On functions, i.e. on $C^\infty(X) = \mathcal{A}_{\mathcal{F}}^0(X)$ instead of $\mathrm{Ker}\,\Delta_{\mathcal{F}}^1$ the distributional trace $\mathrm{tr}(e^{it\sqrt{\Delta}})$ of the operator $e^{it\sqrt{\Delta}}$ has been extensively
studied. By a basic result of Chazarain [Cha] the singular support of
$\mathrm{tr}(e^{it\sqrt{\Delta}})$ consists of $t = 0$ and the numbers $t = ml(\gamma)$ for $m \in \mathbb{Z}, m \neq 0$
and $l(\gamma)$ the lengths of the closed orbits of the geodesic flow. On the
other hand by the analogy with Cramérs theorem we expect that the sin-
gularities of $\mathrm{tr}(e^{it\sqrt{\Delta}} \mid \mathrm{Ker}\,\overline{\Delta}_{\mathcal{F}}^1)$ are contained in the analogous set where
now γ runs over the closed orbits of the flow φ^t.

Based on the work of Hörmander the "big" singularity of $\mathrm{tr}(e^{it\sqrt{\Delta}})$ at
$t = 0$ was analyzed in [DG] proposition 2.1 by an asymptotic expansion
of the Fourier transform of $\mathrm{tr}(e^{it\sqrt{\Delta}})$. An analogue of this expansion for
the Fourier transform of $\mathrm{tr}(e^{it\sqrt{\Delta}} \mid \mathrm{Ker}\,\overline{\Delta}_{\mathcal{F}}^1)$ would correspond well with
asymptotics that can be obtained with some effort from Cramér's the-
ory, except that in Cramér's theory there also appear logarithmic terms
coming from the infinite place. These should also appear in the analysis
of $\mathrm{tr}(e^{it\sqrt{\Delta}} \mid \mathrm{Ker}\,\overline{\Delta}_{\mathcal{F}}^1)$ if one allows the flow to have fixed points. Finally
one should drop the condition that the flow acts isometrically. Then

one has to work with $\mathrm{tr}(e^{it|\theta|} \mid \bar{H}^1_{\mathcal{F},(2)}(X))$ instead of $\mathrm{tr}(e^{it\sqrt{\Delta}} \mid \mathrm{Ker}\,\overline{\Delta}^1_{\mathcal{F}})$. Also one can try to prove a Duistermaat–Guillemin trace formula for $\mathrm{tr}(e^{it|\theta|} \mid \bar{H}^1_{\mathcal{F},(2)}(X))$ and even if the flow has fixed points.

Bibliography

[AK1] J. Álvarez López, Y.A. Kordyukov, Long time behaviour of leafwise heat flow for Riemannian foliations. Compos. Math. **125** (2001), 129–153

[AK2] J. Álvarez López, Y.A. Kordyukov, Distributional Betti numbers of transitive foliations of codimension one. In: Proceedings on Foliations: Geometry and Dynamics, ed. P. Walczak et al. World Scientific, Singapore, 2002, pp. 159–183

[CC] A. Candel, L. Conlon, Foliations I. AMS Graduate studies in Mathematics **23**, 2000.

[Cha] J. Chazarain, Formules de Poisson pour les variétés riemanniennes. Invent. Math. **24** (1974), 65–82

[Ch] J. Cheeger, Analytic torsion and the heat equation. Ann. Math. **109** (1979), 259–322

[C] H. Cramér, Studien über die Nullstellen der Riemannschen Zetafunktion. Math. Z. **4** (1919), 104–130

[DG] J.J. Duistermaat, V.W. Guillemin, The spectrum of positive elliptic operators and periodic bicharacteristics. Invent. Math. **29** (1975), 39–79

[D1] C. Deninger, Motivic L-functions and regularized determinants, (1992), in: Jannsen, Kleiman, Serre (eds.): Seattle conference on motives 1991. Proc. Symp. Pure Math. AMS **55** (1994), Part 1, 707–743

[D2] C. Deninger, Number theory and dynamical systems on foliated spaces. Jber. d. Dt. Math.-Verein. **103** (2001), 79–100

[D3] C. Deninger, On the nature of the "explicit formulas" in analytic number theory – a simple example. In: S. Kanemitsu, C. Jia (eds.), Number Theoretic Methods – Future Trends. In: DEVM "Developments of Mathematics", Kluwer Academic Publ.

[D4] C. Deninger, A dynamical systems analogue of Lichtenbaum's conjectures on special values of Hasse–Weil zeta functions. Preprint arXiv math.NT/0605724

[DS] C. Deninger, W. Singhof, A note on dynamical trace formulas. In: M.L. Lapidus, M. van Frankenhuysen (eds.), Dynamical Spectral and Arithmetic Zeta-Functions. In: AMS Contemp. Math. **290** (2001), 41–55

[DSch] C. Deninger, M. Schröter, A distribution theoretic interpretation of Guinand's functional equation for Cramér's V-function and generalizations. J. London Math. Soc. (2) **52** (1995), 48–60

[Fa] M. Farber, Topology of closed geodesics on hyperbolic manifolds. Mathematical Surveys and Monographs, Vol. **108**, AMS 2004

[Fr1] D. Fried, Homological identities for closed orbits. Invent. Math. **71** (1983), 419–442

[Fr2] D. Fried, Analytic torsion and closed geodesics on hyperbolic manifolds. Invent. Math. **84** (1986), 523–540

[Fr3] D. Fried, Counting circles. In: Dynamical Systems LNM **1342** (1988), 196–215

[Fu] Fuller, An index of fixed point type for periodic orbits. Amer. J. Math. **89** (1967), 133–148

[Ge] T. Geisser, Arithmetic cohomology over finite fields and special values of ζ-functions. Duke Math. J. **133** (2006), 27–57

[Ko] F. Kopei, A remark on a relation between foliations and number theory. ArXiv math.NT/0605184

[K] Y. Kordyukov, The trace formula for transversally elliptic operators on Riemannian foliations. St. Petersburg Math. J. **12** (2001), 407–422

[Le] E. Leichtnam, Scaling group flow and Lefschetz trace formula for laminated spaces with p-adic transversal. To appear in: Bulletin des Sciences Mathématiques

[Li1] S. Lichtenbaum, The Weil-étale topology on schemes over finite fields. Compos. Math. **141** (2005), 689–702

[Li2] S. Lichtenbaum, The Weil étale topology for number rings. ArXiv math.NT/0503604

[M] W. Müller, Analytic torsion and R-torsion of Riemannian manifolds. Adv. Math. **28** (1978), 233–305

[Ra] N. Ramachandran, Values of zeta functions at $s = 1/2$. Int. Math. Res. Not. **25** (2005), 1519–1541

[RS] D. Ray, I. Singer, Analytic torsion. Proc. Symp. Pure Math. XXIII, 167–181

10

Reciprocity Algebras and Branching for Classical Symmetric Pairs

Roger E. Howe

Department of Mathematics
Yale University

Eng-Chye Tan

Department of Mathematics
National University of Singapore

Jeb F. Willenbring

Department of Mathematical Sciences
University of Wisconsin at Milwaukee

ABSTRACT We study branching laws for a classical group G and a symmetric subgroup H. Our approach is by introducing the *branching algebra*, the algebra of covariants for H in the regular functions on the natural torus bundle over the flag manifold for G. We give concrete descriptions of certain subalgebras of the branching algebra using classical invariant theory. In this context, it turns out that the ten classes of classical symmetric pairs (G, H) are associated in pairs, (G, H) and (H', G').

Our results may be regarded as a further development of classical invariant theory as described by Weyl [64], and extended previously in [14]. They show that the framework of classical invariant theory is flexible enough to encompass a wide variety of calculations that have been carried out by other methods over a period of several decades. This framework is capable of further development, and in some ways can provide a more precise picture than has been developed in previous work.

1 Introduction

1.1 The Classical Groups.

Hermann Weyl's book, *The Classical Groups* [64], has influenced many researchers in invariant theory and related fields in the decades since it was written. Written as an updating of "classical" invariant theory, it has itself acquired the patina of a classic. The books [11] and [55] and the references in them give an idea of the extent of the influence. The current authors freely confess to being among those on whom Weyl has had major impact.

The Classical Groups has two main themes: the invariant theory of the classical groups – the general linear groups, the orthogonal groups and the symplectic groups – acting on sums of copies of their standard representations (and, in the case of the general linear groups, copies of the dual representation also); and the description of the irreducible representations of these groups. In invariant theory, the primary results of [64] are what Weyl called the First and Second Fundamental Theorems of invariant theory. The First Fundamental Theorem (FFT) describes a set of "typical basic generators" for the invariants of the selected actions, and the Second Fundamental Theorem (SFT) describes the relations between these generators. The description of the representations culminates in the Weyl Character Formula.

The two themes are not completely integrated. For the first one, Weyl uses the apparatus of classical invariant theory, including the Aronhold polarization operators and the Capelli identity, together with geometrical considerations about orbits, etc. For the second, he abandons polynomial rings and relies primarily on the Schur-Weyl duality, the remarkable connection discovered by I. Schur between representations of the general linear groups and the symmetric groups. This duality takes place on tensor powers of the fundamental representation of GL_n, which of course are finite dimensional. This gives Weyl's description of the irreducible representations more of a combinatorial cast. This combinatorial viewpoint, based around Ferrers-Young diagrams and Young tableaux, has been very heavily developed in the latter half of the twentieth century (see [48], [59], [50], [9], [55] and the references below).

1.2 From Invariants to Covariants.

In [14], it was observed that by combining the results in [64] with another construction of Weyl, namely the *Weyl algebra*, aka the algebra of polynomial coefficient differential operators, it is possible to give a unified treatment of the invariants and the irreducible representations.

Introduction of the Weyl algebra brings several valuable pieces of structure into the picture. A key feature of the Weyl algebra $\mathcal{W}(V)$ associated to a vector space V is that it has a filtration such that the associated graded algebra is commutative, and is canonically isomorphic to the algebra $\mathcal{P}(V \oplus V^*)$ of polynomials on the sum of V with its dual V^*. Moreover, if one extends the natural bilinear pairing between V and V^* to a skew-symmetric bilinear (or *symplectic*) form on $V \oplus V^*$, then the symplectic group $Sp(V \oplus V^*)$ of isometries of this form acts naturally as automorphisms of $\mathcal{W}(V)$, and this action gets carried over to the natural action of $Sp(V \oplus V^*)$ on $\mathcal{P}(V \oplus V^*)$. Also, the natural action of $GL(V)$, the general linear group of V, on $\mathcal{P}(V \oplus V^*)$ embeds $GL(V)$ in $Sp(V \oplus V^*)$. The corresponding action of $GL(V)$ on $\mathcal{W}(V)$ is just the action by conjugation when both $GL(V)$ and $\mathcal{W}(V)$ are regarded as being operators on $\mathcal{P}(V)$. Finally, the Lie algebra $\mathfrak{sp}(V \oplus V^*)$ of $Sp(V \oplus V^*)$ is naturally embedded as a Lie subalgebra of $\mathcal{W}(V)$. The image of $\mathfrak{sp}(V \oplus V^*)$ in $\mathcal{P}(V \oplus V^*)$ consists of the homogeneous polynomials of degree two. The Lie bracket is then given by Poisson bracket with respect to the symplectic form [6].

Given a group $G \subseteq GL(V)$, it is natural in this context to look at $\mathcal{W}(V)^G$, the algebra of polynomial coefficient differential operators invariant under the action of G, or equivalently, of operators that commute with G. One can show in a fairly general context that $\mathcal{W}(V)^G$ provides strong information about the decomposition of $\mathcal{P}(V)$ into irreducible representations for G [10].

In the case of the classical actions considered by Weyl, it turns out that $\mathcal{W}(V)^G$ has an elegant structure. This structure is revealed by considering the centralizer of G inside $Sp(V \oplus V^*)$. Since $G \subseteq GL(V) \subset Sp(V \oplus V^*)$, we can consider $G' = Sp(V \oplus V^*)^G$, the centralizer of G in $Sp(V \oplus V^*)$. The Lie algebra of G' will be $\mathfrak{g}' = \mathfrak{sp}(V \oplus V^*)^G$, the Lie subalgebra of $\mathfrak{sp}(V \oplus V^*)$ consisting of elements that commute with G.

Given a group S and subgroup H, looking at H', the centralizer of H in S, is a construction that has the formal properties of a duality operation, analogous to considering the commutant of a subalgebra in an algebra. It is easy to check that $H'' = (H')'$ contains H, and that $H''' = (H')''$ is again equal to H'. Hence also $H'''' = H''$. Thus H'' constitutes a sort of "closure" of H with respect to the issue of commuting inside S, and the pair (H'', H') constitute a pair of mutually centralizing subgroups of S. In [14] such a pair was termed a *dual pair* of subgroups of S. Dual pairs of subgroups arise naturally in studying the structure of groups. For example, in reductive algebraic groups, the Levi component of a parabolic subgroup and its central torus constitute a dual pair. A subgroup $H \subseteq G$ belongs to a dual pair in G if and only if it is its own double centralizer.

It turns out that, if G is one of the classical groups acting on V

by one of the actions specified by Weyl, and G' is its centralizer in $Sp(V \oplus V^*)$, then (G, G') constitute a dual pair in $Sp(V \oplus V^*)$. That is, $G = G''$. Moreover, by translating Weyl's FFT to the context of the Weyl algebra, one sees that $\mathfrak{g}' = \mathfrak{sp}(V \oplus V^*)^G$ generates $\mathcal{W}(V)^G$ as an associative algebra. Furthermore, the condition that $G = G''$ essentially characterizes the actions considered by Weyl, providing some insight into why the FFT has such a clean statement for these actions, but is known for hardly any other examples. The fact that $\mathfrak{g}'$ generates $\mathcal{W}(V)^G$, together with some simple structural facts about $\mathfrak{g}'$, allows a detailed description of the action of G on $\mathcal{P}(V)$. As part of this picture, one obtains a natural bijection between the irreducible representations of G appearing in $\mathcal{P}(V)$, and certain irreducible representations of $\mathfrak{g}'$.

In the case of when $G = GL_n$ and $V \simeq \mathbb{C}^n \otimes \mathbb{C}^m$, then $G' = GL_m$ acting on the factor $\mathbb{C}^m$ of V. The Lie algebra $\mathfrak{g}' \simeq \mathfrak{gl}_m$ is exactly the span of the Aronhold polarization operators. In this case, the polynomials on V decompose into jointly irreducible representations $\rho \otimes \rho'$ for $GL_n \times GL_m$, and the correspondence $\rho \leftrightarrow \rho'$ is bijective. We refer to this correspondence as (GL_n, GL_m) *duality*. The paper [16] further studied this situation and the foundations of Weyl's Fundamental Theorems, and pointed out that these results could be understood from the point of view of multiplicity free actions. In this development, the FFT for GL_n, Schur-Weyl duality and (GL_n, GL_m) duality appear as three aspects of the same phenomenon. Each can be deduced from either of the others. In particular, the polynomial version of the theory and the combinatorial version are seen as two windows on the same landscape.

1.3 Branching Rules via Invariant Theory.

Already when [64] was being written, work was underway to extend representation theory beyond a basic enumeration of the irreducible representations to describe some aspects of their structure. Part of the motivation for doing this came from quantum mechanics, which also was the inspiration for the Weyl algebra. In [46], Littlewood and Richardson proposed a combinatorial description of the multiplicities of irreducible representations in the tensor product of two irreducible representations of GL_n. These multiplicities are now known as Littlewood-Richardson (LR) coefficients.

The decomposition of tensor products of representations of a group G can be regarded as a *branching problem* – the problem of decomposing the restriction of a representation of some group to a subgroup. Precisely, if ρ_1 and ρ_2 are irreducible representations of G, then the tensor product $\rho_1 \otimes \rho_2$ can be regarded as an <u>irreducible</u> representation of $G \times G$, and all irreducible representations of $G \times G$ are of this form [11, 16].

If we restrict this representation to the diagonal subgroup $\Delta G = \{(g, g) : g \in G\} \subset G \times G$, then we obtain the usual notion of $\rho_1 \otimes \rho_2$ as a representation of G. Note that ΔG is isomorphic to G. Also, the involution $(g_1, g_2) \leftrightarrow (g_2, g_1)$ of $G \times G$ has ΔG as the subgroup of fixed points. Thus, ΔG is a *symmetric subgroup* of $G \times G$ – the fixed point set of an involution (order two automorphism) of $G \times G$. We also call the pair $(G \times G, \Delta(G))$ a *symmetric pair*. In summary, computing tensor products for representations of GL_n can be viewed as studying the branching problem for the symmetric pair $(GL_n \times GL_n, \Delta(GL_n))$.

In the 1940s, Littlewood considered the branching problem from GL_n to the orthogonal group O_n. Note that O_n is the set of fixed points of the involution $g \to (g^{-1})^t$, where A^t denotes the transpose of the $n \times n$ matrix A. Thus, (GL_n, O_n) is a symmetric pair. Since all the representations involved are semisimple (thanks, e.g. to Weyl's *unitarian trick* [64]), to determine them up to isomorphism, it is enough to know the *branching multiplicities* – the multiplicity with which each irreducible representation of O_n occurs in a given irreducible representation of GL_n. Under some restrictions on the representations involved, Littlewood [48, 47] showed how to express the branching multiplicities for the pair (GL_n, O_n), in terms of LR coefficients. Over the decades since [64], these results have been extended in stages, so that one now has a description of the branching multiplicities for any *classical symmetric pair* – a symmetric pair (G, K), where G is a classical group – in terms of LR coefficients [46], [48], [47], [52], [28], [29], [30], [31], [4], [36], [37], [38], [39], [32], [59], [34], [35], [33]. See also [9] and [21].

The goal of this paper is to show how the invariant theory approach can be further developed to encompass much of the work on branching multiplicities cited above. The main ingredient needed for doing this is the notion of *branching algebra*, an idea used by Zhelobenko [65], but relatively little exploited since. The general idea of branching algebra is described in §2.

The concrete description of branching algebras for the classical symmetric pairs in terms of polynomial rings is carried out in §4. More precisely, in §4, certain families of well behaved subalgebras of branching algebras are realized via polynomial rings. We refer to these subalgebras as *partial* branching algebras.

When one realizes partial branching algebras in the context of dual pairs, a lovely reciprocity phenomenon reveals itself. Suppose that a reductive group $G \subset GL(V) \subset Sp(V \oplus V^*)$ belongs to a dual pair (G, G'). Let $K \subset G$ be a symmetric subgroup. It turns out that, if $K' \subset Sp(V \oplus V^*)$ is the centralizer of K in $Sp(V \oplus V^*)$, then (K, K') also form a dual pair in $Sp(V \oplus V^*)$; and furthermore (K', G') is also a classical symmetric pair! (Note that, since $K \subset G$, we clearly will have $K' \supset G'$.) Thus, the dual pairs (G, G') and (K, K') form a *seesaw pair* in the sense of Kudla [41].

Moreover, the partial branching algebra constructed for (G, K) turns out also to be a partial branching algebra for (K', G')! The coincidence implies a reciprocity phenomenon: branching multiplicities for (G, K) also describe branching multiplicities for (K', G'). For this reason, we call these partial branching algebras *reciprocity algebras*. (It should be noted that often some special infinite dimensional representations of K' may be involved in these reciprocity relationships, and sometimes infinite dimensional representations of G' are also involved.) We note that parts of this picture have appeared before. The fact that (K, K') is a dual pair, and that (K', G') is a symmetric pair is implicit in [15]. A numerical version of the reciprocity laws implied by reciprocity algebras was given in [13], and the reciprocity phenomenon for the GL_n tensor product algebra was noted in [16].

The classical symmetric pairs may be sorted into ten infinite families (see Table I in §4). It turns out that if the pair (G, K) is taken from one family, the pair (K', G') is always taken from another family that is determined by the family of (G, K). That is, the seesaw construction applied to classical symmetric pairs, pairs up the ten families into five reciprocal pairs of families. The two families in a pair have many reciprocity laws relating multiplicities of their representations.

Thus, the branching algebra approach to branching rules, when made concrete via classical invariant theory, has some highly attractive formal features. We should keep in mind that they are formal, in the sense that they don't say anything explicit about what certain branching multiplicities <u>are</u>. They just say that branching multiplicities for one pair (G, K) can be expressed in terms of branching multiplicities for another pair (K', G'). To get more specific information, one needs some reasonably explicit description of the partial branching algebras. In this paper we give a <u>relative</u> description, which shows that every reciprocity algebra is related to the GL_n tensor product algebra. This is carried out in §8 for one of the symmetric pairs. The theory works best under some technical restrictions, referred to as the *stable range*. As explained in [21], these relations imply the many formulas in the literature that describe branching multiplicities for symmetric pairs in terms of LR coefficients.

To have a complete theory, one should also give concrete and explicit descriptions of the reciprocity algebras. This paper does not deal with this issue. However, it has been carried out, by the authors and others, in several papers. Using ideas of Grobner/SAGBI theory [56], [51], the paper [22] describes an explicit basis for the basic case, the GL_n tensor product algebra. Analogous bases for most of the other reciprocity algebras are given in [17, 18, 19]. Furthermore, the paper [20] shows how to deduce the Littlewood-Richardson Rule from the results of [22] and representation-theoretic considerations. Also, the paper [12] shows that these branching algebras have toric deformations. More precisely, it shows that they have flat deformations that are semigroup rings of lattice cones, which are explicitly described.

Thus, this paper supplemented by the work just cited shows that the invariant-theory approach using branching algebras provides a uniform framework to deal with a wide variety of issues in representation theory. These computations have been handled in the literature by largely means of combinatorics, as cited above, and more recently by quantum groups and related methods, including the path methods of Littelmann [7], [49], [27], [25], [26], [24], [42], [43], [44], [45]. The branching algebra approach can provide another window on these phenomena.

The branching algebra approach comes with some extra structure attached. The multiplicities are seen not simply as numbers, but intrinsically as cardinalities of integral points in convex sets. (Indeed, one can see this implicitly in the Littlewood-Richardson Rule, and it was made more explicit in [2], [3] and [53]).

Moreover, these points do not simply give the correct count: individual points correspond to <u>specific</u> highest weight vectors in representations. Finally, the fact that all representations are bundled together inside one algebra structure implies relations between the highest weight vectors and the multiplicities of different representations. The authors suspect that this extra structure can be useful in studying certain problems, such as the actions analyzed by Kostant and Rallis [40], and perhaps in understanding the structure of principal series representations of semisimple groups. We hope to return to these themes in future papers.

Acknowledgements: We thank Kenji Ueno and Chen-Bo Zhu for discussions. The second named author also acknowledges the support of NUS grant R-146-000-050-112. The third named author was supported by NSA Grant # H98230-05-1-0078.

2 Branching Algebras

For a reductive complex linear algebraic G, let U_G be a maximal unipotent subgroup of G. The group U_G is determined up to conjugacy in G [5]. Let A_G denote a maximal torus which normalizes U_G, so that $B_G = A_G \cdot U_G$ is a Borel subgroup of G. Also let $\widehat{A}_G^+$ be the set of dominant characters of A_G – the semigroup of highest weights of representations of G. It is well-known [5] [16] and may be thought of as a geometric version of the theory of the highest weight, that the space of regular functions on the coset space G/U_G, denoted by $\mathcal{R}(G/U_G)$, decomposes (under the action of G by left translations) as a direct sum of one copy of each irreducible representation V_ψ (with highest weight ψ) of G (see [60]):

$$\mathcal{R}(G/U_G) \simeq \bigoplus_{\psi \in \widehat{A}_G^+} V_\psi. \qquad (2.1)$$

We note that $\mathcal{R}(G/U_G)$ has the structure of an $\widehat{A}_G^+$-graded algebra, for which the V_ψ are the graded components. To be specific, we note that since A_G normalizes U_G, it acts on G/U_G by right translations, and this action commutes with the action of G by left translations.

Proposition 2.1 (see [63]) *The algebra of regular functions $\mathcal{R}(G/U_G)$ is an $\widehat{A}_G^+$-graded algebra, under the right action of A_G. More precisely, the decomposition (2.1) is the graded algebra decomposition under A_G, where V_ψ is the A_G-eigenspace corresponding to $\varphi \in \widehat{A}_G^+$ with $\varphi = w^*(\psi^{-1})$. Here w is the longest element of the Weyl group with respect to the root system determined by the Borel subgroup B_G.*

Now let $H \subset G$ be a reductive subgroup, and let U_H be a maximal unipotent subgroup of H. We consider the algebra $\mathcal{R}(G/U_G)^{U_H}$, of functions on G/U_G which are invariant under left translations by U_H. Let A_H be a maximal torus of H normalizing U_H, so that $B_H := A_H \cdot U_H$ is a Borel subgroup of H. Then $\mathcal{R}(G/U_G)^{U_H}$ will be invariant under the (left) action of A_H, and we may decompose $\mathcal{R}(G/U_G)^{U_H}$ into eigenspaces for A_H. Since the functions in $\mathcal{R}(G/U_G)^{U_H}$ are by definition (left) invariant under U_H, the (left) A_H-eigenfunctions will in fact be (left) B_H eigenfunctions. In other words, they are highest weight vectors for H. Hence, the characters of A_H acting on (the left of) $\mathcal{R}(G/U_G)^{U_H}$ will all be dominant with respect to B_H, and we may write $\mathcal{R}(G/U_G)^{U_H}$ as a sum of (left) A_H eigenspaces $(\mathcal{R}(G/U_G)^{U_H})^\chi$ for dominant characters χ of H:

$$\mathcal{R}(G/U_G)^{U_H} = \bigoplus_{\chi \in \widehat{A}_H^+} (\mathcal{R}(G/U_G)^{U_H})^\chi. \tag{2.3}$$

Since the spaces V_ψ of decomposition (2.1) are (left) G-invariant, they are a fortiori *left* H-invariant, so we have a decomposition of $\mathcal{R}(G/U_G)^{U_H}$ into *right* A_G-eigenspaces $(\mathcal{R}(G/U_G)^{U_H})_\psi$:

$$\mathcal{R}(G/U_G)^{U_H} = \bigoplus_{\psi \in \widehat{A}_G^+} \mathcal{R}(G/U_G)^{U_H} \cap V_\psi := \bigoplus_{\psi \in \widehat{A}_G^+} \mathcal{R}(G/U_G)_\psi^{U_H}.$$

Combining this decomposition with the decomposition (2.3), we may write

$$\mathcal{R}(G/U_G)^{U_H} = \bigoplus_{\psi \in \widehat{A}_G^+, \ \chi \in \widehat{A}_H^+} (\mathcal{R}(G/U_G)_\psi^{U_H})^\chi. \tag{2.4}$$

To emphasize the key features of this algebra, we note the resulting consequences of decomposition (2.4) in the following proposition.

Proposition 2.2

(a) *The decomposition (2.4) is an $(\widehat{A}_G^+ \times \widehat{A}_H^+)$-graded algebra decomposition of $\mathcal{R}(G/U_G)^{U_H}$.*

(b) *The subspaces $(\mathcal{R}(G/U_G)_\psi^{U_H})^\chi$ tell us the χ highest weight vectors for B_H in the irreducible representation V_ψ of G. Therefore, the decomposition*

$$\mathcal{R}(G/U_G)_\psi^{U_H} = \bigoplus_{\chi \in \widehat{A}_H^+} (\mathcal{R}(G/U_G)_\psi^{U_H})^\chi$$

tells us how V_ψ decomposes as a H-module.

Thus, knowledge of $\mathcal{R}(G/U_G)^{U_H}$ as a $(\widehat{A}_G^+ \times \widehat{A}_H^+)$-graded algebra tell us how representations of G decompose when restricted to H, in other words, it describes the branching rule from G to H.

We will call $\mathcal{R}(G/U_G)^{U_H}$ the (G, H) *branching algebra*. When $G \simeq H \times H$, and H is embedded diagonally in G, the branching algebra describes the decomposition of tensor products of representations of H, and we then call it the *tensor product algebra* for H. More generally, we would like to understand the (G, H) branching algebras for symmetric pairs (G, H).

In the context of regular functions on G/U, such branching algebras are a little too abstract. We shall elaborate later on a more concrete construction of branching algebras, which allows substantial manipulation. We shall come back to the concrete construction at the end of §4, after some preliminaries in §3 and an example in §4.1.

3 Preliminaries and Notations

3.1 Parametrization of Representations

Let G be a classical reductive algebraic group over $\mathbb{C}$: $G = GL_n(\mathbb{C}) = GL_n$, the general linear group; or $G = O_n(\mathbb{C}) = O_n$, the orthogonal group; or $G = Sp_{2n}(\mathbb{C}) = Sp_{2n}$, the symplectic group. We shall explain our notations on irreducible representations of G using integer partitions. In each of these cases, we select a Borel subalgebra of the classical Lie algebra and coordinatize it, as is done in [11]. Consequently, all highest weights are parameterized in the standard way (see [11]).

A non-negative integer *partition* λ, with k parts, is an integer sequence $\lambda_1 \geq \lambda_2 \geq \ldots \geq \lambda_k > 0$. We may sometimes refer to λ as a *Young* or *Ferrers diagram*. We use the same notation for partitions as is done in [50]. For example, we write $\ell(\lambda)$ to denote the *length* (or *depth*) of a partition, i.e., $\ell(\lambda) = k$ for the above partition. Also let $|\lambda| = \sum_i \lambda_i$ be the size of a partition and λ' denote the *transpose* (or *conjugate*) of λ (i.e., $(\lambda')_i = |\{\lambda_j : \lambda_j \geq i\}|$).

GL_n Representations: Given non-negative integers p, q and n such that $n \geq p + q$ and non-negative integer partitions λ^+ and λ^- with p and q parts respectively, let $F_{(n)}^{(\lambda^+, \lambda^-)}$ denote the irreducible rational representation of GL_n with highest weight given by the n-tuple:

$$(\lambda^+, \lambda^-) = \underbrace{\left(\lambda_1^+, \lambda_2^+, \cdots, \lambda_p^+, 0, \cdots, 0, -\lambda_q^-, \cdots, -\lambda_1^-\right)}_{n}$$

If $\lambda^- = (0)$ then we will write $F_{(n)}^{\lambda^+}$ for $F_{(n)}^{(\lambda^+, \lambda^-)}$. Note that if $\lambda^+ = (0)$ then $\left(F_{(n)}^{\lambda^-}\right)^*$ is equivalent to $F_{(n)}^{(\lambda^+, \lambda^-)}$. More generally, $\left(F_{(n)}^{(\lambda^+, \lambda^-)}\right)^*$ is equivalent to $F_{(n)}^{(\lambda^-, \lambda^+)}$.

O_n Representations: The complex orthogonal group has two connected components. Because the group is disconnected we cannot index irreducible representations by highest weights. There is however an analog of Schur-Weyl duality for the case of O_n in which each irreducible rational representation is indexed uniquely by a non-negative integer partition ν such that $(\nu')_1 + (\nu')_2 \leq n$. That is, the sum of the first two columns of the Young diagram of ν is at most n. We will call such a diagram O_n-admissible (see [11] Chapter 10 for details). Let $E^\nu_{(n)}$ denote the irreducible representation of O_n indexed ν in this way.

An irreducible rational representation of SO_n may be indexed by its highest weight. In [11] Section 5.2.2, the irreducible representations of O_n are determined in terms of their restrictions to SO_n (which is a normal subgroup having index 2). We note that if $\ell(\nu) \neq \frac{n}{2}$, then the restriction of $E^\nu_{(n)}$ to SO_n is irreducible. If $\ell(\nu) = \frac{n}{2}$ (n even), then $E^\nu_{(n)}$ decomposes into exactly two irreducible representations of SO_n. See [11] Section 10.2.4 and 10.2.5 for the correspondence between this parametrization and the above parametrization by partitions.

The determinant defines an (irreducible) one-dimensional representation of O_n. This representation is indexed by the length n partition $\zeta = (1, 1, \cdots, 1)$. An irreducible representation of O_n will remain irreducible when tensored by $E^\zeta_{(n)}$, but the resulting representation *may* be inequivalent to the initial representation. We say that a pair of O_n-admissible partitions α and β are *associate* if $E^\alpha_{(n)} \otimes E^\zeta_{(n)} \cong E^\beta_{(n)}$. It turns out that α and β are associate exactly when $(\alpha')_1 + (\beta')_1 = n$ and $(\alpha')_i = (\beta')_i$ for all $i > 1$. This relation is clearly symmetric, and is related to the structure of the underlying SO_n-representations. Indeed, when restricted to SO_n, $E^\alpha_{(n)} \cong E^\beta_{(n)}$ if and only if α and β are either associate or equal.

3.2 Multiplicity-Free Actions

Let G be a complex reductive algebraic group acting on a complex vector space V. We say V is a *multiplicity-free action* if the algebra $\mathcal{P}(V)$ of polynomial functions on V is multiplicity free as a G module. The criterion of Servedio-Vinberg-Kimel'fel'd ([58, 62]) says that V is multiplicity free if and only if a Borel subgroup B of G has a Zariski open orbit in V. In other words, B (and hence G) acts prehomogeneously on V (see [57]). A direct consequence is that B eigenfunctions in $\mathcal{P}(V)$ have a very simple structure. Let $Q_\psi \in \mathcal{P}(V)$ be a B eigenfunction with eigencharacter ψ, normalized so that $Q_\psi(v_0) = 1$ for some fixed v_0 in a Zariski open B orbit in V. Then Q_ψ is completely determined by ψ:

For $v = b^{-1}v_0$ in the Zariski open B orbit,

$$Q_\psi(v) = Q_\psi(b^{-1}v_0) = \psi(b)Q_\psi(v_0) = \psi(b), \qquad b \in B.$$

Q_ψ is then determined on all of V by continuity. Since $B = AU$, and $U = (B, B)$ is the commutator subgroup of B, we can identify a character of B with a character of A. Thus the B eigenfunctions are precisely the G highest weight vectors (with respect to B) in $\mathcal{P}(V)$. Further

$$Q_{\psi_1}Q_{\psi_2} = Q_{\psi_1\psi_2}$$

and so the set of $\widehat{A}^+(V) = \{\psi \in \widehat{A}^+ \mid Q_\psi \neq 0\}$ forms a sub-semigroup of the cone $\widehat{A}^+$ of dominant weights of A.

An element $\psi(\neq 1)$ of a semigroup is *primitive* if it is not expressible as a non-trivial product of two elements of the semigroup. The algebra $\mathcal{P}(V)^U$ has unique factorization (see [23]). The eigenfunctions associated to the primitive elements of $\widehat{A}^+(V)$ are prime polynomials, and $\mathcal{P}(V)^U$ is the polynomial ring on these eigenfunctions. If $\psi = \psi_1\psi_2$, then $Q_\psi = Q_{\psi_1}Q_{\psi_2}$. Thus, if ψ is not primitive, then the polynomial Q_ψ cannot be prime. An element

$$\psi = \Pi_{j=1}^k \psi_j^{c_j}$$

has c_j's uniquely determined, and hence the prime factorization

$$Q_\psi = \Pi_{j=1}^k Q_{\psi_j}^{c_j}.$$

Consider a multiplicity-free action of G on an algebra $\mathcal{W}$. In the general situation, we would like to associate this algebra $\mathcal{W}$ with a sub-algebra of $\mathcal{R}(G/U)$. With this goal in mind, we introduce the following notion:

Definition 7 *Let* $\mathcal{P} = \bigoplus_{\lambda \in \widehat{A}^+} \mathcal{P}_\lambda$ *denote an algebra graded by an abelian semigroup* $\widehat{A}^+$. *If* $\mathcal{W} \subseteq \mathcal{P}$ *is a subalgebra of* $\mathcal{P}$, *then we say that* $\mathcal{W}$ *is a* **total subalgebra of** $\mathcal{P}$ *if*

$$\mathcal{W} = \bigoplus_{\lambda \in Z} \mathcal{P}_\lambda$$

where Z *is a sub-semigroup of* $\widehat{A}^+$, *which we will denote by* $\widehat{A}^+(\mathcal{W}) = Z$. *Note that* $\mathcal{W}$ *is graded by* $\widehat{A}^+(\mathcal{W})$.

In what is to follow, we will usually have $\mathcal{P} = \mathcal{P}(V)$ (polynomial functions on a vector space V) and $\widehat{A}^+$ will denote the dominant chamber of the character group of a maximal torus A of a reductive group G acting on V. In this situation, we introduce an $\widehat{A}^+$-filtration on $\mathcal{W}$ as follows:

$$\mathcal{W}^{(\psi)} = \bigoplus_{\varphi \leq \psi} \mathcal{W}_\varphi \qquad (3.1)$$

where the ordering $\leq$ is the ordering on $\widehat{A}^+$ given by (see [54])

$$\psi_1 \leq \psi_2 \qquad \text{if } \psi_1^{-1}\psi_2 \text{ is expressible as a product of}$$
$$\text{rational powers of positive roots.}$$

Note that positive roots are weights of the adjoint representation of G on its Lie algebra $\mathfrak{g}$. We refer to the abelian group structure on the integral weights multiplicatively. Also, it will turn out that we only need positive *integer* powers of the positive roots.

Next consider the more specific situation where $\mathcal{W}$ which is a G-invariant and G-multiplicity-free subalgebra of a polynomial algebra $\mathcal{P}(V)$. Suppose that $\mathcal{W}^U$ has unique factorization. Then $\mathcal{W}^U$ is a polynomial ring and $\widehat{A}^+(\mathcal{W})$ is a free sub-semigroup in $\widehat{A}^+$ generated by the highest weights corresponding to the non-zero graded components of $\mathcal{W}$. Write the G decomposition as follows:

$$\mathcal{W} = \bigoplus_{\psi \in \widehat{A}^+(\mathcal{W})} \mathcal{W}_\psi$$

noting that $\mathcal{W}_\psi$ is an irreducible G module with highest weight ψ.

If δ occurs with positive multiplicity in the tensor product decomposition

$$\mathcal{W}_\varphi \otimes \mathcal{W}_\psi = \bigoplus_\delta \dim Hom_G(\mathcal{W}_\delta, \mathcal{W}_\varphi \otimes \mathcal{W}_\psi)\, \mathcal{W}_\delta,$$

then $\delta \leq \varphi\psi$. From 3.1 we can see that if

$$\mathcal{W}_\eta \subset \mathcal{W}^{(\varphi)} \quad \text{and} \quad \mathcal{W}_\gamma \subset \mathcal{W}^{(\psi)}, \quad \text{i.e., } \eta \leq \varphi \quad \text{and} \quad \gamma \leq \psi,$$

then it follows that

$$\mathcal{W}_\eta \cdot \mathcal{W}_\gamma \hookrightarrow \mathcal{W}_\eta \otimes \mathcal{W}_\gamma \subset \mathcal{W}^{(\eta\gamma)} \subset \mathcal{W}^{(\varphi\psi)}.$$

Thus

$$\mathcal{W}^{(\varphi)} \cdot \mathcal{W}^{(\psi)} \subset \mathcal{W}^{(\varphi\psi)}.$$

We have now an $\widehat{A}^+$-filtered algebra

$$\mathcal{W} = \bigcup_{\psi \in \widehat{A}^+(\mathcal{W})} \mathcal{W}^{(\psi)},$$

and this filtration is known as the *dominance filtration* [54].

With a filtered algebra, we can form its associated algebra which is $\widehat{A}^+$ graded:

$$\mathrm{Gr}_{\widehat{A}^+}\, \mathcal{W} = \bigoplus_{\psi \in \widehat{A}^+(\mathcal{W})} (\mathrm{Gr}_{\widehat{A}^+}\, \mathcal{W})^\psi$$

where

$$(\mathrm{Gr}_{\widehat{A}^+}\, \mathcal{W})^\psi = \mathcal{W}^{(\psi)} / \left(\bigoplus_{\varphi < \psi} \mathcal{W}^{(\varphi)} \right).$$

Theorem 3.1 *Consider a multiplicity-free G-module $\mathcal{W}$ with a $\widehat{A}^+$-filtered algebra structure such that $\mathcal{W}$ is a unique factorization domain. Assume that the zero degree subspace of $\mathcal{W}$ is $\mathbb{C}$. Then there is a canonical $\widehat{A}^+$-graded algebra injection:*

$$Gr_{\widehat{A}^+}\, \pi \; : \; Gr_{\widehat{A}^+}\, \mathcal{W} \hookrightarrow \mathcal{R}(G/U).$$

Note that this result immediately follows from a more general theorem (see Theorem 5 of [54] and the Appendix to [61]). Moreover, both the assumption on the zero degree subspace and the unique factorization can be removed. We thank the referee for providing both these observations and the references. Below is a proof cast in our present context.

Proof of Theorem 3.1. In [16] it is shown that under the above hypothesis, $\mathcal{W}^U$ is a polynomial ring on a canonical set of generators. Now, $\mathcal{W}^U$ is a $\widehat{A}^+$-graded algebra and therefore, there exists an injective $\widehat{A}^+$-graded algebra homomorphism obtained by sending each generator of the domain to a (indeed any) highest weight vector of the same weight in the codomain:

$$\alpha \; : \; \mathcal{W}^U \hookrightarrow \mathcal{R}(G/U)^U.$$

Note that $\mathcal{W}^U = \mathrm{Gr}_{\widehat{A}^+}(\mathcal{W}^U) = (\mathrm{Gr}_{\widehat{A}^+}\, \mathcal{W})^U$.

There exists a unique G-module homomorphism

$$\overline{\alpha} : \mathrm{Gr}_{\widehat{A}^+}\, \mathcal{W} \hookrightarrow \mathcal{R}(G/U)$$

such that the following diagram commutes:

$$
\begin{array}{ccc}
\alpha : & \mathcal{W}^U & \hookrightarrow & \mathcal{R}(G/U)^U \\
 & \cap & & \cap \\
\overline{\alpha} : & \mathrm{Gr}_{\widehat{A}^+}\, \mathcal{W} & \hookrightarrow & \mathcal{R}(G/U)
\end{array}
$$

We wish to show that $\overline{\alpha}$ is an algebra homomorphism, i.e.,

$$
\begin{array}{ccc}
(\mathrm{Gr}_{\widehat{A}^+}\mathcal{W})^\lambda \ \times\ (\mathrm{Gr}_{\widehat{A}^+}\mathcal{W})^\mu & \xrightarrow[m_\mathcal{W}]{} & (\mathrm{Gr}_{\widehat{A}^+}\mathcal{W})^{\lambda+\mu} \\[2mm]
\overline{\alpha}\downarrow \qquad\qquad \overline{\alpha}\downarrow & & \overline{\alpha}\downarrow \\[2mm]
\mathcal{R}(G/U)^\lambda \ \times\ \mathcal{R}(G/U)^\mu & \xrightarrow[m_{\mathcal{R}(G/U)}]{} & \mathcal{R}(G/U)^{\lambda+\mu}
\end{array}
$$

commutes.

We have two maps:

$$f_i : (\mathrm{Gr}_{\widehat{A}^+}\mathcal{W})^\lambda \otimes (\mathrm{Gr}_{\widehat{A}^+}\mathcal{W})^\mu \to \mathcal{R}(G/U)^{\lambda+\mu}, \qquad i=1,2,$$

defined by: $f_1(v \otimes w) = m_{\mathcal{R}(G/U)}(\overline{\alpha}(v) \otimes \overline{\alpha}(w))$ and $f_2(v \otimes w) = \overline{\alpha}(m_\mathcal{W}(v \otimes w))$.

Each of f_1 and f_2 is G-equivariant and,

$$\dim\left\{\beta \mid \beta : (\mathrm{Gr}_{\widehat{A}^+}\mathcal{W})^\lambda \otimes (\mathrm{Gr}_{\widehat{A}^+}\mathcal{W})^\mu \to \mathcal{R}(G/U)^{\lambda+\mu}\right\} = 1$$

because the Cartan product has multiplicity one in the tensor product of two irreducible G-modules V_λ and V_μ (this is a well known fact that is not difficult to prove, see for example [54]).

Therefore, there exists a constant C such that $f_1 = Cf_2$. We know that $\overline{\alpha}|_{\mathcal{W}^U} = \alpha$ is an algebra homomorphism. So for highest weight vectors $v^\lambda \in \mathcal{W}_\lambda^U$ and $w^\mu \in \mathcal{W}_\mu^U$:

$$
\begin{aligned}
f_1(v^\lambda \otimes w^\mu) &= \overline{\alpha}(v^\lambda)\overline{\alpha}(w^\mu) = \alpha(v^\lambda)\alpha(w^\mu) \\
&= \alpha(v^\lambda w^\mu) = \overline{\alpha}(v^\lambda w^\mu) = f_2(v^\lambda \otimes w^\mu).
\end{aligned}
$$

(Note that $v^\lambda w^\mu$ is a highest weight vector.) Note that $C = 1$. $\square$

3.3 Dual Pairs and Duality Correspondence

The theory of dual pairs will feature prominently in this article. For the treatment of all the branching algebras arising from classical symmetric pairs, we will need to understand dual pairs in three different settings. However, to minimize exposition, we restrict our discussion to just one of the pairs. Details in this section including the more general cases can be found in [11], [14] or [16].

In our context, the theory of dual pairs may be cast in a purely algebraic language. In this section, we will describe the dual pairs

$(O_n, \mathfrak{sp}_{2m})$. Let O_n to be the group of invertible $n \times n$ matrices, g such that $gJg^t = J$ where J is the $n \times n$ matrix:

$$J = \begin{pmatrix} 0 & \cdots & 0 & 1 \\ \vdots & \cdots & 1 & 0 \\ 0 & 1 & \cdots & \vdots \\ 1 & 0 & \cdots & 0 \end{pmatrix}.$$

Let $M_{n,m}$ be the vector space of $n \times m$ complex matrices, and consider the polynomial algebra $\mathcal{P}(M_{n,m})$ over $M_{n,m}$. The group $O_n \times GL_m$ acts on $\mathcal{P}(M_{n,m})$ by $(g, h) \cdot f(x) = f(g^t x h)$, where $g \in O_n$, $h \in GL_m$ and $x \in M_{n,m}$. The derived actions of their Lie algebras act on $\mathcal{P}(M_{n,m})$ by polynomial coefficient differential operators. Using the standard matrices entries as coordinates, we define the following differential operators:

$$\Delta_{ij} := \sum_{s=1}^{n} \frac{\partial^2}{\partial x_{si} \partial x_{n-s+1,j}}, \quad r_{ij}^2 := \sum_{s=1}^{n} x_{si} x_{n-s+1,j}, \text{ and}$$

$$E_{ij} := \sum_{s=1}^{n} x_{si} \frac{\partial}{\partial x_{sj}}.$$

We define three spaces:

$$\mathfrak{sp}_{2m}^{(1,1)} := \text{Span } \left\{ E_{ij} + \tfrac{n}{2} \delta_{i,j} \mid i,j = 1, \ldots, m \right\} \simeq \mathfrak{gl}_m,$$

$$\mathfrak{sp}_{2m}^{(2,0)} := \text{Span } \left\{ r_{ij}^2 \mid 1 \le i \le j \le m \right\}, \quad \text{and} \qquad (3.1)$$

$$\mathfrak{sp}_{2m}^{(0,2)} := \text{Span } \left\{ \Delta_{ij} \mid 1 \le i \le j \le m \right\}.$$

The direct sum, $\mathfrak{g} := \mathfrak{sp}_{2m}^{(2,0)} \oplus \mathfrak{sp}_{2m}^{(1,1)} \oplus \mathfrak{sp}_{2m}^{(0,2)}$, is preserved under the usual operator bracket and is isomorphic, as a Lie algebra, to the rank m complex symplectic Lie algebra, $\mathfrak{sp}_{2m}$. This presentation defines an action of $\mathfrak{sp}_{2m}$ on $\mathcal{P}(M_{n,m})$.

Let $\mathcal{S}^2 \mathbb{C}^m$ be the space of symmetric m by m matrices respectively. If V is a vector space, we denote the symmetric algebra on V by $\mathcal{S}(V)$. For a set S, we shall denote by $\mathbb{C}[S]$ by the algebra generated by elements in the set S.

Theorem 3.2 *Invariants and Harmonics of O_n*

(a) First Fundamental Theorem of Invariant Theory and Separation

of Variables. The invariants

$$\mathcal{J}_{n,m} := \mathcal{P}(M_{n,m})^{O_n} = \mathbb{C}[r_{ij}^2] \quad \left(\cong \mathcal{S}(\mathfrak{sp}_{2m}^{(2,0)}) \cong \mathcal{P}(\mathcal{S}^2\mathbb{C}^m) \ if \ n \geq m \right).$$

Further, if $n \geq 2m$, we have separation of variables

$$\mathcal{P}(M_{n,m}) \simeq \mathcal{H}_{n,m} \otimes \mathcal{J}_{n,m}.$$

where

$$\mathcal{H}_{n,m} = \{ f \in \mathcal{P}(M_{n,m}) \mid \Delta_{ij} f = 0 \ for \ all \ 1 \leq i \leq j \leq m \}$$

denotes the O_n-harmonics in $\mathcal{P}(M_{n,m})$.

(b) *Multiplicity-Free Decomposition under $O_n \times \mathfrak{sp}_{2m}$. We have the decomposition*

$$\mathcal{P}(M_{n,m}) \mid_{O_n \times \mathfrak{sp}_{2m}} = \bigoplus E_{(n)}^\lambda \otimes \widetilde{E}_{(2m)}^\lambda$$

where the sum is over all partitions λ with length at most $\min(n,m)$, and such that $(\lambda')_1 + (\lambda')_2 \leq n$. Furthermore, $\widetilde{E}_{(2m)}^\lambda$ is an irreducible (infinite dimensional) highest weight representation of $\mathfrak{sp}_{2m}$ such that as a representation of GL_m,

$$\widetilde{E}_{(2m)}^\lambda \quad = \mathcal{J}_{n,m} \cdot F_{(m)}^\lambda \qquad \qquad for \ any \ n, m \geq 0,$$
$$\cong \mathcal{S}(\mathcal{S}^2\mathbb{C}^m) \otimes F_{(m)}^\lambda \qquad provided \ n \geq 2m.$$

(c) *Multiplicity-Free Decomposition of Harmonics under $O_n \times \mathfrak{sp}_{2m}^{(1,1)}$. The O_n-harmonics $\mathcal{H}_{n,m}$ is invariant under the action of $O_n \times \mathfrak{sp}_{2m}^{(1,1)}$. Here $\mathfrak{sp}_{2m}^{(1,1)} \simeq \mathfrak{gl}_m$, and as an $O_n \times GL_m$ representation,*

$$\mathcal{P}(M_{n,m})/I(\mathcal{J}_{n,m}^+) \cong \mathcal{H}_{n,m} = \bigoplus E_{(n)}^\lambda \otimes F_{(m)}^\lambda,$$

where the sum is over all partitions λ with length at most $\min(n,m)$ and such that $(\lambda')_1 + (\lambda')_2 \leq n$. Here $I(\mathcal{J}_{n,m}^+)$ refers to the ideal generated by the positive degree O_n invariants in $\mathcal{P}(M_{n,m})$.

4 Reciprocity Algebras

In this paper, we study branching algebras using classical invariant theory. The formulation of classical invariant theory in terms of dual pairs [14] allows one to realize branching algebras for classical symmetric pairs as concrete algebras of polynomials on vector spaces. Furthermore, when realized in this way, the branching algebras have a double interpretation in which they solve two related branching problems simultaneously.

Classical invariant theory also provides a flexible means which allows an inductive approach to the computation of branching algebras, and makes evident natural connections between different branching algebras.

The easiest illustration of the above assertions is the realization of the tensor product algebra for GL_n presented as follows. This example also illustrates the definition of a reciprocity algebra.

4.1 Illustration: Tensor Product Algebra for GL_n

This first example is in [16], which we recall here as it is a model for the other (more involved) constructions of branching algebras as total subalgebras (see Definition 3.1) of GL_n tensor product algebras.

Consider the joint action of $GL_n \times GL_m$ on the $\mathcal{P}(M_{n,m})$ by the rule

$$(g, h) \cdot f(x) = f(g^t x h), \qquad \text{for } g \in GL_n, h \in GL_m, x \in M_{n,m}.$$

For the corresponding action on polynomials, one has the $GL_n \times GL_m$ multiplicity free decomposition (see [16])

$$\mathcal{P}(M_{n,m}) \simeq \bigoplus_\lambda F_{(n)}^\lambda \otimes F_{(m)}^\lambda, \qquad (4.1)$$

of the polynomials into irreducible $GL_n \times GL_m$ representations. Note that the sum is over non-negative partitions λ with length at most $\min(n, m)$.

Let $U_m = U_{GL_m}$ denote the upper triangular unipotent subgroup of GL_m. From decomposition (4.1), we can easily see that

$$\mathcal{P}(M_{n,m})^{U_m} \simeq \left(\bigoplus_\lambda F_{(n)}^\lambda \otimes F_{(m)}^\lambda \right)^{U_m} \simeq \bigoplus_\lambda F_{(n)}^\lambda \otimes (F_{(m)}^\lambda)^{U_m}. \quad (4.2)$$

Since the spaces $(F_{(m)}^\lambda)^{U_m}$ are one-dimensional, the sum in equation (4.2) consists of one copy of each $F_{(n)}^\lambda$. Just as in the discussion of §3.2, the algebra is graded by $\widehat{A}_m^+$, where A_m is the diagonal torus of GL_m, and one sees from (4.2) that the graded components are the $F_{(n)}^\lambda$.

By the arguments in §3.2, $\mathcal{P}(M_{n,m})^{U_m}$ can thus be associated to a graded subalgebra in $\mathcal{R}(GL_n/U_n)$, in particular, this is a total subalgebra as in Definition 3.1. To study tensor products of representations of GL_n, we can take the direct sum of $M_{n,m}$ and $M_{n,\ell}$. We then have an action of $GL_n \times GL_m \times GL_\ell$ on $\mathcal{P}(M_{n,m} \oplus M_{n,\ell})$. Since $\mathcal{P}(M_{n,m} \oplus M_{n,\ell}) \simeq \mathcal{P}(M_{n,m}) \otimes \mathcal{P}(M_{n,\ell})$, we may deduce from (4.1) that

$$\mathcal{P}(M_{n,m} \oplus M_{n,\ell})^{U_m \times U_\ell} \simeq \mathcal{P}(M_{n,m})^{U_m} \otimes \mathcal{P}(M_{n,\ell})^{U_\ell}$$

$$\simeq \bigoplus_{\mu,\nu}(F_{(n)}^{\mu} \otimes F_{(n)}^{\nu}) \otimes \left((F_{(m)}^{\mu})^{U_m} \otimes (F_{(\ell)}^{\nu})^{U_\ell}\right). \qquad (4.3)$$

Thus, this algebra is the sum of one copy of each tensor products $F_{(n)}^{\mu} \otimes F_{(n)}^{\nu}$. Hence, if we take the U_n-invariants, we will get a subalgebra of the tensor product algebra for GL_n. This results in the algebra

$$(\mathcal{P}(M_{n,m} \oplus M_{n,\ell})^{U_m \times U_\ell})^{U_n} \simeq \mathcal{P}(M_{n,m} \oplus M_{n,\ell})^{U_m \times U_\ell \times U_n}.$$

This shows that we can realize the tensor product algebra for GL_n, or more precisely, various total subalgebras of it, as algebras of polynomial functions on matrices, specifically as the algebras $\mathcal{P}(M_{n,m} \oplus M_{n,\ell})^{U_m \times U_\ell \times U_n}$.

However, the algebra $\mathcal{P}(M_{n,m} \oplus M_{n,\ell})^{U_m \times U_\ell \times U_n}$ has a second interpretation, as a different branching algebra. We note that $M_{n,m} \oplus M_{n,\ell} \simeq M_{n,m+\ell}$. On this space we have the action of $GL_n \times GL_{m+\ell}$, which is described by the obvious adaptation of equation (4.1). The action of $GL_n \times GL_m \times GL_\ell$ arises by restriction of the action of $GL_{m+\ell}$ to the subgroup $GL_m \times GL_\ell$ embedded block diagonally in $GL_{m+\ell}$. By (the obvious analog of) decomposition (4.2), we see that

$$\mathcal{P}(M_{n,m+\ell})^{U_n} \simeq \bigoplus_{\lambda}(F_{(n)}^{\lambda})^{U_n} \otimes F_{(m+\ell)}^{\lambda}.$$

This algebra embeds as a subalgebra of $\mathcal{R}(GL_{m+\ell}/U_{m+\ell})$, in particular, this is a total subalgebra as in Definition 3.1. If we then take the $U_m \times U_\ell$ invariants, we find that

$$(\mathcal{P}(M_{n,m+\ell})^{U_n})^{U_m \times U_\ell} \simeq \bigoplus_{\lambda}(F_{(n)}^{\lambda})^{U_n} \otimes (F_{(m+\ell)}^{\lambda})^{U_m \times U_\ell}$$

is (a total subalgebra of) the $(GL_{m+\ell}, GL_m \times GL_\ell)$ branching algebra. Thus, we have established the following result.

Theorem 4.1

(a) *The algebra* $\mathcal{P}(M_{n,m+\ell})^{U_n \times U_m \times U_\ell}$ *is isomorphic to a total subalgebra of the* $(GL_n \times GL_n, GL_n)$ *branching algebra (a.k.a. the* GL_n *tensor product algebra), and to a total subalgebra of the* $(GL_{m+\ell}, GL_m \times GL_\ell)$ *branching algebra.*

(b) *In particular, the dimension of the* $\psi^{\lambda} \times \psi^{\mu} \times \psi^{\nu}$ *homogeneous component for* $A_n \times A_m \times A_\ell$ *of* $\mathcal{P}(M_{n,m+\ell})^{U_n \times U_m \times U_\ell}$ *records simultaneously*

(i) *the multiplicity of* $F_{(n)}^{\lambda}$ *in the tensor product* $F_{(n)}^{\mu} \otimes F_{(n)}^{\nu}$, *and*

(ii) *the multiplicity of $F^\mu_{(m)} \otimes F^\nu_{(\ell)}$ in $F^\lambda_{(m+\ell)}$,*

for partitions μ, ν, λ such that $\ell(\mu) \leq \min(n,m)$, $\ell(\nu) \leq \min(n,\ell)$ and $\ell(\lambda) \leq \min(n, m+\ell)$.

Thus, we can not only realize the GL_n tensor product algebra concretely as an algebra of polynomials, we find that it appears simultaneously in two guises, the second being as the branching algebra for the pair $(GL_{m+\ell}, GL_m \times GL_\ell)$. We emphasize two features of this situation.

First, the pair $(GL_{m+\ell}, GL_m \times GL_\ell)$, as well as the pair $(GL_n \times GL_n, GL_n)$, is a symmetric pair. Hence, both the interpretations of $\mathcal{P}(M_{n,m+\ell})^{U_n \times U_m \times U_\ell}$ are as branching algebras for symmetric pairs.

Second, the relationship between the two situations is captured by the notion of "see-saw pair" of dual pairs [41]. Precisely, a context for understanding the decomposition law (4.1) is provided by observing that GL_n and GL_m (or more correctly, slight modifications of their Lie algebras) are mutual centralizers inside the Lie algebra $\mathfrak{sp}(M_{n,m})$ (of the metaplectic group) of polynomial coefficient differential operators of total degree two on $M_{n,m}$ [14] [16]. We say that they define a *dual pair* inside $\mathfrak{sp}(M_{n,m})$. The decomposition (4.1) then appears as the correspondence of representations associated to this dual pair [14]. Further, the pairs of groups $(GL_n, GL_{m+\ell}) = (G_1, G'_1)$ and $(GL_n \times GL_n, GL_m \times GL_\ell) = (G_2, G'_2)$ both define dual pairs inside the Lie algebra $\mathfrak{sp}(M_{n,m+\ell})$. We evidently have the relations

$$G_1 = GL_n \subset GL_n \times GL_n = G_2, \tag{4.4}$$

and (hence)

$$G'_1 = GL_{m+\ell} \supset GL_m \times GL_\ell = G'_2. \tag{4.5}$$

We refer to a pair of dual pairs related as in inclusions (4.4) and (4.5), a *see-saw pair* of dual pairs.

In these terms, we may think of the symmetric pairs (G_2, G_1) and (G'_1, G'_2) as a *"reciprocal pair"* of symmetric pairs. If we do so, we see that the algebra $\mathcal{P}(M_{n,m+\ell})^{U_n \times U_m \times U_\ell}$ is describable as $\mathcal{P}(M_{n,m+\ell})^{U_{G_1} \times U_{G'_2}}$ – it has a description in terms of the see-saw pair, and in this description the two pairs of the see-saw, or alternatively, the two reciprocal symmetric pairs, enter equivalently into the description of the algebra that describes the branching law for both symmetric pairs. For this reason, we also call this algebra, which describes the branching law for both symmetric pairs, the *reciprocity algebra* of the pair of pairs.

It turns out that any branching algebra associated to a classical symmetric pair, that is, a pair (G, H) in which G is a product of classical groups, has an interpretation as a reciprocity algebra – an algebra that describes a branching law for two reciprocal symmetric pairs simultaneously. Sometimes, however, one of the branching laws involves infinite-dimensional representations.

4.2 Symmetric Pairs and Reciprocity Pairs

In the context of dual pairs, we would like to understand the (G, H) branching of irreducible representations of G to H, for symmetric pairs (G, H). Table I lists the symmetric pairs which we will cover in this paper.

If G is a classical group over $\mathbb{C}$, then G can be embedded as one member of a dual pair in the symplectic group as described in [14]. The resulting pairs of groups are (GL_n, GL_m) or (O_n, Sp_{2m}), each inside Sp_{2nm}, and are called *irreducible* dual pairs. In general, a dual pair of reductive groups in Sp_{2r} is a product of such pairs.

Table I: Classical Symmetric Pairs

Description	G	H
Diagonal	$GL_n \times GL_n$	GL_n
Diagonal	$O_n \times O_n$	O_n
Diagonal	$Sp_{2n} \times Sp_{2n}$	Sp_{2n}
Direct Sum	GL_{n+m}	$GL_n \times GL_m$
Direct Sum	O_{n+m}	$O_n \times O_m$
Direct Sum	$Sp_{2(n+m)}$	$Sp_{2n} \times Sp_{2m}$
Polarization	O_{2n}	GL_n
Polarization	Sp_{2n}	GL_n
Bilinear Form	GL_n	O_n
Bilinear Form	GL_{2n}	Sp_{2n}

Proposition 4.2 *Let G be a classical group, or a product of two copies of a classical group. Let G belong to a dual pair (G, G') in a symplectic group Sp_{2m}. Let $H \subset G$ be a symmetric subgroup, and let H' be the centralizer of H in Sp_{2m}. Then (H, H') is also a dual pair in Sp_{2m}, and G' is a symmetric subgroup inside H'.*

Proof: This can be shown by fairly easy case-by-case checking. The basic reason that (H, H') form a dual pair is that, for any classical symmetric pair (G, H), the restriction of the standard module of G, or its dual, to H is a sum of standard modules of H, or their duals [14]. This is very easy to check on a case-by-case basis. The see-saw relationship of symmetric pairs organizes the 10 series of symmetric pairs as given in Table I into five pairs of series. These are shown in Table II.   □

Table II: Reciprocity Pairs

SymmetricPair($\mathbf{G}, \mathbf{H}$)	$(\mathbf{H}, \mathfrak{h}')$	$(\mathbf{G}, \mathfrak{g}')$
$(GL_n \times GL_n, GL_n)$	$(GL_n, \mathfrak{gl}_{m+\ell})$	$(GL_n \times GL_n, \mathfrak{gl}_m \oplus \mathfrak{gl}_l)$
$(O_n \times O_n, O_n)$	$(O_n, \mathfrak{sp}_{2(m+\ell)})$	$(O_n \times O_n, \mathfrak{sp}_{2m} \oplus \mathfrak{sp}_{2l})$
$(Sp_{2n} \times Sp_{2n}, Sp_{2n})$	$(Sp_{2n}, \mathfrak{so}_{2(m+\ell)})$	$(Sp_{2n} \times Sp_{2n}, \mathfrak{so}_{2m} \oplus \mathfrak{so}_{2\ell})$
$(GL_{n+m}, GL_n \times GL_m)$	$(GL_n \times GL_m, \mathfrak{gl}_\ell \oplus \mathfrak{gl}_\ell)$	$(GL_{n+m}, \mathfrak{gl}_\ell)$
$(O_{n+m}, O_n \times O_m)$	$(O_n \times O_m, \mathfrak{sp}_{2\ell} \oplus \mathfrak{sp}_{2\ell})$	$(O_{n+m}, \mathfrak{sp}_{2\ell})$
$(Sp_{2(n+m)}, Sp_{2n} \times Sp_{2m})$	$(Sp_{2n} \times Sp_{2m}, \mathfrak{so}_{2\ell} \oplus \mathfrak{so}_{2\ell})$	$(Sp_{2(n+m)}, \mathfrak{so}_{2\ell})$
(O_{2n}, GL_n)	$(GL_n, \mathfrak{gl}_{2m})$	$(O_{2n}, \mathfrak{sp}_{2m})$
(Sp_{2n}, GL_n)	$(GL_n, \mathfrak{gl}_{2m})$	$(Sp_{2n}, \mathfrak{so}_{2m})$
(GL_n, O_n)	$(O_n, \mathfrak{sp}_{2m})$	$(GL_n, \mathfrak{gl}_m)$
(GL_{2n}, Sp_{2n})	$(Sp_{2n}, \mathfrak{so}_{2m})$	$(GL_{2n}, \mathfrak{gl}_m)$

Remark: Table II also amounts to another point of view on the structure on which [15] is based.

As alluded to in §2, we need a more concrete realization of branching algebras. With this goal in mind, we shall introduce the general definition of a reciprocity algebra through the following sequence of steps:

Step 1 Consider a symmetric pair (G, H).

Step 2 Use the theory of dual pairs to construct a multiplicity free $G \times K$ variety $\mathcal{V}$, for a group K associated to a dual pair involving G. Analogues to the Theory of Spherical Harmonics (see Theorem 3.3) allow us to consider a dual pair $(G, \mathfrak{g}')$, which has the Lie algebra of K as $\mathfrak{g}'^{(1,1)}$ (similar to the space $\mathfrak{sp}_{2m}^{(1,1)}$ in (3.1)). We note that $\mathfrak{g}'$ forms a *family* of Lie algebras, and each choice of $\mathfrak{g}'$ determines the type of G irreducible representations involved.

Step 3 Consider the coordinate ring of $\mathcal{V}$, which we denote by $\mathbb{C}[\mathcal{V}]$. Note that $\mathbb{C}[\mathcal{V}]$ is either polynomial algebra or a quotient of a polynomial algebra, depending on which dual pair we are considering. If U_K is a maximal unipotent subgroup of K, then $\mathbb{C}[\mathcal{V}]^{U_K}$ is a *partial model* of G, in other words, a collection of irreducible representations of G appearing once and only once in $\mathbb{C}[\mathcal{V}]^{U_K}$.

Step 4 Taking U_H covariants, the algebra $\mathbb{C}[\mathcal{V}]^{U_K \times U_H}$ will be our candidate. We will abuse our terminology and still call it a "branching algebra". This is because $\mathbb{C}[\mathcal{V}]^{U_K \times U_H}$ sits in $\mathcal{R}(G/U_G)^{U_H}$ as a total subalgebra (see Definition 3.1). We hasten to add, as pointed out at the end of Step 2, that we have a family of total subalgebras in $\mathcal{R}(G/U_G)^{U_H}$. Further, each total subalgebra relates two branching phenomena, and thus we call it a reciprocity algebra.

In the following table we provide the ingredients for the special cases as well as the stability range for the classical branching formula involving Littlewood-Richardson coefficients (see [21]).

Table III: Stability Range

Sym. Pair, (G,H)	K	Rep. of $G \times K$	Stability Range
(GL_k, O_k)	$GL_p \times GL_q$	$M_{k,p} \oplus M_{k,q}^*$	$k \geq 2(p+q)$
(GL_{2k}, Sp_{2k})	$GL_p \times GL_q$	$M_{2k,p} \oplus M_{2k,q}^*$	$k \geq p+q$
(O_{2k}, GL_k)	GL_{2n}	$M_{2k,2n}$	$k \geq 2n$
(Sp_{2k}, GL_k)	GL_{2n}	$M_{2k,2n}$	$k \geq 2n$
$(GL_{k+\ell}, GL_k \times GL_\ell)$	$GL_p \times GL_q$	$M_{k+\ell,p} \oplus M_{k+\ell,q}^*$	$\min(k,l) \geq p+q$
$(O_{k+\ell}, O_k \times O_\ell)$	GL_n	$M_{k+\ell,n}$	$\min(k,l) \geq 2n$
$(Sp_{2k+2\ell}, Sp_{2k} \times Sp_{2\ell})$	GL_n	$M_{2(k+\ell),n}$	$\min(k,l) \geq n$
$(GL_k \times GL_k, GL_k)$	$GL_p \times GL_q \times$ $GL_r \times GL_s$	$M_{k,p} \oplus M_{k,q} \oplus$ $M_{k,r}^* \oplus M_{k,s}^*$	$k \geq p+q+r+s$
$(O_k \times O_k, O_k)$	$GL_n \times GL_m$	$M_{k,n+m}$	$k \geq 2(n+m)$
$(Sp_{2k} \times Sp_{2k}, Sp_{2k})$	$GL_n \times GL_m$	$M_{2k,n+m}$	$k \geq (n+m)$

With the above general construction in mind, we begin with two of the reciprocity algebras in the next two sections. We have chosen the examples more to illustrate the subtleties and the general framework.

5 Branching from GL_n to O_n

Consider the problem of restricting irreducible representations of GL_n to the orthogonal group O_n. We consider the symmetric see-saw pair (GL_n, O_n) and (Sp_{2m}, GL_m). As in the discussion of §4.1, we can realize (a total subalgebra of) the coordinate ring of the flag manifold GL_n/U_n as the algebra of U_m-invariants on $\mathcal{P}(M_{n,m})$. If we then look at the U_{O_n}-invariants in this algebra, then we will have (a certain total subalgebra of) the (GL_n, O_n) branching algebra. Thus, we are interested in the algebra

$$\mathcal{P}(M_{n,m})^{U_{O_n} \times U_m}.$$

We note that, in analogy with the situation of §4.1, this is the algebra of invariants for the unipotent subgroups of the smaller member of each symmetric pair.

Let us investigate what this algebra appears to be if we first take invariants with respect to U_{O_n}. We have a decomposition of $\mathcal{P}(M_{n,m})$ as a joint $O_n \times \mathfrak{sp}_{2m}$-module (see Theorem 3.3(b)):

$$\mathcal{P}(M_{n,m}) \simeq \bigoplus_\mu E^\mu_{(n)} \otimes \widetilde{E}^\mu_{(2m)}. \tag{5.1}$$

Recall that the sum runs through the set of all non-negative integer partitions μ such that $\ell(\mu) \leq \min(n, m)$ and $(\mu')_1 + (\mu')_2 \leq n$. Here $E^\mu_{(n)}$ denotes the irreducible O_n representation parameterized by μ. Recall from (4.1), the multiplicity free $GL_n \times GL_m$ decomposition $\mathcal{P}(M_{n,m}) \simeq \bigoplus_\mu F^\mu_{(n)} \otimes F^\mu_{(m)}$. The module $E^\mu_{(n)}$ is generated by the GL_n highest weight vector in $F^\mu_{(n)}$. Further, $\widetilde{E}^\mu_{(2m)}$ is an irreducible infinite-dimensional representation of $\mathfrak{sp}_{2m}$ with lowest $\mathfrak{gl}_m$-type $F^\mu_{(m)}$.

Theorem 5.1 *Assume $n > 2m$.*

(a) *The algebra $\mathcal{P}(M_{n,m})^{U_{O_n} \times U_m}$ is isomorphic to a total subalgebra of the (GL_n, O_n) branching algebra, and to a total subalgebra of the $(\mathfrak{sp}_{2m}, GL_m)$ branching algebra.*

(b) *In particular, the dimension of the $\varphi^\mu \times \psi^\lambda$ homogeneous component for $A_{O_n} \times A_m$ of $\mathcal{P}(M_{n,m})^{U_{O_n} \times U_m}$ records simultaneously*

  (i) *the multiplicity of $E^\mu_{(n)}$ in the representation $F^\lambda_{(n)}$, and*

  (ii) *the multiplicity of $F^\lambda_{(m)}$ in $\widetilde{E}^\mu_{(2m)}$.*

  for partitions μ, λ such that $\ell(\mu) \leq m$, and $\ell(\lambda) \leq m$.

Proof. Taking the U_{O_n}-invariants for the decomposition (5.1), we find that

$$\mathcal{P}(M_{n,m})^{U_{O_n}} \simeq \bigoplus_{\mu} (E^\mu_{(n)})^{U_{O_n}} \otimes \widetilde{E}^\mu_{(2m)}, \qquad (5.2)$$

where the sum is over partitions μ such that $\ell(\mu) \leq \min(n,m)$ and $(\mu')_1 + (\mu')_2 \leq n$. Note that the stability condition $n > 2m$ guarantees the latter inequality. The space $(E^\mu_{(n)})^{U_{O_n}}$ is the space of highest weight vectors for $(E^\mu_{(n)})^{U_{O_n}}$. We would like to say that it is one-dimensional, so that $\mathcal{P}(M_{n,m})^{U_{O_n}}$ would consist of one copy of each of the irreducible representations $\widetilde{E}^\mu_{(2m)}$. But, owing to the disconnectedness of O_n, this is not quite true, and when it is true, the highest weight may not completely determine $E^\mu_{(n)}$.

However, if $n > 2m$, then $(E^\mu_{(n)})^{U_{O_n}}$ is one-dimensional, and does single out $E^\mu_{(n)}$ among the representations which appear in the sum (5.1). Hence, let us make this restriction for the present discussion. Taking the U_m invariants in the sum (5.2), we find that

$$(\mathcal{P}(M_{n,m})^{U_{O_n}})^{U_m} \simeq \bigoplus_{\mu} (E^\mu_{(n)})^{U_{O_n}} \otimes (\widetilde{E}^\mu_{(2m)})^{U_m}. \qquad (5.3)$$

Note that the sum is over all partitions μ such that $\ell(\mu) \leq m$ (since $n > 2m$). The space $(\widetilde{E}^\mu_{(2m)})^{U_m}$ describes how the representation $\widetilde{E}^\mu_{(2m)}$ of $\mathfrak{sp}_{2m}$ decomposes as a $\mathfrak{gl}_m$ module, or equivalently, as a GL_m-module. In other words, $(\widetilde{E}^\mu_{(2m)})^{U_m}$ describes the branching rule from $\mathfrak{sp}_{2m}$ to $\mathfrak{gl}_m$ for the module $\widetilde{E}^\mu_{(2m)}$.

We know (thanks to our restriction to $n > 2m$) that the space $(E^\mu_{(n)})^{U_{O_n}}$ is one-dimensional. Let φ^μ be the A_{O_n} weight of $(E^\mu_{(n)})^{U_{O_n}}$. Thus, φ^μ is the restriction to the diagonal maximal torus A_{O_n} of the character ψ^μ of the group A_n of diagonal $n \times n$ matrices. Our assumption further implies that φ^μ determines $E^\mu_{(n)}$. Therefore, for a given dominant A_m weight ψ^λ, corresponding to the partition λ, where $\ell(\lambda) \leq m$, the ψ^λ-eigenspace in $(\widetilde{E}^\mu_{(2m)})^{U_m}$ tells us the multiplicity of $F^\lambda_{(m)}$ in the restriction of $\widetilde{E}^\mu_{(2m)}$ to $\mathfrak{gl}_m$. This is the same as the dimension of the joint $(\varphi^\mu \times \psi^\lambda)$ eigenspace in

$$(\mathcal{P}(M_{n,m})^{U_{O_n}})^{U_m} \simeq \mathcal{P}(M_{n,m})^{U_{O_n} \times U_m} \simeq (\mathcal{P}(M_{n,m})^{U_m})^{U_{O_n}}.$$

But we have already seen that this eigenspace describes the multiplicity of $E^\mu_{(n)}$ in $F^\lambda_{(n)}$. Thus, again the $A_{O_n} \times A_m$ homogeneous components of $\mathcal{P}(M_{n,m})^{U_{O_n} \times U_m}$ have a simultaneous interpretation, one for a branching law associated to each of the two symmetric pairs composing the symmetric see-saw pair.   □

In this case, one of the branching laws involves infinite-dimensional representations. However, they are highest weight representations, which are the most tractable of infinite-dimensional representations, from an algebraic point of view.

6 Tensor Product Algebra for O_n

Using the symmetric see-saw pair

$$\left((O_n \times O_n, O_n), (Sp_{2(m+\ell)}, Sp_{2m} \times Sp_{2\ell})\right),$$

we can construct (total subalgebras of) the tensor product algebra for O_n. To prepare for this, we should explicate the decomposition (5.1) further.

Let us recall the basic setup as in §3.3. Recall that $\mathcal{J}_{n,m} = \mathcal{P}(M_{n,m})^{O_n}$ is the algebra of O_n-invariant polynomials. Theorem 3.3(a) implies that $\mathcal{J}_{n,m}$ is a quotient of $\mathcal{S}(\mathfrak{sp}_{2m}^{(2,0)})$, the symmetric algebra on $\mathfrak{sp}_{2m}^{(2,0)}$.

The natural mapping

$$\mathcal{H}_{n,m} \to \mathcal{P}(M_{n,m})/I(\mathcal{J}_{n,m}^+)$$

is a linear $O_n \times GL_m$-module isomorphism. Further, the $O_n \times GL_m$ structure of $\mathcal{H}_{n,m}$ is as follows (see Theorem 3.3(c)):

$$\mathcal{H}_{n,m} \simeq \bigoplus_\mu E_{(n)}^\mu \otimes F_{(m)}^\mu.$$

Here μ ranges over the same diagrams as in (5.1).

From Theorem 3.3(b),

$$\widetilde{E}_{(2m)}^\mu \simeq F_{(m)}^\mu \cdot \mathcal{J}_{n,m} \simeq \mathcal{S}(\mathfrak{sp}_{2m}^{(2,0)}) \cdot F_{(m)}^\mu, \qquad (6.1)$$

and it follows that

$$\widetilde{E}_{(2m)}^\mu / (\mathfrak{sp}_{2m}^{(2,0)} \cdot \widetilde{E}_{(2m)}^\mu) \simeq F_{(m)}^\mu.$$

In other words, we can detect the $\mathfrak{sp}_{2m}$ isomorphism class of the module $\widetilde{E}_{(2m)}^\mu$ by the GL_m isomorphism class of the quotient $\widetilde{E}_{(2m)}^\mu / (\mathfrak{sp}_{2m}^{(2,0)} \cdot \widetilde{E}_{(2m)}^\mu)$. Also, if $W \subset \mathcal{P}(M_{n,m})$ is any $\mathfrak{sp}_{2m}$-invariant subspace, then

$$W/(\mathfrak{sp}_{2m}^{(2,0)} \cdot W) \simeq W \cap \mathcal{H}_{n,m},$$

and this subspace also reveals the $\mathfrak{sp}_{2m}$ isomorphism type of W.

We can use the above to find a total subalgebra of the tensor product algebra of O_n. One consequence of the above discussion is that

$$\left(\mathcal{P}(M_{n,m})/I(\mathcal{J}_{n,m}^+)\right)^{U_m} \simeq \bigoplus_\mu E_{(n)}^\mu \otimes (F_{(m)}^\mu)^{U_m}$$

consists of one copy of each irreducible representation $E_{(n)}^\mu$.

If we repeat the above discussion for $M_{n,\ell}$, and combine the results, we find that

$$\left(\mathcal{P}(M_{n,m})/I(\mathcal{J}_{n,m}^+)\right)^{U_m} \otimes \left(\mathcal{P}(M_{n,\ell})/I(\mathcal{J}_{n,\ell}^+)\right)^{U_\ell}$$

$$\simeq \bigoplus_{\mu,\nu} \left(E_{(n)}^\mu \otimes E_{(n)}^\nu\right) \otimes \left((F_{(m)}^\mu)^{U_m} \otimes (F_{(\ell)}^\nu)^{U_\ell}\right) \qquad (6.2)$$

is a direct sum of one copy of each possible tensor product of an $E_{(n)}^\mu$ with an $E_{(n)}^\nu$. At this point, we make the assumption that $n > 2(m+\ell)$, as in this range the O_n constituents of decomposition are irreducible when restricted to the connected component of the identity in O_n. If we now take the U_{O_n}-invariants in equation (6.2), we will have (a total subalgebra of) the tensor product algebra of O_n:

$$\left(\left(\mathcal{P}(M_{n,m})/I(\mathcal{J}_{n,m}^+)\right)^{U_m} \otimes \left(\mathcal{P}(M_{n,\ell})/I(\mathcal{J}_{n,\ell}^+)\right)^{U_\ell}\right)^{U_{O_n}}$$

$$\simeq \bigoplus_{\mu,\nu} \left(E_{(n)}^\mu \otimes E_{(n)}^\nu\right)^{U_{O_n}} \otimes \left((F_{(m)}^\mu)^{U_m} \otimes (F_{(\ell)}^\nu)^{U_\ell}\right).$$

We can describe this algebra in another way. Begin with the observation that $\mathcal{P}(M_{n,m}) \otimes \mathcal{P}(M_{n,\ell}) \simeq \mathcal{P}(M_{n,m+\ell})$, and

$$\mathcal{P}(M_{n,m})/I(\mathcal{J}_{n,m}^+) \otimes \mathcal{P}(M_{n,\ell})/I(\mathcal{J}_{n,\ell}^+) \simeq \mathcal{P}(M_{n,m+\ell})/I(\mathcal{J}_{n,m}^+ \oplus \mathcal{J}_{n,\ell}^+).$$

Thus

$$\left(\mathcal{P}(M_{n,m})/I(\mathcal{J}_{n,m}^+)\right)^{U_m} \otimes \left(\mathcal{P}(M_{n,\ell})/I(\mathcal{J}_{n,\ell}^+)\right)^{U_\ell}$$

$$\simeq \left(\mathcal{P}(M_{n,m+\ell})/I(\mathcal{J}_{n,m}^+ \oplus \mathcal{J}_{n,\ell}^+)\right)^{U_m \times U_\ell},$$

and taking U_{O_n} invariants of the above, we get

$$\left(\left(\mathcal{P}(M_{n,m})/I(\mathcal{J}_{n,m}^+)\right)^{U_m} \otimes \left(\mathcal{P}(M_{n,\ell})/I(\mathcal{J}_{n,\ell}^+)\right)^{U_\ell}\right)^{U_{O_n}}$$

$$\simeq \left(\left(\mathcal{P}(M_{n,m+\ell})/I(\mathcal{J}_{n,m}^+ \oplus \mathcal{J}_{n,\ell}^+)\right)^{U_m \times U_\ell}\right)^{U_{O_n}}$$

$$\simeq \left(\left(\mathcal{P}(M_{n,m+\ell})/I(\mathcal{J}_{n,m}^+ \oplus \mathcal{J}_{n,\ell}^+)\right)^{U_{O_n}}\right)^{U_m \times U_\ell}.$$

Theorem 6.1 *Given positive integers n, m and ℓ with $n > 2(m + \ell)$ we have:*

(a) *The algebra*

$$\left((\mathcal{P}(M_{n,m})/I(\mathcal{J}_{n,m}^+))^{U_m} \otimes (\mathcal{P}(M_{n,\ell})/I(\mathcal{J}_{n,\ell}^+))^{U_\ell} \right)^{U_{O_n}}$$

is isomorphic to a total subalgebra of the $(O_n \times O_n, O_n)$ branching algebra (a.k.a. the O_n tensor product algebra), and to a total subalgebra of the $(\mathfrak{sp}_{2(m+\ell)}, \mathfrak{sp}_{2m} \oplus \mathfrak{sp}_{2\ell})$ branching algebra.

(b) *Specifically, the dimension of the $(\varphi^\lambda \times \psi^\mu \times \psi^\nu)$-eigenspace for $A_{O_n} \times A_m \times A_\ell$ of $\left((\mathcal{P}(M_{n,m+\ell})/I(\mathcal{J}_{n,m}^+ \oplus \mathcal{J}_{n,\ell}^+))^{U_{O_n}} \right)^{U_m \times U_\ell}$ records simultaneously*

(i) *the multiplicity of $E_{(n)}^\lambda$ in $E_{(n)}^\mu \otimes E_{(n)}^\nu$, as well as*

(ii) *the multiplicity of $\widetilde{E}_{(2m)}^\mu \otimes \widetilde{E}_{(2\ell)}^\nu$ in the restriction of $\widetilde{E}_{(2(m+\ell))}^\lambda$.*

Here the partitions μ, ν, λ satisfy the following conditions: $\ell(\mu) \leq \min(n,m)$, $\ell(\nu) \leq \min(n,\ell)$, and $\ell(\lambda) \leq \min(n, m + \ell)$.

Proof. Let us now compute the ring expressed in this way. From Theorem 3.3(b), we know that

$$\mathcal{P}(M_{n,m})^{U_{O_n}} \simeq \left(\bigoplus_\mu E_{(n)}^\mu \otimes \widetilde{E}_{(2m)}^\mu \right)^{U_{O_n}} \simeq \bigoplus_\mu (E_{(n)}^\mu)^{U_{O_n}} \otimes \widetilde{E}_{(2m)}^\mu.$$

Note that within the range $n > 2(m + \ell)$ we have $\dim(E_{(n)}^\mu)^{U_{O_n}} = 1$ since the O_n-representations $E_{(n)}^\mu$ remain irreducible when restricted to SO_n.

Now repeat this with m replaced by $m + \ell$:

$$\mathcal{P}(M_{n,m+\ell})^{U_{O_n}} \simeq \left(\bigoplus_\mu E_{(n)}^\mu \otimes \widetilde{E}_{(2(m+\ell))}^\mu \right)^{U_{O_n}}$$

$$\simeq \bigoplus_\mu (E_{(n)}^\mu)^{U_{O_n}} \otimes \widetilde{E}_{(2(m+\ell))}^\mu.$$

Hence

$$(\mathcal{P}(M_{n,m+\ell})/I(\mathcal{J}_{n,m}^+ \oplus \mathcal{J}_{n,\ell}^+))^{U_{O_n}}$$

$$\simeq \left(\left(\bigoplus_\lambda E_{(n)}^\lambda \otimes \widetilde{E}_{(2(m+\ell))}^\lambda \right) /I(\mathcal{J}_{n,m}^+ \oplus \mathcal{J}_{n,\ell}^+) \right)^{U_{O_n}}$$

$$\simeq \bigoplus_\lambda (E_{(n)}^\lambda)^{U_{O_n}} \otimes \left(\widetilde{E}_{(2(m+\ell))}^\lambda /(\mathfrak{sp}_{2m}^{(2,0)} \oplus \mathfrak{sp}_{2\ell}^{(2,0)}) \cdot \widetilde{E}_{(2(m+\ell))}^\lambda \right).$$

From this we finally get

$$\left((\mathcal{P}(M_{n,m+\ell})/I(\mathcal{J}_{n,m}^+ \oplus \mathcal{J}_{n,\ell}^+))^{U_{O_n}} \right)^{U_m \times U_\ell}$$

$$\simeq \bigoplus_\lambda (E_{(n)}^\lambda)^{U_{O_n}} \otimes \left(\widetilde{E}_{(2(m+\ell))}^\lambda / (\mathfrak{sp}_{2m}^{(2,0)} \oplus \mathfrak{sp}_{2\ell}^{(2,0)}) \cdot \widetilde{E}_{(2(m+\ell))}^\lambda \right)^{U_m \times U_\ell}.$$

From the discussion following equation (6.1), we see that the factor

$$\left(\widetilde{E}_{(2(m+\ell))}^\lambda / (\mathfrak{sp}_{2m}^{(2,0)} \oplus \mathfrak{sp}_{2\ell}^{(2,0)}) \cdot \widetilde{E}_{(2(m+\ell))}^\lambda \right)^{U_m \times U_\ell}$$

tells us the $\mathfrak{sp}_{2m} \oplus \mathfrak{sp}_{2\ell}$ decomposition of $\widetilde{E}_{(2(m+\ell))}^\lambda$. $\square$

Hence, again the algebra has a double interpretation, one in terms of decomposing tensor products of O_n representations, and one in terms of branching from $\mathfrak{sp}_{2(m+\ell)}$ to $\mathfrak{sp}_{2m} \oplus \mathfrak{sp}_{2\ell}$ (although the second branching law involves infinite-dimensional representations).

7 The Stable Range and Relations Between Reciprocity Algebras

Let us summarize our discussions this far. Given any classical symmetric pair, we can embed it in a (family of) see-saw symmetric pair(s). Doing this, we find that (a total subalgebra of) the branching algebra for the pair can equally well be interpreted as the branching algebra for a dual family of representations of the dual symmetric pair. The representations of the dual symmetric pair will frequently be infinite dimensional, but they are always highest weight modules.

An immediate consequence of this isomorphism of algebras is the isomorphisms of intertwining spaces and hence equality of multiplicities, which we have collectively described as *reciprocity laws*. These reciprocity laws are of the same nature as Frobenius Reciprocity for induced representations of groups.

From §4.2, we see that the see-saw symmetric pairs actually come in two parameter families. If one of the pairs involves many more variables than the other, then certain features of the discussions above become simpler.

Take the results of Theorem 4.1 as an illustration: Let n, m and ℓ denote positive integers. Now suppose that λ, μ and ν are partitions such that the length of λ (resp. μ, resp. ν) is at most $\min(n, m + \ell)$ (resp. $\min(n, m)$, resp. $\min(n, \ell)$). Then the Littlewood-Richardson coefficient

$$c_{\mu\nu}^\lambda = \dim \operatorname{Hom}_{GL_n}(F_{(n)}^\lambda, F_{(n)}^\mu \otimes F_{(n)}^\nu)$$

while at the same time,

$$c_{\mu\nu}^\lambda = \dim \operatorname{Hom}_{GL_m \times GL_\ell}(F_{(m)}^\mu \otimes F_{(\ell)}^\nu, F_{(m+\ell)}^\lambda).$$

Thus, for fixed λ, μ and ν, we have two distinct interpretations of the Littlewood-Richardson coefficients for sufficiently large n, m and ℓ.

Consider another example: branching from GL_n to O_n. If we let these groups act on $\mathcal{P}(M_{n,m})$, we get the see-saw pairs $(O_n, \mathfrak{sp}_{2m})$ and (GL_n, GL_m). The branching coefficients d_λ^μ from GL_n to O_n can be described as follows:

$$F_{(n)}^\lambda \mid_{O_n} = \sum_\mu d_\lambda^\mu \, E_{(n)}^\mu$$

where

$$\begin{aligned}
d_\lambda^\mu &= \dim \operatorname{Hom}_{O_n}(E_{(n)}^\mu, F_{(n)}^\lambda) \\
&= \dim \operatorname{Hom}_{GL_m}\left(F_{(m)}^\lambda, F_{(m)}^\mu \otimes \mathcal{S}(\mathfrak{sp}_{2m}^{(2,0)})\right) \\
&= \dim \operatorname{Hom}_{GL_m}\left(F_{(m)}^\lambda, F_{(m)}^\mu \otimes \mathcal{S}(S^2\mathbb{C}^m)\right)
\end{aligned}$$

is independent of n, if $n \geq m$, and only depends on the diagrams λ and μ. This allows one to create a theory of "stable characters" for O_n. Similar considerations apply to GL_n and Sp_{2n} and this idea has been actively pursued by [39], amongst others.

These are all instances of stability laws. The well-known one-step branching from GL_n to GL_{n-1} is another instance. More precisely, the dominant weights of GL_n (resp. GL_{n-1}) may be indexed by partitions as in Section 3.1. Relative to this parametrization, the only requirement is that n be sufficiently large when compared to the lengths of the partitions. In other words, a single partition indexes a highest weight of a representation of GL_n for all sufficiently large n. Thus, this branching can be described entirely by diagrams, with no mention of the size n, if n is large. Iteration of this branching also shows that when n is large, the weight multiplicities of dominant weights of an irreducible GL_n representation are independent of n, in a similar sense. See [1] for the other classical groups, which don't share this stability property.

In the last two sections that follow, we will illustrate the simplifications that occur in the stable range, highlighting certain specific see-saw pairs. In all these cases, we show that the branching algebras associated to symmetric pairs can all be described by use of suitable branching algebras associated to the general linear groups. Thus, if we can have control of the solution in the general linear group case, we will have some control of the other classical groups. The other non-trivial examples will be important extensions of this work, and we will see them in further papers, for example, [21], [22], [17], [18] and [19].

8 Stability for Branching from GL_n to O_n

We begin with a detailed discussion of the case of

$$(GL_n, O_n) \quad \text{and} \quad (\mathfrak{sp}_{2m}, GL_m).$$

Here we have already encountered the stable range, without the name. It is when $n > 2m$. Several things happen in the stable range:

(a) The representations $E^\mu_{(n)}$ of the orthogonal group remain irreducible when restricted to the special orthogonal group SO_n, and furthermore, no two of them are equivalent.

(b) Recall the algebra $\mathcal{J}_{n,m}$ of O_n-invariant polynomials on $M_{n,m}$ generated by the quadratic invariants, which is the abelian subalgebra $\mathfrak{sp}_{2m}^{(2,0)}$ of $\mathfrak{sp}_{2m}$. In the stable range (in fact it holds true whenever $n \geq m$), the natural surjective homomorphism

$$\mathcal{S}(\mathfrak{sp}_{2m}^{(2,0)}) \to \mathcal{J}_{n,m}$$

is an isomorphism. See Theorem 3.3(a).

(c) In the stable range, the multiplication map

$$\mathcal{H}_{n,m} \otimes \mathcal{J}_{n,m} \simeq \mathcal{H}_{n,m} \otimes \mathcal{S}(\mathfrak{sp}_{2m}^{(2,0)}) \to \mathcal{P}(M_{n,m})$$

is also an isomorphism of $O_n \times GL_m$-modules. See Theorem 3.3(a) and 3.3(b).

Of course, the subspace $\mathcal{H}_{n,m}$ of harmonic polynomials is not an algebra – it is not closed under multiplication. This is quite clear, since $\mathcal{H}_{n,m}$ contains all the linear functions, which generate the whole polynomial ring. However, to form the reciprocity algebra associated to the symmetric see-saw pairs (GL_n, O_n) and $(\mathfrak{sp}_{2m}, GL_m)$, we need to take the U_{O_n}-invariants. Thus, our reciprocity algebra is a subalgebra of

$$(\mathcal{H}_{n,m} \otimes \mathcal{S}(\mathfrak{sp}_{2m}^{(2,0)}))^{U_{O_n}} = \mathcal{H}_{n,m}^{U_{O_n}} \otimes \mathcal{S}(\mathfrak{sp}_{2m}^{(2,0)})$$

$$\simeq \left(\bigoplus_\mu (E^\mu_{(n)})^{U_{O_n}} \otimes F^\mu_{(m)} \right) \otimes \mathcal{S}(\mathfrak{sp}_{2m}^{(2,0)}). \quad (8.1)$$

Theorem 8.1 *When $n > 2m$, the space $\mathcal{H}_{n,m}^{U_{O_n}}$ is a subalgebra of $\mathcal{P}(M_{n,m})$. Hence, the algebra $\mathcal{P}(M_{n,m})^{U_{O_n}}$ is isomorphic to a tensor product*

$$\mathcal{P}(M_{n,m})^{U_{O_n}} \simeq \mathcal{H}_{n,m}^{U_{O_n}} \otimes \mathcal{S}(\mathfrak{sp}_{2m}^{(2,0)})$$

of the algebras $\mathcal{H}_{n,m}^{U_{O_n}}$ and $\mathcal{S}(\mathfrak{sp}_{2m}^{(2,0)})$. Furthermore, the algebra $\mathcal{H}_{n,m}^{U_{O_n}}$ is isomorphic (as a representation) to the subalgebra $\mathcal{R}^+(GL_m/U_m)$ of $\mathcal{R}(GL_m/U_m)$ defined by the polynomial representations.

Proof. Note that $\mathcal{H}_{n,m}^{U_{O_n}}$ can be identified with a subalgebra $\mathcal{R}^+(GL_m/U_m)$ of $\mathcal{R}(GL_m/U_m)$ defined by the polynomial representations, from our discussion in §3.2. Consider the space of polynomials belonging to the sum in the last expression of equation (8.1). Let $\{x_{jk} \mid j = 1,\dots,n, k = 1,\dots,m\}$ be the standard matrix entries on $M_{n,m}$. In order to make the unipotent group U_{O_n} of O_n maximally compatible with (in fact, contained in) the unipotent subgroup U_n of GL_n, we choose the inner product on $\mathbb{C}^n$ as in Section 3.3. By this choice, joint $O_n \times GL_m$ harmonic highest weight vectors are monomials in the determinants

$$\delta_j = \det \begin{bmatrix} x_{11} & x_{12} & \cdots & x_{1j} \\ x_{21} & x_{22} & \cdots & x_{1j} \\ \vdots & \vdots & \vdots & \vdots \\ x_{j1} & x_{j2} & \cdots & x_{jj} \end{bmatrix} \qquad \text{for } j = 1,\dots,m.$$

From this, we can see that the space $\sum_{\mu}(E_{(n)}^{\mu})^{U_{O_n}} \otimes F_{(m)}^{\mu}$ is spanned by the monomials in the determinants

$$\det \begin{bmatrix} x_{1,b_1} & x_{1,b_2} & \cdots & x_{1,b_j} \\ x_{2,b_1} & x_{2,b_2} & \cdots & x_{2,b_j} \\ \vdots & \vdots & \vdots & \vdots \\ x_{j,b_1} & x_{j,b_2} & \cdots & x_{j,b_j} \end{bmatrix} \qquad (8.2)$$

as $\{b_1, b_2, b_3, \dots, b_j\}$ ranges over all j-tuples of integers from 1 to m. Indeed, the span of such monomials is clearly invariant under $\mathfrak{gl}_m$, and consists of highest weight vectors for O_n. Finally, we see that these monomials will all be harmonic, because the partial Laplacians spanning $\mathfrak{sp}_{2m}^{(0,2)}$ have the form

$$\Delta_{ab} = \sum_{j=1}^{n} \frac{\partial^2}{\partial x_{j,a} \partial x_{n+1-j,b}}.$$

Since every term of Δ_{ab} involves differentiating with respect to a variable x_{jk} with $j > n/2$, and the determinants (8.2) do not depend on these variables, we see that they will be annihilated by the Δ_{ab}, which means that they are harmonic. This shows that $\mathcal{H}_{n,m}^{U_{O_n}}$ is a subalgebra of $\mathcal{P}(M_{n,m})$.

We have thus completed the proof of the theorem. □

We can use the description in Theorem 8.1 of $\mathcal{P}(M_{n,m})^{U_{O_n}}$ to relate the branching algebra $\mathcal{P}(M_{n,m})^{U_{O_n} \times U_m}$ to the tensor product algebra for GL_m. As a GL_m-module, the space $\mathfrak{sp}_{2m}^{(2,0)}$ is isomorphic to $\mathcal{S}^2\mathbb{C}^m$, the space of symmetric $m \times m$ matrices. It is well known that the symmetric algebra $\mathcal{S}(\mathcal{S}^2\mathbb{C}^m)$ is multiplicity-free as a representation of GL_m, and decomposes into a sum of one copy of each polynomial representation corresponding to a diagram with rows of even length (or a partition of even parts):

$$\mathcal{S}(\mathcal{S}^2\mathbb{C}^m) \simeq \bigoplus_\nu F_{(m)}^{2\nu}.$$

(Note that this result is in several places in the literature. See [11] and [16] for example.)

As a GL_m-module, $\mathcal{S}(\mathcal{S}^2\mathbb{C}^m)$ could be embedded in $\mathcal{R}(GL_m/U_m)$, but the algebra structures on these two algebras are quite different.

Using the dominance filtration (see §3.2), we have a canonical $\widehat{A}^+$-algebra filtration on $\mathcal{S}(\mathcal{S}^2\mathbb{C}^m)$. If we form the associated graded algebra, then Theorem 3.2 says that it will be isomorphic to the subalgebra of $R(GL_m/U_m)$ spanned by the representations attached to diagrams with even length rows.

Let us denote the associated graded algebra of $\mathcal{S}(\mathcal{S}^2\mathbb{C}^m)$ by

$$\mathrm{Gr}_{\widehat{A}_m^+} \mathcal{S}(\mathcal{S}^2\mathbb{C}^m).$$

Let us denote the subalgebra of $\mathcal{R}(GL_m/U_m)$ spanned by the representations attached to diagrams with even length rows by $\mathcal{R}^{+2}(GL_m/U_m)$.

We can filter the tensor product $\mathcal{H}_{n,m}^{U_{O_n}} \otimes \mathcal{S}(\mathfrak{sp}_{2m}^{(2,0)})$ by means of the filtration on $\mathcal{S}(\mathfrak{sp}_{2m}^{(2,0)})$. The associated graded algebra will then be $\mathcal{H}_{n,m}^{U_{O_n}} \otimes \mathrm{Gr}_{\widehat{A}_m^+} \mathcal{S}(\mathfrak{sp}_{2m}^{(2,0)})$. This discussion has indicated that the following result holds.

Theorem 8.2 *When $n > 2m$, the associated graded algebra of $\mathcal{P}(M_{n,m})^{U_{O_n}}$ with respect to the dominance filtration on the factor $\mathcal{J}_{n,m}$ is isomorphic to the tensor product of the graded subalgebras $\mathcal{R}^+(GL_m/U_m)$ and $\mathcal{R}^{+2}(GL_m/U_m)$ of $\mathcal{R}(GL_m/U_m)$:*

$$Gr_{\widehat{A}_m^+}(\mathcal{P}(M_{n,m})^{U_{O_n}}) \simeq \mathcal{R}^+(GL_m/U_m) \otimes \mathcal{R}^{+2}(GL_m/U_m).$$

Of course, $\mathrm{Gr}_{\widehat{A}_m^+}\left(\mathcal{P}(M_{n,m})^{U_{O_n}}\right)$ is isomorphic as a GL_m-module to $\mathcal{P}(M_{n,m})^{U_{O_n}}$ in an obvious way, by construction. Also $\mathrm{Gr}_{\widehat{A}_m^+}\left(\mathcal{P}(M_{n,m})^{U_{O_n}}\right)$ inherits the $\widehat{A}_{O_n}^+$ grading from $\mathcal{P}(M_{n,m})^{U_{O_n}}$ – it becomes identified with the $\widehat{A}_m^+$ grading on the first factor $\mathcal{R}^+(GL_m/U_m)$ in the tensor product of Theorem 8.2. On the other hand, the second factor is also $\widehat{A}_m^+$-graded in the obvious way, since it is the factor which defines the associated graded. When we take the U_m invariants, we get another grading by $\widehat{A}_m^+$, associated to the A_m action on the U_m invariants. This triply $\widehat{A}_m^+$-graded algebra is evidently a total subalgebra of the tensor product algebra of GL_m.

On the other hand, we could take the U_m invariants inside $\mathcal{P}(M_{n,m})^{U_{O_n}}$, and then pass to the associated graded. It is not hard to convince oneself that these two processes commute with each other. Hence, we finally have:

Corollary 8.3 *When $n > 2m$, the associated graded algebra of U_m invariants in $\mathcal{P}(M_{n,m})^{U_{O_n}}$,*

$$Gr_{\widehat{A}_m^+}\left(\left(\mathcal{P}(M_{n,m})^{U_{O_n}}\right)^{U_m}\right) \simeq \left(Gr_{\widehat{A}_m^+}\left(\mathcal{P}(M_{n,m})^{U_{O_n}}\right)\right)^{U_m}$$

$$\simeq \left(\mathcal{R}^+(GL_m/U_m) \otimes \mathcal{R}^{+2}(GL_m/U_m)\right)^{U_m}$$

is a triply-graded total subalgebra of the tensor product algebra of GL_m. The restrictions on the gradings which define $Gr_{\widehat{A}_m^+}\left(\left(\mathcal{P}(M_{n,m})^{U_{O_n}}\right)^{U_m}\right)$ are:

(a) *the weight on the first factor of $(\mathcal{R}(GL_m/U_m) \otimes \mathcal{R}(GL_m/U_m))^{U_m}$ should correspond to a partition (i.e., it should be a polynomial weight), and*

(b) *the weight on the second factor should correspond to a partition with even parts.*

Remark: The content of Corollary 8.3 in terms of multiplicities is the Littlewood Restriction Formula [8], [21]; see formula (2.4.1), [28]; see (5.7) with (4.19), [39]; see Theorem 1.5.3 and 2.3.1, [47] and [48]. With this result it is possible to compute a basis of the reciprocity algebra for (GL_n, O_n) using [21]; see [18].

9 Tensor Products for O_n

According to Theorem 6.1, we can compute tensor products for the orthogonal group via the algebra

$$\left(\left(\mathcal{P}(M_{n,m})/I(\mathcal{J}_{n,m}^+) \right)^{U_m} \otimes \left(\mathcal{P}(M_{n,\ell})/I(\mathcal{J}_{n,\ell}^+) \right)^{U_\ell} \right)^{U_{O_n}}.$$

Here the stable range is $n > 2(m + \ell)$. Then we have

$$\mathcal{P}(M_{n,m+\ell}) \simeq \mathcal{H}_{n,m+\ell} \otimes \mathcal{S}(\mathfrak{sp}_{2(m+\ell)}^{(2,0)}).$$

Furthermore,

$$\mathcal{S}(\mathfrak{sp}_{2(m+\ell)}^{(2,0)}) = \mathcal{S}(\mathfrak{sp}_{2m}^{(2,0)} \oplus \mathfrak{sp}_{2\ell}^{(2,0)} \oplus (\mathbb{C}^m \otimes \mathbb{C}^\ell))$$

$$\simeq \mathcal{S}(\mathfrak{sp}_{2m}^{(2,0)}) \otimes \mathcal{S}(\mathfrak{sp}_{2\ell}^{(2,0)}) \otimes \mathcal{S}(\mathbb{C}^m \otimes \mathbb{C}^\ell)$$

Since $\mathcal{J}_{n,m} \simeq \mathcal{S}(\mathfrak{sp}_{2m}^{(2,0)})$ and $\mathcal{J}_{n,\ell} \simeq \mathcal{S}(\mathfrak{sp}_{2\ell}^{(2,0)})$, we see that

$$\mathcal{P}(M_{n,m})/I(\mathcal{J}_{n,m}^+) \otimes \mathcal{P}(M_{n,\ell})/I(\mathcal{J}_{n,\ell}^+) \simeq \mathcal{P}(M_{n,m} \oplus M_{n,\ell})/I(\mathcal{J}_{n,m}^+ \oplus \mathcal{J}_{n,\ell}^+)$$

$$\simeq \mathcal{H}_{n,m+\ell} \otimes \mathcal{S}(\mathfrak{sp}_{2(m+\ell)}^{(2,0)})/I(\mathfrak{sp}_{2m}^{(2,0)} \oplus \mathfrak{sp}_{2\ell}^{(2,0)}) \tag{9.1}$$

$$\simeq \mathcal{H}_{n,m+\ell} \otimes \mathcal{S}(\mathbb{C}^m \otimes \mathbb{C}^\ell).$$

Thus, using equation (9.1), we see that

$$\left(\mathcal{P}(M_{n,m})/I(\mathcal{J}_{n,m}^+) \otimes \mathcal{P}(M_{n,\ell})/I(\mathcal{J}_{n,\ell}^+) \right)^{U_{O_n}}$$

$$\simeq \left(\mathcal{H}_{n,m+\ell} \otimes \mathcal{S}(\mathbb{C}^m \otimes \mathbb{C}^\ell) \right)^{U_{O_n}} \simeq \mathcal{H}_{n,m+\ell}^{U_{O_n}} \otimes \mathcal{S}(\mathbb{C}^m \otimes \mathbb{C}^\ell)$$

$$\simeq \left(\bigoplus_\lambda E_{(n)}^\lambda \otimes F_{(m+\ell)}^\lambda \right)^{U_{O_n}} \otimes \mathcal{S}(\mathbb{C}^m \otimes \mathbb{C}^\ell)$$

$$\simeq \left(\bigoplus_\lambda (E_{(n)}^\lambda)^{U_{O_n}} \otimes F_{(m+\ell)}^\lambda \right) \otimes \mathcal{S}(\mathbb{C}^m \otimes \mathbb{C}^\ell)$$

$$\simeq \left(\bigoplus_\lambda (F_{(n)}^\lambda)^{U_n} \otimes F_{(m+\ell)}^\lambda \right) \otimes \mathcal{S}(\mathbb{C}^m \otimes \mathbb{C}^\ell).$$

Note that $F_{(n)}^\lambda$ is the GL_n representation generated by the highest weight of the O_n representation $E_{(n)}^\lambda$ and both $(F_{(n)}^\lambda)^{U_n}$ and $(E_{(n)}^\lambda)^{O_n}$ are one dimensional.

Hence, finally we get

$$\left(\left(\mathcal{P}(M_{n,m})/I(\mathcal{J}_{n,m}^+) \right)^{U_m} \otimes \left(\mathcal{P}(M_{n,\ell})/I(\mathcal{J}_{n,\ell}^+) \right)^{U_\ell} \right)^{U_{O_n}} .$$

$$\simeq \left(\left(\mathcal{P}(M_{n,m})/I(\mathcal{J}_{n,m}^+) \otimes \mathcal{P}(M_{n,\ell})/I(\mathcal{J}_{n,\ell}^+) \right)^{U_{O_n}} \right)^{U_m \times U_\ell}$$

$$\simeq \left(\left(\bigoplus_\lambda (F_{(n)}^\lambda)^{U_n} \otimes F_{(m+\ell)}^\lambda \right) \otimes \mathcal{S}(\mathbb{C}^m \otimes \mathbb{C}^\ell) \right)^{U_m \times U_\ell} . \qquad (9.2)$$

We can interpret this algebra in term of tensor product algebras for general linear groups. Now consider the (polynomial) tensor product algebras

$$(\mathcal{R}^+(GL_k/U_k) \otimes \mathcal{R}^+(GL_k/U_k))^{U_k} \simeq \bigoplus_{\lambda,\mu} \left(F_{(k)}^\lambda \otimes F_{(k)}^\mu \right)^{U_k}$$

for $k = n, m$ and ℓ. If we form the tensor product of these, we get

$$(\mathcal{R}^+(GL_n/U_n) \otimes \mathcal{R}^+(GL_n/U_n))^{U_n} \otimes (\mathcal{R}^+(GL_m/U_m) \otimes \mathcal{R}^+(GL_m/U_m))^{U_m}$$

$$\otimes \ (\mathcal{R}^+(GL_\ell/U_\ell) \otimes \mathcal{R}^+(GL_\ell/U_\ell))^{U_\ell}$$

$$\simeq \bigoplus_{\alpha,\beta,\delta,\lambda,\mu,\nu} \left(F_{(n)}^\alpha \otimes F_{(n)}^\beta \right)^{U_n} \otimes \left(F_{(m)}^\delta \otimes F_{(m)}^\lambda \right)^{U_m} \otimes \left(F_{(\ell)}^\mu \otimes F_{(\ell)}^\nu \right)^{U_\ell}$$

Let us denote this algebra by $\mathbb{T}_{n,m,\ell}$. The algebra $\mathbb{T}_{n,m,\ell}$ is $(\widehat{A}_n^+)^3 \times (\widehat{A}_m^+)^3 \times (\widehat{A}_\ell^+)^3$-graded. If we require that $\lambda = \alpha$, or that $\mu = \beta$, or that $\nu = \delta$, then we obtain total subalgebras of $\mathbb{T}_{n,m,\ell}$. If $\delta = \alpha$, we will denote it by $\Delta_{1,3}\mathbb{T}_{n,m,\ell}$, and so forth. The subalgebra obtained by requiring that all three diagonal conditions occur at once will be denoted by using all three Δ's. Thus we will write

$$\Delta_{1,3}\Delta_{2,5}\Delta_{4,6}\mathbb{T}_{n,m,\ell}$$

$$= \sum_{\alpha,\beta,\delta} \left(F_{(n)}^\alpha \otimes F_{(n)}^\beta \right)^{U_n} \otimes \left(F_{(m)}^\alpha \otimes F_{(m)}^\delta \right)^{U_m} \otimes \left(F_{(\ell)}^\beta \otimes F_{(\ell)}^\delta \right)^{U_\ell}$$

It takes a similar argument by expanding (9.2) (as in §8) to see that $\Delta_{1,3}\Delta_{2,5}\Delta_{4,6}\mathbb{T}_{n,m,\ell}$ and

$$\left(\left(\mathcal{P}(M_{n,m})/I(\mathcal{J}_{n,m}^+) \otimes \mathcal{P}(M_{n,\ell})/I(\mathcal{J}_{n,\ell}^+) \right)^{U_{O_n}} \right)^{U_m \times U_\ell}$$

are isomorphic as multigraded vector spaces. They may not be isomorphic as algebras, because

$$\left(\left(\mathcal{P}(M_{n,m})/I(\mathcal{J}_{n,m}^+) \otimes \mathcal{P}(M_{n,\ell})/I(\mathcal{J}_{n,\ell}^+) \right)^{U_{O_n}} \right)^{U_m \times U_\ell}$$

is not graded, while we see that $\Delta_{1,3}\Delta_{2,5}\Delta_{4,6}I\!\!T_{n,m,\ell}$ is. However, if we pass to the associated graded of $\mathcal{S}(\mathbb{C}^m \otimes \mathbb{C}^\ell)$, then the two algebras do become isomorphic. We record this fact.

Theorem 9.1 *Assume the stable range $n > 2(m + \ell)$. We have the following isomorphisms of $(\widehat{A}_n^+)^3 \times (\widehat{A}_m^+)^3 \times (\widehat{A}_\ell^+)^3$-graded algebras:*

$$Gr_{(\widehat{A}_n^+)^3 \times (\widehat{A}_m^+)^3 \times (\widehat{A}_\ell^+)^3} \left(\left(\left(\mathcal{P}(M_{n,m})/I(\mathcal{J}_{n,m}^+) \right)^{U_m} \otimes \left(\mathcal{P}(M_{n,\ell})/I(\mathcal{J}_{n,\ell}^+) \right)^{U_\ell} \right)^{U_{O_n}} \right)$$

$$\simeq \Delta_{1,3}\Delta_{2,5}\Delta_{4,6}I\!\!T_{n,m,\ell}.$$

Remark: The content of Theorem 9.1 in terms of multiplicities can be found in [21]; see formula (2.1.2), [32]; see Theorem 4.1 and [52].

Bibliography

[1] Benkart, G. M., Britten, D. J., Lemire, F. W., Stability in modules for classical Lie algebras—a constructive approach, Mem. Amer. Math. Soc., vol.85, (1990), **430**, vi+165, MR1010997 (90m:17012),

[2] Berenstein, A. D., Zelevinsky, A. V., Triple multiplicities for sl(r + 1) and the spectrum of the exterior algebra of the adjoint representation, J. Algebraic Combin., vol.1, (1992), **1**, 7–22, MR1162639 (93h:17012),

[3] Berenstein, Arkady, Zelevinsky, Andrei, Tensor product multiplicities, canonical bases and totally positive varieties, Invent. Math., vol.143, (2001), **1**, 77–128, MR1802793 (2002c:17005),

[4] Black, G. R. E., King, R. C., Wybourne, B. G., Kronecker products for compact semisimple Lie groups, J. Phys. A, vol.16, (1983), **8**, 1555–1589, MR708193 (85e:22020),

[5] Borel, Armand, Linear algebraic groups, Graduate Texts in Mathematics, vol.126, edition=2, Springer-Verlag, New York, (1991), xii+288, MR1102012 (92d:20001),

[6] Cannas da Silva, Ana, Lectures on symplectic geometry, Lecture Notes in Mathematics, vol.1764, Springer-Verlag, Berlin, (2001), xii+217, MR1853077 (2002i:53105),

[7] Drinfel'd, V. G., Quantum groups, Proceedings of the International Congress of Mathematicians, Vol. 1, 2 (Berkeley, Calif., 1986), 798–820, Amer. Math. Soc., Providence, RI, (1987), MR934283 (89f:17017),

[8] Enright, Thomas J., Willenbring, Jeb F., Hilbert series, Howe duality and branching for classical groups, Ann. of Math. (2), vol.159, (2004), **1**, 337–375, MR2052357 (2005d:22013),

[9] Fulton, William, Young tableaux, London Mathematical Society Student Texts, vol.35, With applications to representation theory and geometry, Cambridge University Press, Cambridge, (1997), x+260, MR1464693 (99f:05119),

[10] Goodman, Roe, Multiplicity-free spaces and Schur-Weyl-Howe duality, Representations of real and p-adic groups, , Lect. Notes Ser. Inst. Math. Sci. Natl. Univ. Singap., vol.2, Singapore Univ. Press, Singapore, , (2004), 305–415, MR2090873 (2005j:20053),

[11] Goodman, Roe, Wallach, Nolan R., Representations and invariants of the classical groups, Encyclopedia of Mathematics and its Applications, vol.68, Cambridge University Press, Cambridge, (1998), xvi+685, MR1606831 (99b:20073),

[12] Howe, Roger, Jackson, Steve, Lee, Soo Teck, Tan, Eng-Chye, Willenbring, JebF. Toric Deformation of Branching Algebras, in preparation

[13] Howe, Roger, Reciprocity laws in the theory of dual pairs, Representation theory of reductive groups (Park City, Utah, 1982), Progr. Math., vol.40, 159– 175, Birkhäuser Boston, Boston, MA, (1983), MR733812 (85k:22033),

[14] Howe, Roger, Remarks on classical invariant theory, Trans. Amer. Math. Soc., vol.313, (1989), **2**, 539– 570, MR986027 (90h:22015a),

[15] Howe, Roger, Transcending classical invariant theory, J. Amer. Math. Soc., vol.2, (1989), **3**, 535– 552, MR985172 (90k:22016),

[16] Howe, Roger, Perspectives on invariant theory: Schur duality, multiplicity-free actions and beyond, The Schur lectures (1992) (Tel Aviv), Israel Math. Conf. Proc., vol.8, 1– 182, Bar-Ilan Univ., Ramat Gan, (1995), MR1321638 (96e:13006),

[17] Howe, Roger, Lee, Soo Teck, Bases for some reciprocity algebras. I, Trans. Amer. Math. Soc., vol.359, (2007), **9**, 4359–4387 (electronic), MR2309189,

[18] Howe, Roger, Lee, Soo Teck, Bases for some reciprocity algebras. II, Adv. Math., vol.206, (2006), **1**, 145–210, MR2261753,

[19] Howe, Roger, Lee, Soo Teck, Bases for some reciprocity algebras. III, Compos. Math., vol.142, (2006), **6**, 1594–1614, MR2278762,

[20] Howe, Roger, Lee, Soo Teck, Why should the Littlewood-Richardson Rule be true?, in preparation.

[21] Howe, Roger, Tan, Eng-Chye, Willenbring, Jeb F., Stable branching rules for classical symmetric pairs, Trans. Amer. Math. Soc., vol.357, (2005), **4**, 1601– 1626 (electronic), MR2115378,

[22] Howe, Roger E., Tan, Eng-Chye, Willenbring, Jeb F., A basis for the GL_n tensor product algebra, Adv. Math., vol.196, (2005), **2**, 531–564, MR2166314 (2006h:20062),

[23] Howe, Roger, Umeda, Tōru, The Capelli identity, the double commutant theorem, and multiplicity-free actions, Math. Ann., vol.290, (1991), **3**, 565– 619, MR1116239 (92j:17004),

[24] Jantzen, Jens Carsten, Introduction to quantum groups, Representations of reductive groups, Publ. Newton Inst., 105– 127, Cambridge Univ. Press, Cambridge, (1998), MR1714152 (2000j:17021),

[25] Jimbo, Michio, Solvable lattice models and quantum groups, Proceedings of the International Congress of Mathematicians, Vol. I, II (Kyoto, 1990), 1343– 1352, Math. Soc. Japan, Tokyo, (1991), MR1159318 (93g:82036),

[26] Joseph, Anthony, Quantum groups and their primitive ideals, Ergeb-

nisse der Mathematik und ihrer Grenzgebiete (3) [Results in Mathematics and Related Areas (3)], vol.29, Springer-Verlag, Berlin, (1995), x+383, MR1315966 (96d:17015),

[27] Kashiwara, Masaki, Crystalizing the q-analogue of universal enveloping algebras, Comm. Math. Phys., vol.133, (1990), **2**, 249– 260, MR1090425 (92b:17018),

[28] King, R. C., Modification rules and products of irreducible representations of the unitary, orthogonal, and symplectic groups, J. Mathematical Phys., vol.12, (1971), 1588– 1598, MR0287816 (44 #5019),

[29] King, R. C., Branching rules using tensor methods, Group theoretical methods in physics (Proc. Third Internat. Colloq., Centre Phys. Théor., Marseille, 1974), Vol. 2, 400– 408, Centre Nat. Recherche Sci., Centre Phys. Théor., Marseille, (1974), MR0480644 (58 #800),

[30] King, R. C., Branching rules for GL(N) $\supset_m$ and the evaluation of inner plethysms, J. Mathematical Phys., vol.15, (1974), 258– 267, MR0331999 (48 #10331),

[31] King, R. C., Branching rules for classical Lie groups using tensor and spinor methods, J. Phys. A, vol.8, (1975), 429– 449, MR0411400 (53 #15136),

[32] King, R. C., S-functions and characters of Lie algebras and superalgebras, Invariant theory and tableaux (Minneapolis, MN, 1988), IMA Vol. Math. Appl., vol.19, 226– 261, Springer, New York, (1990), MR1035497 (90k:17014),

[33] King, R, C., Branching rules and weight multiplicities for simple and affine Lie algebras, Algebraic methods in physics (Montréal, QC, 1997), CRM Ser. Math. Phys., 121– 133, Springer, New York, (2001), MR1847252 (2002g:17013),

[34] King, R. C., Wybourne, B. G., Analogies between finite-dimensional irreducible representations of SO($2n$) and infinite-dimensional irreducible representations of Sp($2n$, **R**). I. Characters and products, J. Math. Phys., vol.41, (2000), **7**, MR1765836 (2001f:20090a),

[35] King, R. C., Wybourne, B. G., Analogies between finite-dimensional irreducible representations of SO($2n$) and infinite-dimensional irreducible representations of Sp($2n$, **R**). II. Plethysms, J. Math. Phys., vol.41, (2000), **8**, MR1770978 (2001f:20090b),

[36] Koike, Kazuhiko, Terada, Itaru, Littlewood's formulas and their application to representations of classical Weyl groups, Commutative algebra and combinatorics (Kyoto, 1985), Adv. Stud. Pure Math., vol.11, 147– 160, North-Holland, Amsterdam, (1987), MR951200 (89i:20028),

[37] Koike, Kazuhiko, Terada, Itaru, Young-diagrammatic methods for the representation theory of the groups Sp and SO, The Arcata Conference on Representations of Finite Groups (Arcata, Calif., 1986), Proc. Sympos. Pure Math., vol.47, 437– 447, Amer. Math. Soc., Providence, RI, (1987), MR933432 (89b:20086),

[38] Koike, Kazuhiko, Terada, Itaru, Young-diagrammatic methods for the representation theory of the classical groups of type B_n, C_n, D_n, J. Algebra, vol.107, (1987), **2**, 466– 511, MR885807 (88i:22035),

[39] Koike, Kazuhiko, Terada, Itaru, Young diagrammatic methods for the restriction of representations of complex classical Lie groups to reductive subgroups of maximal rank, Adv. Math., vol.79, (1990), **1**, 104– 135, MR1031827 (91a:22013),

[40] Kostant, B., Rallis, S., Orbits and representations associated with symmetric spaces, Amer. J. Math., vol.93, (1971), 753– 809, MR0311837 (47 #399),

[41] Kudla, Stephen S., Seesaw dual reductive pairs, Automorphic forms of several variables (Katata, 1983), Progr. Math., vol.46, 244– 268, Birkhäuser Boston, Boston, MA, (1984), MR763017 (86b:22032),

[42] Littelmann, Peter, A Littlewood-Richardson rule for symmetrizable Kac-Moody algebras, Invent. Math., vol.116, (1994), 1-3, 329– 346, MR1253196 (95f:17023),

[43] Littelmann, Peter, The path model for representations of symmetrizable Kac-Moody algebras, Proceedings of the International Congress of Mathematicians, Vol. 1, 2 (Zürich, 1994), 298– 308, Birkhäuser, Basel, (1995), MR1403930 (97h:17024),

[44] Littelmann, Peter, The path model for representations of symmetrizable Kac-Moody algebras, Proceedings of the International Congress of Mathematicians, Vol. 1, 2 (Zürich, 1994), 298– 308, Birkhäuser, Basel, (1995), MR1403930 (97h:17024),

[45] Littelmann, Peter, Paths and root operators in representation theory, Ann. of Math. (2), vol.142, (1995), 3, 499– 525, MR1356780 (96m:17011),

[46] Littlewood, D. E., Richardson, A. R., Group characters and algebra, Philos. Trans. Roy. Soc. London. Ser. A., vol.233, (1934), 99– 142

[47] Littlewood, D. E., On invariant theory under restricted groups, Philos. Trans. Roy. Soc. London. Ser. A., vol.239, (1944), 387– 417, MR0012299 (7,6e),

[48] Littlewood, D. E., The Theory of Group Characters and Matrix Representations of Groups, Oxford University Press, New York, (1940), viii+292, MR0002127 (2,3a),

[49] Lusztig, G., Canonical bases arising from quantized enveloping algebras, J. Amer. Math. Soc., vol.3, (1990), 2, 447– 498, MR1035415 (90m:17023),

[50] Macdonald, I. G., Symmetric functions and Hall polynomials, Oxford Mathematical Monographs, With contributions by A. Zelevinsky; Oxford Science Publications, The Clarendon Press Oxford University Press, New York, (1995), x+475, MR1354144 (96h:05207),

[51] Miller, Ezra, Sturmfels, Bernd, Combinatorial commutative algebra, Graduate Texts in Mathematics, vol.227, Springer-Verlag, New York, (2005), xiv+417, MR2110098 (2006d:13001),

[52] Newell, M. J., Modification rules for the orthogonal and symplectic groups, Proc. Roy. Irish Acad. Sect. A., vol.54, (1951), 153– 163, MR0043093 (13,204e),

[53] Pak, Igor, Vallejo, Ernesto, Combinatorics and geometry of Littlewood-Richardson cones, European J. Combin., vol.26, (2005), 6, MR2143205 (2006e:05187),

[54] Popov, V. L., Contractions of actions of reductive algebraic groups, Mat. Sb. (N.S.), vol.130(172), (1986), 3, 310– 334, 431, MR865764 (88c:14065),

[55] Procesi, Claudio, Lie groups, Universitext, An approach through invariants and representations, Springer, New York, (2007), xxiv+596, MR2265844 (2007j:22016),

[56] Robbiano, Lorenzo, Sweedler, Moss, Subalgebra bases, Commutative algebra (Salvador, 1988), Lecture Notes in Math., vol.1430, 61– 87, Springer, Berlin, (1990), MR1068324 (91f:13027),

[57] Sato, M., Kimura, T., A classification of irreducible prehomogeneous vec-

tor spaces and their relative invariants, Nagoya Math. J., vol.65, (1977), 1– 155, MR0430336 (55 #3341),

[58] Servedio, Frank J., Prehomogeneous vector spaces and varieties, Trans. Amer. Math. Soc., vol.176, (1973), 421– 444, MR0320173 (47 #8712),

[59] Sundaram, Sheila, Tableaux in the representation theory of the classical Lie groups, bookInvariant theory and tableaux (Minneapolis, MN, 1988), IMA Vol. Math. Appl., vol.19, 191– 225, Springer, New York, (1990), MR1035496 (91e:22022),

[60] Towber, Jacob, Two new functors from modules to algebras, J. Algebra, vol.47, (1977), 1, 80– 104, MR0469955 (57 #9735),

[61] Vinberg, È. B., Complexity of actions of reductive groups, Funktsional. Anal. i Prilozhen., vol.20, (1986), 1, 1– 13, 96, MR831043 (87j:14077),

[62] Vinberg, È. A., Kimel'fel'd, B. N., Homogeneous domains on flag manifolds and spherical subsets of semisimple Lie groups, Funktsional. Anal. i Prilozhen., vol.12, (1978), 3, 12–19, 96, MR509380 (82e:32042),

[63] Vinberg, È. B., Popov, V. L., A certain class of quasihomogeneous affine varieties, Izv. Akad. Nauk SSSR Ser. Mat., vol.36, (1972), 749–764, MR0313260 (47 #1815),

[64] Weyl, Hermann, The classical groups, Princeton Landmarks in Mathematics, note–Their invariants and representations; Fiftcenth printing Princeton Paperbacks, Princeton University Press, Princeton, NJ, (1997), xiv+320, MR1488158 (98k:01049),

[65] Želobenko, D. P., Compact Lie groups and their representations, note=Translated from the by Israel Program for Scientific Translations; Translations of Mathematical Monographs, Vol. 40, American Mathematical Society, Providence, R.I., (1973), viii+448, MR0473098 (57 #12776b),

11
Character formulae from
Hermann Weyl to the present

Jens Carsten Jantzen
Mathematics Institute
Aarhus Universitet

Introduction

In 1926 Hermann Weyl published a paper that contains his character formula for irreducible finite dimensional complex representations of complex and real semi-simple Lie groups and their Lie algebras. It can also
be interpreted as a character formula for connected compact groups and
for semi-simple algebraic groups in characteristic 0. (Here I am using
modern terminology; when Weyl wrote his paper, terms like "Lie groups"
were not yet in use.)

When we look at Weyl's character formula as a statement for Lie
algebras, then it is a theorem on purely algebraic objects. However,
Weyl used analytic methods to prove it. Not surprisingly, people looked
for algebraic proofs. These attempts were finally successful and led also
to useful reformulations of Weyl's formula. This development will be
described in the first section of this survey.

The other topic to be discussed will be the search for analogues to
Weyl's formula in more general cases. To start with, a finite dimensional complex semi-simple Lie algebra has an abundance of irreducible
representations that are infinite dimensional. It was natural to look for
character formulae for at least some families of representations sharing
features of the finite dimensional ones — for example those generated
by a highest weight vector.

Furthermore, it was also natural to go beyond finite dimensional complex semi-simple Lie algebras. There are several algebraic objects that
share many structural features with these Lie algebras and that have
similar representation theories. Let me mention those that we look at
here:

In the 1950s it was proved that semi-simple algebraic groups over an algebraically closed field of prime characteristic are classified in the same way as semi-simple algebraic groups over $\mathbf{C}$ and also that the classification of simple modules by their highest weights generalises from $\mathbf{C}$ to prime characteristic. In the late 1960s Kac–Moody algebras were constructed as generalisations of finite dimensional semi-simple Lie algebras, and in the 1980s quantum groups were introduced as deformations of universal enveloping algebras. Also for these new objects simple modules with a highest weight play an important role.

In all three cases the quest for character formulae has been a central activity over the years. It turned out that there were strong connections between the different set-ups. And in all three cases the strongest results cannot be proved with algebraic methods alone (at least so far).

I have to admit that a quite different story could be told under the title that I have given here. Weyl's 1926 paper was also the starting point for harmonic analysis on semi-simple Lie groups. A survey of this line of development —looking, for example, at the monumental work of Harish-Chandra including his character formula for discrete series representations — would certainly require as much space as what I have done here. Given the limited time I had for my lecture in Bielefeld I had to restrict myself and thus decided to select the topics closest to my own work and area of expertise.

1 Semi-simple complex Lie algebras

Weyl's character formula can be regarded as a statement about representations of Lie algebras or about representations of groups (Lie groups or algebraic groups). In either case it requires a certain amount of notation. We first develop the Lie algebra version. For unexplained terms we refer to [Hu1].

1.1. Let $\mathfrak{g}$ be a finite dimensional semi-simple complex Lie algebra and $\mathfrak{h}$ a Cartan subalgebra of $\mathfrak{g}$. For any $\mathfrak{g}$–module V and any $\lambda \in \mathfrak{h}^*$ we denote by

$$ V_\lambda = \{\, v \in V \mid h\,v = \lambda(h)\,v \text{ for all } h \in \mathfrak{h} \,\} $$

the *weight space* of V for the weight λ. The set of all $\lambda \in \mathfrak{h}^*$ with $V_\lambda \neq 0$ is called the set of *weights* of V.

The set of non-zero weights of $\mathfrak{g}$ (considered as a $\mathfrak{g}$–module under the adjoint action) is called the *root system* of $\mathfrak{g}$ with respect to $\mathfrak{h}$; this set will be denoted by R. One can choose a *base* $\{\alpha_1, \alpha_2, \ldots, \alpha_r\}$ for the root system R. This base determines a partial ordering on $\mathfrak{h}^*$ such that $\lambda \leq \mu$ if and only if there exist non-negative integers $m_1, m_2, \ldots, m_r$ such that $\mu - \lambda = \sum_{i=1}^{r} m_i \alpha_i$.

Set $R^+ = \{ \alpha \in R \mid \alpha > 0 \}$. Then R is the disjoint union of R^+ and $-R^+$. Let $\mathfrak{n}^+$ be the direct sum of all $\mathfrak{g}_\alpha$ with $\alpha \in R^+$, let $\mathfrak{n}^-$ be the direct sum of all $\mathfrak{g}_\alpha$ with $\alpha \in -R^+$. Then $\mathfrak{g}$ has the triangular decomposition $\mathfrak{g} = \mathfrak{n}^- \oplus \mathfrak{h} \oplus \mathfrak{n}^+$.

For each $\alpha \in R$ there exists a unique $h_\alpha \in [\mathfrak{g}_\alpha, \mathfrak{g}_{-\alpha}] \subset \mathfrak{h}$ such that $\alpha(h_\alpha) = 2$. Then the reflection $s_\alpha : \mathfrak{h}^* \to \mathfrak{h}^*$ given by $s_\alpha(\mu) = \mu - \mu(h_\alpha)\alpha$ permutes the root system R. The subgroup of $\mathrm{GL}(\mathfrak{h}^*)$ generated by all s_α with $\alpha \in R$ is called the *Weyl group* of $\mathfrak{g}$ with respect to $\mathfrak{h}$; it will be denoted by W.

We call

$$X := \{ \lambda \in \mathfrak{h}^* \mid \lambda(h_\alpha) \in \mathbf{Z} \text{ for all } \alpha \in R \}$$

the set of *integral weights*. It is a free abelian group of rank equal to $\dim \mathfrak{h}$. The Weyl group W permutes X; under this action each element of X is conjugate to exactly one element in

$$X_+ := \{ \lambda \in X \mid \lambda(h_\alpha) \geq 0 \text{ for all } \alpha \in R^+ \}.$$

Elements in X_+ are called *dominant* integral weights.

1.2. If V is a finite dimensional $\mathfrak{g}$–module, then all weights of V are integral and V is the direct sum of its weight spaces: $V = \bigoplus_{\lambda \in X} V_\lambda$. The *formal character* of V is then defined as an element in the group ring $\mathbf{Z}[X]$:

$$\mathrm{ch}\, V = \sum_{\lambda \in X} \dim V_\lambda \, e(\lambda)$$

where $(e(\lambda) \mid \lambda \in X)$ is the standard basis of $\mathbf{Z}[X]$. This character is invariant under the action of the Weyl group W given by $w \cdot e(\lambda) = e(w(\lambda))$ for all $w \in W$ and $\lambda \in X$.

If V is a simple finite dimensional $\mathfrak{g}$–module, then the set of weights of V has a unique maximal weight, called the *highest weight* of V. This highest weight is dominant; conversely each dominant integral weight is the highest weight of exactly one (up to isomorphism) simple finite dimensional $\mathfrak{g}$–module. In other words, mapping V to its highest weight induces a bijection from the set of isomorphism classes of simple finite dimensional $\mathfrak{g}$–modules onto the set X_+. For each $\lambda \in X_+$ let $V(\lambda)$ denote a simple finite dimensional $\mathfrak{g}$–module with highest weight λ.

Set ρ equal to half the sum of all positive roots, i.e., of all elements in R^+. It turns out that $\rho \in X$, in fact that $\rho(h_{\alpha_i}) = 1$ for all simple roots α_i.

Now *Weyl's character formula* states that

$$\text{ch}\, V(\lambda) = \frac{\sum_{w \in W} \det(w)\, e(w(\lambda + \rho))}{\sum_{w \in W} \det(w)\, e(w(\rho))}. \tag{1}$$

More precisely one should say that here the denominator is non-zero and divides the numerator in the integral domain $\mathbf{Z}[X]$; so the quotient is a well defined element in $\mathbf{Z}[X]$. In many cases it is useful to write Weyl's formula in the form

$$\text{ch}\, V(\lambda) = \frac{\sum_{w \in W} \det(w)\, e(w(\lambda + \rho) - \rho)}{\sum_{w \in W} \det(w)\, e(w(\rho) - \rho)}, \tag{2}$$

i.e., to multiply the numerator and the denominator by $e(-\rho)$.

From the character formula one deduces *Weyl's dimension formula*:

$$\dim V(\lambda) = \prod_{\alpha \in R^+} \frac{(\lambda + \rho)(h_\alpha)}{\rho(h_\alpha)}. \tag{3}$$

There is a ring homomorphism $\varphi \colon \mathbf{Z}[X] \to \mathbf{Z}$ mapping each $e(\mu)$ to 1. One has then $\varphi(\text{ch}\, V) = \dim V$ for any finite dimensional $\mathfrak{g}$–module V. Unfortunately a direct application of φ to (1) leads to a fraction of the form $0/0$. Therefore a more complicated argument is needed that involves the *denominator formula*

$$\sum_{w \in W} \det(w)\, e(w\,\rho - \rho) = \prod_{\alpha \in R^+} (1 - e(-\alpha)), \tag{4}$$

see [W3], page 386, line 7.

1.3. Let me now state how to regard Weyl's character formula as a result for algebraic groups. (See [Bo1], [Hu2] or [Sp3] for unexplained terms and background.)

Consider a connected semi-simple algebraic group G over $\mathbf{C}$ with Lie $G = \mathfrak{g}$. Then we can find a *maximal torus* H in G with Lie $H = \mathfrak{h}$. Here "maximal torus" is to be interpreted in the algebraic group sense as a direct product of multiplicative groups $\mathbf{C}^\times$.

Let $X(H)$ denote the group of all *characters* of H, i.e., of all homomorphisms $H \to \mathbf{C}^\times$ of algebraic groups. Taking the tangent map at the identity is an injective group homomorphism $d \colon X(H) \to \mathfrak{h}^*$. The image $d\, X(H)$ is a subgroup of finite index in the lattice X of all integral weights. We have $d\, X(H) = X$ if G is simply connected; for arbitrary G the image contains with finite index the subgroup $\sum_{\alpha \in R} \mathbf{Z}\, \alpha$ generated by all roots.

If V is a finite dimensional G–module (so by definition the representation $G \to \mathrm{GL}(V)$ is a homomorphism of algebraic groups), then V becomes a $\mathfrak{g}$–module when we take the tangent map at the identity of $G \to \mathrm{GL}(V)$. In the decomposition $V = \bigoplus_{\mu \in X} V_\mu$ into weight spaces for $\mathfrak{h}$, only $\mu \in d\,X(H)$ can contribute non-zero summands. Any $h \in H$ acts on any $V_{d\,\nu}$ with $\nu \in X(H)$ as multiplication by $\nu(h)$. Given $g \in G$ let $\chi_V(g)$ denote the trace of g acting on V. We get then for all $h \in H$

$$\chi_V(h) = \sum_{\nu \in X(H)} \nu(h)\,\dim V_{d\,\nu}.$$

An arbitrary $g \in G$ has a *Jordan decomposition* $g = g_s g_u$ into a a unipotent factor g_u and a semisimple factor g_s. One gets then $\chi_V(g) = \chi_V(g_s)$ since the representation $G \to \mathrm{GL}(V)$ preserves the Jordan decomposition. Furthermore, g_s is conjugate in G to some $h_g \in H$ and we get $\chi_V(g) = \chi_V(g_s) = \chi_V(h_g)$. This shows that the formal character $\mathrm{ch}\,V$ determines the character χ_V. (One can avoid the use of the Jordan decomposition and observe that the semisimple elements in G are dense in G and that χ_V is continuous.)

If V is a simple G–module, then also the derived $\mathfrak{g}$–module is simple, hence isomorphic to some $V(\lambda)$ with $\lambda \in X_+ \cap d\,X(H)$. One can show conversely that each $V(\lambda)$ with $\lambda \in X_+ \cap d\,X(H)$ lifts uniquely to a G–module. Therefore these $V(\lambda)$ are a system of representatives for the isomorphism classes of simple G–modules. So Weyl's character formula determines the characters of all simple G–modules.

We can regard G also as a complex Lie group. The simple G–modules above yield then also the irreducible holomorphic finite dimensional representations of G. Furthermore, if $G_\mathbf{R}$ is a real form of G, then we get also the irreducible complex representations of the real Lie group $G_\mathbf{R}(\mathbf{R})$. (These representations correspond to irreducible complex representations of the Lie algebra $\mathfrak{g}_\mathbf{R} = \mathrm{Lie}\,G_\mathbf{R}$, hence to simple modules over $\mathfrak{g}_\mathbf{R} \otimes_\mathbf{R} \mathbf{C} \simeq \mathfrak{g}$.)

1.4. Weyl proved his character formula in a series of three papers [W1], [W2], [W3] from 1925/26 where he first deals with special linear groups, then with symplectic and special orthogonal groups[1] before finally dealing with the general case.

His main tool is the integration over a compact real form of G. The proof uses that the orthogonality relations for irreducible characters, first proved by Frobenius and Schur for finite groups, extend to compact groups. (This had already been observed by Schur.) Another important

---

1 then called Komplexgrupppen and Drehungsgruppen in German

ingredient in the proof is the connection between the representations of the group and its Lie algebra; one exploits the fact that a formal character ch V is a $\mathbf{Z}$–linear combination of sums of the form $\sum_{\nu \in W\mu} e(\nu)$ with $\mu \in X_+$. For a version of this proof in modern terminology see [BtD], chap. VI, §1. See also Slodowy's survey [Sl] discussing the use of the integration technique by Weyl and in earlier work.

Weyl's results confirm the character formulae found by Schur in [S1] (with completely different methods) for the irreducible polynomial representations of general linear groups as well as Schur's results in [S2] for the compact real orthogonal groups. When comparing with Schur's results, Weyl underlines that he (Weyl) can deal with all orthogonal groups (real and complex), not only with the compact ones as Schur did.

Side results of Weyl's three papers are the introduction of Weyl groups and the proof of the complete reducibility of finite dimensional representations of semi-simple complex Lie algebras.

1.5. As stated in the preceding subsection Weyl used methods from analysis to prove his character formula which clearly is a result on algebraic objects. The earlier work by Schur showed that a purely algebraic approach existed for special linear groups, while Brauer achieved this for orthogonal groups in [Br1]. But an algebraic proof working for all semi-simple complex Lie algebras was first found by Freudenthal in his paper [F] from 1954.

In order to state Freudenthal's crucial result we need one extra bit of notation. The Killing form on $\mathfrak{g}$ is non-degenerate on $\mathfrak{h}$ and thus induces an isomorphism $\mathfrak{h} \xrightarrow{\sim} \mathfrak{h}^*$ of vector spaces. We use this isomorphism to transport the Killing form over to $\mathfrak{h}^*$. We get thus a W–invariant bilinear form $(\,,\,)$ on $\mathfrak{h}^*$ that is positive definite on $\sum_{\mu \in X} \mathbf{R}\mu$.

Now *Freudenthal's formula* states for all $\lambda \in X_+$ and $\mu \in X$ that

$$((\lambda + \rho, \lambda + \rho) - (\mu + \rho, \mu + \rho)) \dim V(\lambda)_\mu$$

$$= 2 \sum_{\alpha \in R^+} \sum_{i=1}^{\infty} \dim V(\lambda)_{\mu + i\alpha} (\mu + i\alpha, \alpha). \qquad (1)$$

This formula allows an inductive calculation of all $\dim V(\lambda)_\mu$ starting with the general fact that $\dim V(\lambda)_\lambda = 1$. Then one has to know that $(\mu + \rho, \mu + \rho) < (\lambda + \rho, \lambda + \rho)$ for all $\mu \in X$ with $V(\lambda)_\mu \neq 0$ and $\mu \neq \lambda$. Note that the sum on the right hand side has only finitely many non-zero terms because for any $\alpha \in R^+$ there are only finitely many $i \in \mathbf{N}$ with $\mu + i\alpha \leq \lambda$.

The proof of (1) involves a reduction to the case $\mathfrak{g} = \mathfrak{sl}(2, \mathbf{C})$ and a comparison of two different calculations of the action of the Casimir element for $\mathfrak{g}$ on $V(\lambda)$.

Once one has (1) one can deduce Weyl's character formula using some clever tricks. However, today the importance of Freudenthal's formula comes from another direction: It yields the fastest known algorithm for the calculation of the dimensions of all $V(\lambda)_\mu$. It is therefore used in computer programs (such as the package LiE) for this purpose. Regrettably analogous formulae are missing for the characters of more general modules that we shall discuss later on.

1.6. Another important version of the character formula was found by Kostant in his paper [Ko] from 1959. It involves the *Kostant partition function* P that associates to any $\mu \in X$ the number of R^+–tuples $(n_\alpha)_{\alpha \in R^+}$ of non-negative integers such that $\mu = \sum_{\alpha \in R^+} n_\alpha \alpha$. So $P(\mu)$ is non-zero if and only if $\mu \in \sum_{\alpha \in R^+} \mathbf{N}\alpha$.

Now Kostant's formula states for all $\lambda \in X_+$ and $\mu \in X$ that

$$\dim V(\lambda)_\mu = \sum_{w \in W} \det(w)\, P(w(\lambda + \rho) - (\mu + \rho)). \qquad (1)$$

This equation can be easily deduced from the denominator formula 1.2(4). In fact, using 1.2(4) one checks that (1) and 1.2(2) are equivalent, cf. [Hu1], 24.3.

Actually, Kostant used a more complicated approach to deduce his formula from Weyl's one. But soon Cartier (see [Cr]) and Steinberg discovered the simpler argument.

Kostant's formula is not as useful for practical calculations as Freudenthal's formula because of the large sum over all Weyl group elements and because of the many cancellations in the sum. Note that for dominant μ all summands $(\mu + i\alpha, \alpha)$ in 1.5(1) are positive. So there are no cancellations in Freudenthal's formula in this case, and the invariance of ch $V(\lambda)$ under W shows that it suffices to compute $\dim V(\lambda)_\mu$ for dominant μ.

1.7. One gets a better insight into Kostant's formula if one works with certain infinite dimensional $\mathfrak{g}$–modules. For any Lie algebra $\mathfrak{a}$ denote by $U(\mathfrak{a})$ its universal enveloping algebra. For each $\mu \in \mathfrak{h}^*$ let $\mathbf{C}_\mu$ denote the one dimensional $(\mathfrak{h} \oplus \mathfrak{n}^+)$–module on which any $h \in \mathfrak{h}$ acts as $\mu(h)$ and any $x \in \mathfrak{n}^+$ acts as 0. The induced $\mathfrak{g}$–module

$$M(\mu) := U(\mathfrak{g}) \otimes_{U(\mathfrak{h} \oplus \mathfrak{n}^+)} \mathbf{C}_\mu \qquad (1)$$

is now usually called the *Verma module* with highest weight μ.

The map $u \mapsto u \otimes 1$ is an isomorphism $U(\mathfrak{n}^-) \xrightarrow{\sim} M(\mu)$ of vector spaces. It takes a weight vector of weight ν in $U(\mathfrak{n}^-)$ to a weight vector of weight $\nu + \mu$ in $M(\mu)$. Now a look at a PBW basis for $U(\mathfrak{n}^-)$ constructed with root vectors in $\mathfrak{n}^-$ yields that $M(\mu)$ is the direct sum of its weight spaces and that

$$\operatorname{ch} M(\mu) = \sum_{\nu \in \mathfrak{h}^*} P(\mu - \nu) \, e(\nu). \tag{2}$$

For this to make sense we have to extend the definition of the formal character as in 1.2 to infinite dimensional $\mathfrak{g}$–modules that have finite dimensional weight spaces and are the direct sums of these weight spaces; these general formal characters live then in a completion of $\mathbf{Z}[\mathfrak{h}^*]$ and will be written as infinite sums of all $e(\nu)$ with $\nu \in \mathfrak{h}^*$. These more general characters need not be invariant under the action of the Weyl group.

Now Kostant's formula can be rewritten as

$$\operatorname{ch} V(\lambda) = \sum_{w \in W} \det(w) \operatorname{ch} M(w \bullet \lambda) \qquad \text{for all } \lambda \in X_+ \tag{3}$$

where we use the "dot notation"

$$w \bullet \mu = w(\mu + \rho) - \rho \qquad \text{for all } w \in W \text{ and } \mu \in \mathfrak{h}^*. \tag{4}$$

Actually, something stronger is true: There is for each $\lambda \in X_+$ an exact sequence of $\mathfrak{g}$–modules (called the *BGG resolution*)

$$0 \to M(w_0 \bullet \lambda) \to \cdots \tag{5}$$
$$\to \bigoplus_{l(w)=2} M(w \bullet \lambda) \to \bigoplus_{l(w)=1} M(w \bullet \lambda) \to M(\lambda) \to V(\lambda) \to 0.$$

Here $l(w)$ denotes for any $w \in W$ the minimal length of an expression $w = s_{\alpha_{i_1}} s_{\alpha_{i_2}} \dots s_{\alpha_{i_r}}$ as product of reflections with respect to simple roots α_j. Furthermore w_0 is the unique element in W of maximal length, equal to $\dim \mathfrak{n}^+$; it is also characterised as the unique element in W with $w_0(R^+) = -R^+$. So each term in (5) is the direct sum of all $M(w \bullet \lambda)$ with $w \in W$ and $l(w) = i$ for some i with $0 \le i \le \dim \mathfrak{n}^+$ — except for the terms $V(\lambda)$ and 0, of course. Since each reflection has determinant -1, we get $\det(w) = (-1)^{l(w)}$ for all $w \in W$. Therefore (3) follows easily from (5) taking formal characters.

The surjection from $M(\lambda)$ onto $V(\lambda)$ appears already in the work of Chevalley [Ch] and Harish-Chandra [HC] who use this map to give an algebraic construction of the finite dimensional modules for $\mathfrak{g}$ that does not proceed case-by-case. From Harish-Chandra's work one also can deduce that the kernel of this surjection is the image of the direct sum of all $M(w{\bullet}\lambda)$ with $l(w) = 1$. The existence of the whole exact sequence was then proved by Bernštein, Gel'fand, and Gel'fand in [BGG2]. This article appeared in 1975, but had been available for a few years before that date. Their proof uses detailed knowledge about homomorphisms between Verma modules and about the Bruhat–Chevalley ordering on the Weyl group; it involves a bit of homological algebra.

Note that each $M(\mu)$ is isomorphic to $U(\mathfrak{n}^-)$ as an $\mathfrak{n}^-$–module. Therefore (5) is a free resolution of $V(\lambda)$ considered as a $U(\mathfrak{n}^-)$–module and we can compute the Ext-groups $\mathrm{Ext}^{\bullet}_{\mathfrak{n}^-}(V(\lambda), \mathbf{C})$ as the cohomology groups of the complex $\mathrm{Hom}_{\mathfrak{n}^-}(M^{\bullet}, \mathbf{C})$ where $\mathbf{C}$ is the trivial one dimensional $\mathfrak{n}^-$–module and where M^i is the direct sum of all $M(w{\bullet}\lambda)$ with $l(w) = i$. Note that this is a complex of $\mathfrak{h}$–modules.

For any $\mathfrak{n}^-$–module V we have a natural identification

$$\mathrm{Hom}_{\mathfrak{n}^-}(V, \mathbf{C}) \xrightarrow{\sim} (V/\mathfrak{n}^- V)^*.$$

For any $\mu \in \mathfrak{h}^*$ the $\mathfrak{h}$–module $M(\mu)/\mathfrak{n}^- M(\mu)$ is isomorphic to $\mathbf{C}_\mu$. It follows that the i-th term in the complex $\mathrm{Hom}_{\mathfrak{n}^-}(M^{\bullet}, \mathbf{C})$ is isomorphic to the direct sum of all $\mathbf{C}_{-w{\bullet}\lambda}$ with $l(w) = i$. The $w{\bullet}\lambda$ with $w \in W$ are pairwise distinct; therefore all maps in the complex are 0. It follows that

$$\mathrm{Ext}^i_{\mathfrak{n}^-}(V(\lambda), \mathbf{C}) \simeq \bigoplus_{l(w)=i} \mathbf{C}_{-w{\bullet}\lambda} = \bigoplus_{l(w)=i} \mathbf{C}_{-w(\lambda+\rho)+\rho}. \qquad (6)$$

Exchanging R^+ with $-R^+$ this implies

$$\mathrm{Ext}^i_{\mathfrak{n}^+}(V(\lambda), \mathbf{C}) \simeq \bigoplus_{l(w)=i} \mathbf{C}_{-w(w_0\lambda-\rho)-\rho} = \bigoplus_{l(w)=i} \mathbf{C}_{w{\bullet}(-w_0\,\lambda)}. \qquad (7)$$

(One has to replace ρ by $-\rho$ and the highest weight λ by the lowest weight $w_0\,\lambda$.) The isomorphisms $\mathrm{Ext}^i_{\mathfrak{n}^+}(V(\lambda), \mathbf{C}) \simeq \mathrm{Ext}^i_{\mathfrak{n}^+}(\mathbf{C}, V(\lambda)^*) = H^i(\mathfrak{n}^+, V(\lambda)^*)$ yield now

$$H^i(\mathfrak{n}^+, V(\lambda)) = \bigoplus_{l(w)=i} \mathbf{C}_{w{\bullet}\lambda} \qquad (8)$$

since $V(\lambda)^* \simeq V(-w_0\,\lambda)$. We get in particular *Bott's theorem* on the dimensions of these cohomology groups.

1.8. The fastest way to prove Kostant's (or Weyl's) character formula algebraically is not to use the resolution 1.7(5). Instead one uses some elementary properties of the Verma modules. Any $M(\mu)$ with $\mu \in \mathfrak{h}^*$ has a unique simple factor module that we denote by $L(\mu)$. For $\lambda \in X_+$

we have $L(\lambda) = V(\lambda)$. Using Harish-Chandra's description of the centre of $U(\mathfrak{g})$ one shows that each $M(\mu)$ has finite length with all composition factors of the form $L(w{\bullet}\mu)$ with $w \in W$ and $w{\bullet}\mu \leq \mu$; the simple module $L(\mu)$ occurs with multiplicity 1 in a composition series. This implies that there exist integers $a_{\mu,w} \geq 0$ with

$$\operatorname{ch} M(\mu) = \operatorname{ch} L(\mu) + \sum_{w{\bullet}\mu < \mu} a_{\mu,w} \operatorname{ch} L(w{\bullet}\mu). \tag{1}$$

From this we get integers $b_{\mu,w} \in \mathbf{Z}$ with

$$\operatorname{ch} L(\mu) = \sum_{w{\bullet}\mu \leq \mu} b_{\mu,w} \operatorname{ch} M(w{\bullet}\mu) \tag{2}$$

such that $b_{\mu,1} = 1$.

Now in order to prove Kostant's formula in the form 1.7(3) we just have to show for all $\lambda \in X_+$ that $b_{\lambda,w} = \det w$ for all $w \in W$. This, however, is an easy consequence of the W–invariance of $\operatorname{ch} V(\lambda)$.

This proof was discovered by Bernštein, Gel'fand, and Gel'fand in their 1971 paper [BGG1]. It can be found in [Hu1], 24.2 with a simplification in an appendix.

We have here looked at algebraic proofs of Weyl's formula that use the point of view of Lie algebras. There are also algebraic proofs that use the point of view of algebraic groups. Let me mention papers by Springer [Sp1], Demazure [De], Donkin [Do], and Andersen [A1].

1.9. Another way of looking at Weyl's character formula is *Littelmann's path model* developed in his papers [Li1] and [Li2] from 1994/95. Here a path is a piecewise linear, continuous map π from the interval $[0,1]_{\mathbf{Q}} := \{t \in \mathbf{Q} \mid 0 \leq t \leq 1\}$ to $X_{\mathbf{Q}} := \sum_{\nu \in X} \mathbf{Q}\nu \subset \mathfrak{h}^*$ such that $\pi(0) = 0$. One identifies paths π and π' if there exists a piecewise linear, surjective, non-decreasing, continuous map $\varphi : [0,1]_{\mathbf{Q}} \to [0,1]_{\mathbf{Q}}$ with $\pi' = \pi \circ \varphi$. These identifications ensure that the concatenation of paths is associative. Let $\mathcal{P}$ denote the set of all paths modulo these identifications.

In the path model one associates to each dominant weight $\lambda \in X_+$ a subset $\mathcal{P}_\lambda$ of $\mathcal{P}$ such that

$$\operatorname{ch} V(\lambda) = \sum_{\pi \in \mathcal{P}_\lambda} e(\pi(1)). \tag{1}$$

In other words, for each μ there are $\dim V(\lambda)_\mu$ paths in $\mathcal{P}_\lambda$ ending at μ.

According to [Li2] there are several possible choices for $\mathcal{P}_\lambda$. In [Li1] a special choice was made, the set of Lakshmibai–Seshadri paths (short:

L–S paths), cf. Section 4 in [Li2]. One of these paths is π_λ given by $\pi_\lambda(t) = t\lambda$ for all t, $0 \le t \le 1$. The other L–S paths arise then by applying Littelmann's *root operators*: To each simple root α_i he associates two root operators, e_i and f_i, that map $\mathcal{P}$ to $\mathcal{P} \cup \{0\}$, see Section 1 in [Li2]. If $f_i(\pi) \ne 0$ for some $\pi \in \mathcal{P}$, then $f_i(\pi)(1) = \pi(1) - \alpha_i$; if $e_i(\pi) \ne 0$, then $e_i(\pi)(1) = \pi(1) + \alpha_i$. Now $\mathcal{P}_\lambda$ consists of all paths of the form $f_{i_1} \circ f_{i_2} \circ \cdots \circ f_{i_s}(\pi_\lambda)$.

One can associate to $\mathcal{P}_\lambda$ an oriented coloured graph whose vertices are the elements of $\mathcal{P}_\lambda$ and which has an arrow $\pi \overset{i}{\longrightarrow} \pi'$ if and only if $\pi' = f_i(\pi)$. This graph turns out to be isomorphic to the crystal graph associated to $V(\lambda)$ by Kashiwara using the corresponding quantum group, cf. [Jo2], 6.4.27.

One application of the path model is a formula for the decomposition of any tensor product $V(\lambda) \otimes V(\lambda')$ with $\lambda, \lambda' \in X$ into a direct sum of simple submodules. This formula has the great advantage that the multiplicities are sums of non-negative terms, not alternating sums as in the formulae by Brauer in [Br2] (see also note 22 to chap. VII in [W4]) or Steinberg in [St].

1.10. We now leave the consideration of finite dimensional $\mathfrak{g}$–modules and turn to character formulae for more general, but similar objects. A first example involves the in general infinite dimensional simple highest weight modules $L(\mu)$ from 1.8. The characters of these modules are of interest not only in themselves, but also because they determine characters for simple Harish-Chandra modules for the complex Lie group G considered as a Lie group over the real numbers, cf. [BG], [E], or [Jo1].

For the sake of simplicity let us concentrate on modules of the form $L(w{\scriptstyle\bullet}\lambda)$ with $\lambda \in X_+$ and $w \in W$. There are by 1.8(2) integers $b_{w,x}$ with

$$\operatorname{ch} L(w{\scriptstyle\bullet}\lambda) = \sum_{x{\scriptstyle\bullet}\lambda \le w{\scriptstyle\bullet}\lambda} b_{w,x} \operatorname{ch} M(x{\scriptstyle\bullet}\lambda). \tag{1}$$

Here the coefficients $b_{w,x}$ are actually independent of λ; this was proved in [Ja1].

In their 1979 paper [KL1] Kazhdan and Lusztig came up with a conjecture on the coefficients $b_{w,x}$ that was then proved by Brylinski and Kashiwara in [BK] and independently by Beilinson and Bernstein, see [BB]. The Kazhdan–Lusztig conjecture (now a theorem) states that

$$b_{w,x} = \det(xw)\, P_{xw_0,ww_0}(1) \tag{2}$$

where $P_{x,w}$ is the *Kazhdan–Lusztig polynomial* attached to w and x.

These polynomials are constructed using the Hecke algebra of W. The construction leads to a recursion formula that determines the polynomials uniquely and allows (in principle) their calculation. For another description of these polynomials and some variations of them, see [So3].

Soon after stating their conjecture in [KL1], Kazhdan and Lusztig gave in [KL2] a geometric interpretation of their polynomials. It involves the flag variety $\mathcal{B} := G/B$ where B is the Borel subgroup of G with Lie $B = \mathfrak{h} \oplus \mathfrak{n}^+$. For each $w \in W$ let $\mathcal{B}_w$ denote the Bruhat cell $\mathcal{B}_w = BwB/B$. On the closure $\overline{\mathcal{B}}_w$ one considers now the intersection cohomology complex $\mathbf{IC}^\bullet(\overline{\mathcal{B}}_w)$. This is an object in the derived category $D^b_{cs}(\overline{\mathcal{B}}_w)$ of sheaves on $\overline{\mathcal{B}}_w$ with constructible cohomology sheaves; its restriction to the open subset $\mathcal{B}_w$ is the constant sheaf $\mathbf{C}$ put into degree $-l(w) = -\dim \mathcal{B}_w$. Now the result in [KL2] says for any $x \in W$ with $\mathcal{B}_x \subset \overline{\mathcal{B}}_w$ that

$$P_{x,w}(t^2) = t^{l(w)} \sum_i \dim \mathcal{H}^i(\mathbf{IC}^\bullet(\overline{\mathcal{B}}_w))_x \, t^i. \tag{3}$$

Here $\mathcal{H}^i(\mathbf{IC}^\bullet(\overline{\mathcal{B}}_w))$ denotes the i-th cohomology sheaf of the complex $\mathbf{IC}^\bullet(\overline{\mathcal{B}}_w)$ and the index x indicates that we take the stalk at any point of $\mathcal{B}_x$; the dimension in (3) does not depend on the choice of this point since the restriction of $\mathcal{H}^i(\mathbf{IC}^\bullet(\overline{\mathcal{B}}_w))$ to $\mathcal{B}_x$ turns out to be a constant sheaf. An alternative proof of (3) due to MacPherson can be found in [Sp2].

In case $\mathcal{B}_x \not\subset \overline{\mathcal{B}}_w$ one usually does not define $P_{x,w}$. However, we set $P_{x,w} = 0$ in this case in order to simplify the formulation of (2).

1.11. Let me describe the main ingredients of the proof of 1.10(2). For more details see [Sp2].

Given 1.10(3) it appears reasonable to find some connection between complexes of sheaves on $\mathcal{B}$ and modules of the form $L(w \bullet \lambda)$ and $M(w \bullet \lambda)$. This can be done using the sheaf $\mathcal{D}_\mathcal{B}$ of (algebraic) differential operators on $\mathcal{B}$.

The action of G on $\mathcal{B} = G/B$ by left translation induces an algebra homomorphism from the universal enveloping algebra $U(\mathfrak{g})$ to the algebra $\mathcal{D}_\mathcal{B}(\mathcal{B})$ of global sections of $\mathcal{D}_\mathcal{B}$. This homomorphism turns out to be surjective; its kernel is the ideal $I_0 U(\mathfrak{g})$ generated by the annihilator I_0 of the trivial one dimensional $\mathfrak{g}$–module $\mathbf{C} = L(0)$ in the centre of $U(\mathfrak{g})$. Setting $U^0 = U(\mathfrak{g})/I_0 U(\mathfrak{g})$ we get an isomorphism $U^0 \xrightarrow{\sim} \mathcal{D}_\mathcal{B}(\mathcal{B})$, hence an equivalence between $\{U^0\text{–modules}\}$ and $\{\mathcal{D}_\mathcal{B}(\mathcal{B})\text{–modules}\}$. All $L(w \bullet 0)$ and $M(w \bullet 0)$ with $w \in W$ are U^0–modules.

Next one shows that $\{\mathcal{D}_\mathcal{B}(\mathcal{B})\text{–modules}\}$ and $\{\mathcal{D}_\mathcal{B}\text{–modules}\}$ are equivalent categories: One maps any $\mathcal{D}_\mathcal{B}$–module $\mathcal{M}$ to the $\mathcal{D}_\mathcal{B}(\mathcal{B})$–module $\mathcal{M}(\mathcal{B})$ of its global sections, while any $\mathcal{D}_\mathcal{B}(\mathcal{B})$–module M is sent to the

$\mathcal{D}_\mathcal{B}$–module $\mathcal{D}_\mathcal{B} \otimes_{\mathcal{D}_\mathcal{B}(\mathcal{B})} M$. Combining this with the earlier result one has now an equivalence between { U^0–modules } and { $\mathcal{D}_\mathcal{B}$–modules }. For any $w \in W$ let $\mathcal{L}_w$ denote the $\mathcal{D}_\mathcal{B}$–module corresponding to $L(ww_0{\scriptstyle\bullet}0)$ and $\mathcal{M}_w$ the one corresponding to $M(ww_0{\scriptstyle\bullet}0)$. (Note the factor w_0.)

All $\mathcal{L}_w$ and $\mathcal{M}_w$ belong to a special class of $\mathcal{D}_\mathcal{B}$–modules, the *holonomic modules with regular singularities*. On this class the functor $\mathbf{R}\,\mathcal{H}om_{\mathcal{D}_\mathcal{B}}(\mathcal{O}_\mathcal{B}, \ )$ takes values in $D^b_{cs}(\mathcal{B})$; in fact it induces an equivalence of derived categories. (This is the *Riemann–Hilbert correspondence*.)

One shows that $\mathbf{R}\,\mathcal{H}om_{\mathcal{D}_\mathcal{B}}(\mathcal{O}_\mathcal{B}, \mathcal{L}_w)$ is the intersection cohomology complex $\mathbf{IC}^\bullet(\overline{\mathcal{B}}_w)$ extended by 0 outside $\overline{\mathcal{B}}_w$ and shifted in degree by $\dim \mathcal{B}$, whereas $\mathbf{R}\,\mathcal{H}om_{\mathcal{D}_\mathcal{B}}(\mathcal{O}_\mathcal{B}, \mathcal{M}_w)$ is the constant sheaf $\mathbf{C}_{\mathcal{B}_w}$ on $\mathcal{B}_w$ extended by 0 outside $\mathcal{B}_w$ and put in degree $\dim \mathcal{B} - l(w)$.

Let us write $[M]$ for the class of an object M in a suitable Grothendieck group. Now 1.10(1) can be rewritten $[L(w{\scriptstyle\bullet}\lambda)] = \sum_x b_{w,x}\,[M(x{\scriptstyle\bullet}\lambda)]$. The equivalence between { U^0–modules } and { $\mathcal{D}_\mathcal{B}$–modules } yields then $[\mathcal{L}_w] = \sum_x b_{ww_0,xw_0}\,[\mathcal{M}_x]$. And since $\mathbf{R}\,\mathcal{H}om_{\mathcal{D}_\mathcal{B}}(\mathcal{O}_\mathcal{B}, \ )$ is an equivalence on the subcategory involved, one gets

$$(-1)^{\dim \mathcal{B}}[\mathbf{IC}^\bullet(\overline{\mathcal{B}}_w)] = \sum_x b_{ww_0,xw_0}\,(-1)^{\dim \mathcal{B} - l(x)}[\mathbf{C}_{\mathcal{B}_x}]. \qquad (1)$$

(Note that a shift in degree by d in $D^b_{cs}(\mathcal{B})$ corresponds to a multiplication by $(-1)^d$ in the Grothendieck group.)

For any $z \in \mathcal{B}$ the map taking any X in $D^b_{cs}(\mathcal{B})$ to $\sum_i (-1)^i \dim \mathcal{H}^i(X)_z$ factors through the Grothendieck group of $D^b_{cs}(\mathcal{B})$. Applying it to (1) with $z \in \mathcal{B}_x$ for some $x \in W$ we get

$$\sum_i (-1)^i \dim \mathcal{H}^i(\mathbf{IC}^\bullet(\overline{\mathcal{B}}_w))_x = (-1)^{l(x)} b_{ww_0,xw_0}. \qquad (2)$$

By 1.10(3) the left hand side in (2) is equal to $(-1)^{l(w)} P_{x,w}((-1)^2) = \det(w) P_{x,w}(1)$. This now yields 1.10(2) since $w_0^2 = 1$.

Finally, I have to admit to some cheating above. In order to get the Riemann–Hilbert correspondence one has to consider $\mathcal{B}$ as an analytic manifold. In the functor $\mathbf{R}\,\mathcal{H}om_{\mathcal{D}_\mathcal{B}}(\mathcal{O}_\mathcal{B}, \ )$ we have to replace $\mathcal{O}_\mathcal{B}$ by the sheaf $\mathcal{O}_\mathcal{B}^{an}$ of analytic functions on $\mathcal{B}$ and similarly $\mathcal{D}_\mathcal{B}$ by the sheaf $\mathcal{D}_\mathcal{B}^{an}$ of differential operators with analytic coefficients. If $\mathcal{M}$ is a $\mathcal{D}_\mathcal{B}$–module, then this functor has to be applied to $\mathcal{M}^{an} := \mathcal{O}_\mathcal{B}^{an} \otimes_{\mathcal{O}_\mathcal{B}} \mathcal{M}$.

1.12. In 1.10/11 we have restricted to $L(\mu)$ with $\mu \in W{\scriptstyle\bullet}\lambda$ for some $\lambda \in X_+$. An arbitrary $\mu \in X$ can be written in the form $\mu = w{\scriptstyle\bullet}\lambda$ with $\lambda \in -\rho + X_+$; here λ is uniquely determined by μ, but w is not except for $\lambda \in X_+$. However, we can make w unique by requiring that $l(w)$ is maximal among all $l(w')$ with $\mu = w' {\scriptstyle\bullet} \lambda$. If we do this, then 1.10(2) extends to this case and we get $\operatorname{ch} L(\mu) = \sum_x \det(wx) P_{xw_0, ww_0}(1) \operatorname{ch} L(x {\scriptstyle\bullet} \lambda)$; this follows from [Ja1].

For arbitrary $\mu \in \mathfrak{h}^*$ one has to consider $R_\mu := \{\, \alpha \in R \mid \mu(h_\alpha) \in \mathbf{Z} \,\}$ and $W_\mu := \{\, w \in W \mid w\mu - \mu \in \mathbf{Z}R \,\}$. One checks that R_μ is a root system with Weyl group W_μ. Given μ one can find $w \in W_\mu$ and $\lambda \in \mathfrak{h}^*$ with $\mu = w {\scriptstyle\bullet} \lambda$ and $(\lambda + \rho)(h_\alpha) \geq 0$ for all $\alpha \in R_\mu \cap R^+$. If here $(\lambda + \rho)(h_\alpha) > 0$ for all $\alpha \in R_\mu \cap R^+$, then one gets

$$\operatorname{ch} L(\mu) = \sum_x \det(wx) P^\lambda_{xw_\lambda, ww_\lambda}(1) \operatorname{ch} L(x {\scriptstyle\bullet} \lambda) \qquad (1)$$

where we sum over $x \in W_\mu = W_\lambda$, with w_λ the longest element in W_λ, and where $P^\lambda_{xw_\lambda, ww_\lambda}$ denotes a Kazhdan–Lusztig polynomial for the reflection group W_λ. If $(\lambda + \rho)(h_\alpha) = 0$ for some $\alpha \in R_\mu \cap R^+$, then (1) extends provided one chooses w of maximal length (taken in W_λ) with $\mu = w {\scriptstyle\bullet} \lambda$. One gets (1) from the equivalence of categories constructed in [So1] though this is not explicitly stated there.

2 Kac–Moody algebras

We now turn to Kac–Moody algebras. Their structure is very similar to that of the finite dimensional semi-simple Lie algebras from Section 1. We take the notation from that classical case and add a sub- or superscript "KM" to denote the corresponding object in the Kac–Moody case. For more background on Kac–Moody algebras one may consult [Kc2], [MP], [Wa], or [Ca].

2.1. A Kac–Moody algebra is a Lie algebra over $\mathbf{C}$ associated to a *generalised Cartan matrix*, i.e., an $I \times I$–matrix $A = (a_{ij})$ for some finite index set I such that all entries in A belong to $\mathbf{Z}$, such that $a_{ii} = 2$ for all i and $a_{ij} \leq 0$ whenever $i \neq j$; in addition one requires that $a_{ij} = 0$ if and only if $a_{ji} = 0$. This last condition is always satisfied when A is *symmetrisable*, which means that there exist integers $d_i > 0$ such that $d_i a_{ij} = d_j a_{ji}$ for all i and j. A Kac–Moody algebra associated to a symmetrisable generalised Cartan matrix is called symmetrisable.

The Kac–Moody algebra $\mathfrak{g}_{\mathrm{KM}}$ associated to A has a triangular decomposition $\mathfrak{g}_{\mathrm{KM}} = \mathfrak{n}^-_{\mathrm{KM}} \oplus \mathfrak{h}_{\mathrm{KM}} \oplus \mathfrak{n}^+_{\mathrm{KM}}$. The subalgebra $\mathfrak{h}_{\mathrm{KM}}$ is commutative and finite dimensional. As in the classical case $\mathfrak{g}_{\mathrm{KM}}$ is the direct sum of weight spaces for $\mathfrak{h}_{\mathrm{KM}}$ under the adjoint action. The 0–weight space is $\mathfrak{h}_{\mathrm{KM}}$ itself; the non-zero weights form the root system $R_{\mathrm{KM}} \subset \mathfrak{h}^*_{\mathrm{KM}}$

of $\mathfrak{g}_{\mathrm{KM}}$. It is the disjoint union of R^+_{KM} and $-R^+_{\mathrm{KM}}$ where R^+_{KM} consists of the $\alpha \in R_{\mathrm{KM}}$ such that the root space $\mathfrak{g}_{\mathrm{KM},\alpha}$ is contained in $\mathfrak{n}^+_{\mathrm{KM}}$. Then $\mathfrak{n}^-_{\mathrm{KM}}$ is the direct sum of all $\mathfrak{g}_{\mathrm{KM},\alpha}$ with $\alpha \in -R^+_{\mathrm{KM}}$. Any root space $\mathfrak{g}_{\mathrm{KM},\alpha}$ is finite dimensional; its dimension may well be greater than 1.

There are linearly independent elements $(h_i)_{i \in I}$ in $\mathfrak{h}_{\mathrm{KM}}$ and linearly independent elements $(\alpha_i)_{i \in I}$ in $\mathfrak{h}^*_{\mathrm{KM}}$ such that $\alpha_i(h_j) = a_{ji}$ for all i and j. Usually these families are not bases for their respective spaces; the only exception is the case $\det A \neq 0$. Each α_i is a positive root; one has $R^+_{\mathrm{KM}} \subset \sum_{i \in I} \mathbf{N}\alpha_i$. As in 1.1 one uses the α_i to define a partial ordering $\leq$ on $\mathfrak{h}^*_{\mathrm{KM}}$.

For each $i \in I$ let s_i denote the reflection on $\mathfrak{h}^*_{\mathrm{KM}}$ defined by $s_i(\lambda) = \lambda - \lambda(h_i)\alpha_i$. The group W_{KM} generated by all s_i is called the Weyl group of $\mathfrak{g}_{\mathrm{KM}}$. The root system R_{KM} is stable under W_{KM}. A root is called real if it belongs to some $W_{\mathrm{KM}}\alpha_i$ with $i \in I$; all other roots are called imaginary.

If A is the Cartan matrix of the Lie algebra $\mathfrak{g}$ from 1.1, then $\mathfrak{g}_{\mathrm{KM}}$ is isomorphic to $\mathfrak{g}$. If A is not the Cartan matrix of some finite dimensional complex semi-simple Lie algebra, then $\mathfrak{g}_{\mathrm{KM}}$ is infinite dimensional, in fact R_{KM} and W_{KM} are infinite.

If $\mathfrak{g}_{\mathrm{KM}}$ is symmetrisable, then there exists a non-degenerate bilinear form $(,)$ on $\mathfrak{h}^*_{\mathrm{KM}}$ invariant under W_{KM}. It satisfies $(\alpha_i, \alpha_i) \neq 0$ and $\lambda(h_i) = 2(\lambda, \alpha_i)/(\alpha_i, \alpha_i)$ for all i and all $\lambda \in \mathfrak{h}^*_{\mathrm{KM}}$. When we use such a form, we shall assume that it is normalised so that each (α_i, α_i) is a positive real number. In this case a root α is imaginary if and only if $(\alpha, \alpha) \leq 0$.

2.2. As in 1.7 any $\mu \in \mathfrak{h}^*_{\mathrm{KM}}$ defines a one dimensional module $\mathbf{C}_\mu$ over $\mathfrak{h}_{\mathrm{KM}} \oplus \mathfrak{n}^+_{\mathrm{KM}}$ and then an induced Verma module

$$M_{\mathrm{KM}}(\mu) := U(\mathfrak{g}_{\mathrm{KM}}) \otimes_{U(\mathfrak{h}_{\mathrm{KM}} \oplus \mathfrak{n}^+_{\mathrm{KM}})} \mathbf{C}_\mu. \tag{1}$$

It is a direct sum of finite dimensional weight spaces. So one can define its character $\operatorname{ch} M_{\mathrm{KM}}(\mu)$ in a suitable completion of the group ring $\mathbf{Z}[\mathfrak{h}^*_{\mathrm{KM}}]$ and gets

$$\operatorname{ch} M_{\mathrm{KM}}(\mu) = e(\mu) \prod_{\alpha \in R^+_{\mathrm{KM}}} (1 - e(-\alpha))^{-m_\alpha} \tag{2}$$

where $m_\alpha = \dim \mathfrak{g}_{\mathrm{KM},\alpha} = \dim \mathfrak{g}_{\mathrm{KM},-\alpha}$.

Any $M_{\mathrm{KM}}(\mu)$ has a unique simple quotient that we denote by $L_{\mathrm{KM}}(\mu)$. In contrast to the classical case Verma modules usually have infinite length. However, given $\nu \in \mathfrak{h}^*_{\mathrm{KM}}$ one can find a submodule N of $M_{\mathrm{KM}}(\mu)$

such that ν is not a weight of N and such that $M_{\mathrm{KM}}(\mu)/N$ has finite length with all composition factors of the form $L_{\mathrm{KM}}(\nu')$ with $\nu' \leq \mu$. It follows that there are uniquely determined integers $a_{\mu\nu} \geq 0$ with $a_{\mu\mu} = 1$ and

$$\mathrm{ch}\, M_{\mathrm{KM}}(\mu) = \sum_{\nu \leq \mu} a_{\mu\nu}\, \mathrm{ch}\, L_{\mathrm{KM}}(\nu). \tag{3}$$

This sum usually involves infinitely many non-zero terms. However, for any given ν' there are only finitely many ν with $\nu' \leq \nu$ and $a_{\mu\nu} \neq 0$; so $e(\nu')$ occurs with a non-zero coefficient only for finitely many summands on the right hand side of (3).

One can now invert the matrix of all $a_{\mu\nu}$; one gets integers $b_{\mu\nu}$ with

$$\mathrm{ch}\, L_{\mathrm{KM}}(\mu) = \sum_{\nu \leq \mu} b_{\mu\nu}\, \mathrm{ch}\, M_{\mathrm{KM}}(\nu) \tag{4}$$

and $b_{\mu\mu} = 1$.

2.3. Set

$$X^{\mathrm{KM}} := \{\, \mu \in \mathfrak{h}_{\mathrm{KM}}^* \mid \mu(h_i) \in \mathbf{Z} \text{ for all } i \,\}$$

and

$$X_+^{\mathrm{KM}} := \{\, \mu \in X^{\mathrm{KM}} \mid \mu(h_i) \geq 0 \text{ for all } i \,\}.$$

Choose an arbitrary $\rho_{\mathrm{KM}} \in X_+^{\mathrm{KM}}$ with $\rho_{\mathrm{KM}}(h_i) = 1$ for all i. As in 1.7(4) we introduce the dot notation $w{\boldsymbol{\cdot}}\mu = w(\mu+\rho_{\mathrm{KM}})-\rho_{\mathrm{KM}}$ for all $w \in W_{\mathrm{KM}}$ and $\mu \in \mathfrak{h}_{\mathrm{KM}}^*$; it does not depend on the choice of ρ_{KM}.

In his 1974 paper [Kc1] Kac generalised Weyl's character formula to symmetrisable Kac–Moody algebras. He showed first for all $\lambda \in X_+^{\mathrm{KM}}$ that

$$\mathrm{ch}\, L_{\mathrm{KM}}(\lambda) = \sum_{w \in W_{\mathrm{KM}}} \det(w)\, \mathrm{ch}\, M_{\mathrm{KM}}(w{\boldsymbol{\cdot}}\lambda). \tag{1}$$

The simple module $\mathrm{ch}\, L_{\mathrm{KM}}(0)$ is the one dimensional trivial module $\mathbf{C}$. Now a comparison with 2.2(2) yields the *denominator formula*

$$\prod_{\alpha \in R_{\mathrm{KM}}^+} (1 - e(-\alpha))^{m_\alpha} = \sum_{w \in W_{\mathrm{KM}}} \det(w)\, e(w{\boldsymbol{\cdot}}0). \tag{2}$$

Using this we can rewrite (1) as

$$\mathrm{ch}\, L_{\mathrm{KM}}(\lambda) = \frac{\sum_{w \in W_{\mathrm{KM}}} \det(w)\, e(w{\boldsymbol{\cdot}}\lambda)}{\sum_{w \in W_{\mathrm{KM}}} \det(w)\, e(w{\boldsymbol{\cdot}}0)} \tag{3}$$

in perfect analogy to 1.2(2).

The proof in [Kc1] follows to a large extent the ideas in [BGG1], cf. 1.8. However, at one point something new was needed: In order to show that $\operatorname{ch} L_{\mathrm{KM}}(\lambda)$ is a linear combination of all $\operatorname{ch} L_{\mathrm{KM}}(w \boldsymbol{\cdot} \lambda)$ with $w \in W_{\mathrm{KM}}$, Kac could not use the centre of the enveloping algebra. Instead he introduced Casimir operators that act on the modules, but do not come from the enveloping algebra.

Actually (3) holds also when $\mathfrak{g}_{\mathrm{KM}}$ is not symmetrisable. This was proved more than ten years later, independently by Kumar in [Ku2] and by Mathieu in [M].

Let me add that Littelmann's path model (see 1.9) works equally well for the modules $\operatorname{ch} L_{\mathrm{KM}}(\lambda)$ with $\lambda \in X_+^{\mathrm{KM}}$ over symmetrisable Kac–Moody algebras.

2.4. Assume from now on that $\mathfrak{g}_{\mathrm{KM}}$ is symmetrisable. In this case Kac and Kazhdan determined in [KK] all composition factors of all Verma modules $M_{\mathrm{KM}}(\mu)$, i.e., all ν with $a_{\mu\nu} \neq 0$ in 2.2(3). In contrast to the classical case they could not conclude that $a_{\mu\nu} \neq 0$ implies $\nu \in W_{\mathrm{KM}} \boldsymbol{\cdot} \mu$. However, if we exclude certain weights, then the situation gets better.

For each positive imaginary root α set

$$H_\alpha := \{ \lambda \in \mathfrak{h}_{\mathrm{KM}}^* \mid 2(\lambda + \rho_{\mathrm{KM}}, \alpha) = (\alpha, \alpha) \}. \tag{1}$$

These H_α are called the *critical hyperplanes* in $\mathfrak{h}_{\mathrm{KM}}^*$. The complement $\mathfrak{h}_{\mathrm{KM}}^* \setminus \bigcup_\alpha H_\alpha$ of the union of the critical hyperplanes is stable under the dot action of W_{KM}. For each μ in this complement the equation 2.2(4) has the form

$$\operatorname{ch} L_{\mathrm{KM}}(\mu) = \sum_{\nu \in W_{\mathrm{KM}} \boldsymbol{\cdot} \mu} b_{\mu\nu} \operatorname{ch} M_{\mathrm{KM}}(\nu). \tag{2}$$

This was shown in [Ku1] extending partial results in [DGK].

A weight $\lambda \in X_+^{\mathrm{KM}}$ does not belong to any critical hyperplane because we have $(\lambda + \rho_{\mathrm{KM}}, \alpha) \geq (\rho_{\mathrm{KM}}, \alpha) > 0$ for any positive root α, but $(\alpha, \alpha) \leq 0$ for any imaginary root α. Therefore (2) can be applied to any $\mu = w \boldsymbol{\cdot} \lambda$ with $w \in W_{\mathrm{KM}}$. In this case [DGK] contains a conjecture for the $b_{\mu\nu}$ (or rather for the inverse matrix of the $b_{\mu\nu}$). This conjecture was then proved in the two papers [Ks] by Kashiwara and [KT1] by Kashiwara and Tanisaki; it was independently proved by Casian in [Cs]. The result is that

$$\operatorname{ch} L_{\mathrm{KM}}(w \boldsymbol{\cdot} \lambda) = \sum_{x \in W_{\mathrm{KM}}} \det(wx)\, Q_{w,x}^{\mathrm{KM}}(1) \operatorname{ch} M_{\mathrm{KM}}(x \boldsymbol{\cdot} \lambda) \tag{3}$$

for all $\lambda \in X_+^{\mathrm{KM}}$ and $w \in W_{\mathrm{KM}}$.

Here the $Q_{w,x}^{\mathrm{KM}}$ are the *inverse Kazhdan–Lusztig polynomials* for W_{KM}. The construction of the Kazhdan–Lusztig polynomials works for all Coxeter groups, hence also for W_{KM}; we denote here the Kazhdan–Lusztig polynomials by $P_{w,x}^{\mathrm{KM}}$. These polynomials are usually only introduced when $w \preccurlyeq x$ where $\preccurlyeq$ is the Bruhat-Chevalley order on W_{KM}; we set here $P_{w,x}^{\mathrm{KM}} = 0$ in all other cases.

The inverse Kazhdan–Lusztig polynomials $Q_{w,x}^{\mathrm{KM}}$ are zero unless $w \preccurlyeq x$ and are defined by the equations

$$\sum_{y \in W_{\mathrm{KM}}} \det(xy)\, Q_{x,y}^{\mathrm{KM}}\, P_{y,z}^{\mathrm{KM}} = \delta_{x,z} \tag{4}$$

for all $x, z \in W_{\mathrm{KM}}$. (Note that only y with $x \preccurlyeq y \preccurlyeq z$ can contribute a non-zero term to this sum; there are finitely many y with this property.)

If A is the Cartan matrix of the Lie algebra $\mathfrak{g}$ from 1.1, then $P_{x,w}^{\mathrm{KM}}$ is the polynomial $P_{x,w}$ from 1.10. In this case one can show that $Q_{w,x}^{\mathrm{KM}} = P_{xw_0,ww_0}^{\mathrm{KM}}$. Therefore (3) is a generalisation of the result in 1.10, the Kazhdan–Lusztig conjecture.

2.5. In the classical case one gets easily the character formula for all $L(\mu)$ with $\mu \in X$ once one knows the character formula for all $L(\mu)$ with $\mu \in W{\scriptstyle\bullet}X_+$. This does not extend to general Kac–Moody algebras. Besides the problem with the critical hyperplanes, also the fact that X^{KM} is in general not equal to $W_{\mathrm{KM}}X_+^{\mathrm{KM}}$ makes trouble. Let us look at a crucial special case, the *affine Kac–Moody algebras*.

Consider $\mathfrak{g}$ as in 1.1 and suppose in addition that $\mathfrak{g}$ is simple. Denote again the simple roots by $\alpha_1, \alpha_2, \ldots, \alpha_r$. Let $\beta \in R^+ \cap X_+$ be a positive root that is also a dominant weight. There are one or two choices for β depending on whether all roots have the same length or not. If we write $\beta = \sum_{i=1}^r n_i \alpha_i$ with $n_i \in \mathbf{N}$, then $n_i > 0$ for all i.

We now associate to $\mathfrak{g}$ and β a generalised Cartan matrix A with index set $I = \{0, 1, \ldots, r\}$. For $i, j \geq 1$ set $a_{ij} = 2(\alpha_i, \alpha_j)/(\alpha_i, \alpha_i)$; so this part yields the usual Cartan matrix for $\mathfrak{g}$. To this add $a_{0i} = -2(\beta, \alpha_i)/(\beta, \beta)$ and $a_{i0} = -2(\beta, \alpha_i)/(\alpha_i, \alpha_i)$ for all $i > 0$ as well as $a_{00} = 2$. Note that A is symmetrisable.

Assume now that $\mathfrak{g}_{\mathrm{KM}}$ is the Kac–Moody algebra associated to A. Any Kac–Moody algebra constructed this way is called an affine Kac–Moody algebra.

Set $\delta = \alpha_0^{\mathrm{KM}} + \sum_{i=1}^r n_i \alpha_i^{\mathrm{KM}} \in \mathfrak{h}_{\mathrm{KM}}^*$. In this subsection I am writing α_i^{KM} for the simple roots in $\mathfrak{h}_{\mathrm{KM}}^*$ in order to avoid confusion. Now an elementary calculation shows that $\delta(h_i) = 0$ for all i, $0 \leq i \leq r$. This implies $s_i(\delta) = \delta$ for all i, hence $w(\delta) = \delta$ for all $w \in W_{\mathrm{KM}}$.

As mentioned in 2.1 there is a W_{KM}–invariant non-degenerate bilinear form on $\mathfrak{h}_{\mathrm{KM}}^*$. Now $\delta(h_i) = 0$ implies $(\delta, \alpha_i^{\mathrm{KM}}) = 0$ for all i, hence $(\delta, \delta) = 0$. Any $\lambda \in X_+^{\mathrm{KM}}$ satisfies $(\lambda + \rho_{\mathrm{KM}}, \alpha_i^{\mathrm{KM}}) > 0$ for all i, hence $(\lambda + \rho_{\mathrm{KM}}, \delta) > 0$. This implies for all $w \in W_{\mathrm{KM}}$

$$(w{\scriptstyle\bullet}\lambda + \rho_{\mathrm{KM}}, \delta) = (w(\lambda + \rho_{\mathrm{KM}}), w(\delta)) = (\lambda + \rho_{\mathrm{KM}}, \delta) > 0.$$

So we can never reach any μ with $(\mu + \rho_{\mathrm{KM}}, \delta) < 0$.

It can be shown that δ belongs to the root system R_{KM}. In fact, all $m\delta$ with $m \in \mathbf{Z}$, $m \neq 0$ are roots; they are precisely the imaginary roots of $\mathfrak{g}_{\mathrm{KM}}$. It follows that there is only one critical hyperplane in this case, equal to $H_\delta = \{\, \mu \in \mathfrak{h}_{\mathrm{KM}}^* \mid (\mu + \rho_{\mathrm{KM}}, \delta) = 0 \,\}$.

There is actually one more family of affine Kac–Moody algebras not covered by the construction above. If $\mathfrak{g}$ is of type C_r with $r \geq 1$, then one takes above as β also half the largest root in R. Then one has to write $2\beta = \sum_{i=1}^r n_i \alpha_i$ and to set $\delta = 2\alpha_0^{\mathrm{KM}} + \sum_{i=1}^r n_i \alpha_i^{\mathrm{KM}}$; otherwise no change is needed.

2.6. Keep the assumptions on $\mathfrak{g}_{\mathrm{KM}}$ from 2.5. We could have defined X_+^{KM} in 1.2 also as the set of all $\lambda \in X^{\mathrm{KM}}$ with $(\lambda + \rho_{\mathrm{KM}})(h_i) > 0$ for all i since $\rho_{\mathrm{KM}}(h_i) = 1$ for all i. We now set

$$X_-^{\mathrm{KM}} := \{\, \lambda \in X^{\mathrm{KM}} \mid (\lambda + \rho_{\mathrm{KM}})(h_i) < 0 \text{ for all } i \,\}. \tag{1}$$

Any $\lambda \in X_-^{\mathrm{KM}}$ satisfies $(\lambda + \rho_{\mathrm{KM}}, \delta) < 0$, hence $(\mu + \rho_{\mathrm{KM}}, \delta) < 0$ for all $\mu \in W_{\mathrm{KM}}{\scriptstyle\bullet}\lambda$. Therefore $\mathrm{ch}\, L_K(\mu)$ is certainly not determined by the results from 2.4. However, one can apply 2.4(1) to any $\mu = w{\scriptstyle\bullet}\lambda$ with $w \in W_{\mathrm{KM}}$. The coefficients $b_{\mu\nu}$ in this case were determined by Kashiwara and Tanisaki in the paper [KT2] from 1995; they showed that

$$\mathrm{ch}\, L_{\mathrm{KM}}(w{\scriptstyle\bullet}\lambda) = \sum_{x \in W_{\mathrm{KM}}} \det(wx)\, P_{x,w}^{\mathrm{KM}}(1)\, \mathrm{ch}\, M_{\mathrm{KM}}(x{\scriptstyle\bullet}\lambda). \tag{2}$$

This confirms a conjecture by Lusztig from [Lu3] for the case where all roots in R have the same length.

One gets from (2) and 2.4(3) character formulae for all $L_{\mathrm{KM}}(\mu)$ with $\mu \in X$ and $(\mu + \rho_{\mathrm{KM}}, \delta) \neq 0$. If $(\mu + \rho_{\mathrm{KM}}, \delta) > 0$, then one can find $\lambda \in -\rho_{\mathrm{KM}} + X_+^{\mathrm{KM}}$ and $w \in W_{\mathrm{KM}}$ with $\mu = w{\scriptstyle\bullet}\lambda$; if w has maximal length for this property, then 2.4(3) extends to this case. If $(\mu + \rho_{\mathrm{KM}}, \delta) < 0$, then one can find $\lambda \in \rho_{\mathrm{KM}} + X_-^{\mathrm{KM}}$ and $w \in W_{\mathrm{KM}}$ with $\mu = w{\scriptstyle\bullet}\lambda$; if w has minimal length for this property, then (2) extends to this case. This follows using translation functors, cf. [Ku3] for the second case.

For $\mu \in \mathfrak{h}_{\mathrm{KM}}^*$ with $\mu \notin X$ and $(\mu + \rho_{\mathrm{KM}}, \delta) \neq 0$ one gets similar results when one replaces W_{KM} with its subgroup of all $w \in W_{\mathrm{KM}}$ such that $w\mu - \mu \in \mathbf{Z}R_{\mathrm{KM}}$. This was proved by Kashiwara and Tanisaki in a sequence of papers ([KT4], [KT5], and [KT6]). Alternatively one can now use Fiebig's paper [Fi] where he generalises Soergel's approach for $\mathfrak{g}$ mentioned in 1.12.

One can express the condition $(\mu + \rho_{\mathrm{KM}}, \delta) \neq 0$ differently. There exists a unique element $c \in h_0 + \sum_{i=1}^{r} \mathbf{C}h_i \subset \mathfrak{h}_{\mathrm{KM}}$ such that $\alpha(c) = 0$ for all $\alpha \in R_{\mathrm{KM}}$. Then the one dimensional subspace $\mathbf{C}c$ is the centre of the Lie algebra $\mathfrak{g}_{\mathrm{KM}}$. The element c acts as scalar multiplication by $\mu(c)$ on any $M_{\mathrm{KM}}(\mu)$ and $L_{\mathrm{KM}}(\mu)$. Now $(\mu + \rho_{\mathrm{KM}}, \delta) = 0$ is equivalent to $\mu + \rho_{\mathrm{KM}} \in \sum_{i=0}^{r} \mathbf{C}\alpha_i$, hence to $(\mu + \rho_{\mathrm{KM}})(c) = 0$. So the simple modules $L_{\mathrm{KM}}(\mu)$ not covered by the results above are exactly those on which c acts as multiplication by $-\rho_{\mathrm{KM}}(c)$. This value is often called the critical level; the number $\rho_{\mathrm{KM}}(c)$ is a positive integer called the *dual Coxeter number* of $\mathfrak{g}_{\mathrm{KM}}$.

2.7. The proofs given by Kashiwara and Tanisaki for the character formulae 2.4(3) and 2.6(2) have many features in common with the proof of the original Kazhdan–Lusztig conjecture sketched in 1.11. Again flag varieties and $\mathcal{D}$–modules are involved.

Let us exclude the finite dimensional case. One associates to $\mathfrak{g}_{\mathrm{KM}}$ a Kac–Moody group G_{KM} and subgroups B_{KM}^+, B_{KM}^- that play the role of Borel subgroups. These groups can be regarded as group schemes or as infinite dimensional complex Lie groups.

Set $\mathcal{B}_{\mathrm{KM}} := G_{\mathrm{KM}}/B_{\mathrm{KM}}^+$; this is the (infinite dimensional) flag variety for G_{KM}. It contains Schubert cells $\mathcal{B}_{\mathrm{KM},w} := B_{\mathrm{KM}}^+ w B_{\mathrm{KM}}^+/B_{\mathrm{KM}}^+$ with $w \in W_{\mathrm{KM}}$. Each $\mathcal{B}_{\mathrm{KM},w}$ is isomorphic to an affine space of dimension $l(w)$ and its closure $\overline{\mathcal{B}}_{\mathrm{KM},w}$ is the finite union of all $\mathcal{B}_{\mathrm{KM},x}$ with $x \preccurlyeq w$.

One can also consider $\mathcal{B}_{\mathrm{KM}}^w := B_{\mathrm{KM}}^- w B_{\mathrm{KM}}^+/B_{\mathrm{KM}}^+$. Each $\mathcal{B}_{\mathrm{KM}}^w$ is isomorphic to an infinite dimensional affine space. It has codimension $l(w)$; its closure is the infinite union of all $\mathcal{B}_{\mathrm{KM}}^x$ with $w \preccurlyeq x$.

Now the proof of the character formula 2.6(2) — where we can restrict the sum to x with $x \preccurlyeq w$ — involves the geometry of $\overline{\mathcal{B}}_{\mathrm{KM},w}$ whereas the proof of 2.4(3) — where we can restrict the sum to x with $w \preccurlyeq x$ — involves the geometry of $\overline{\mathcal{B}_{\mathrm{KM}}^w}$.

The link between $\mathfrak{g}_{\mathrm{KM}}$–modules and the geometry of $\mathcal{B}_{\mathrm{KM}}$ is provided by suitable categories of $\mathcal{D}$–modules on $\mathcal{B}_{\mathrm{KM}}$. The construction of these categories together with the appropriate functors to $\mathfrak{g}_{\mathrm{KM}}$–modules and to $D_{\mathrm{cs}}^n(\mathcal{B}_{\mathrm{KM}})$ is quite lengthy and requires some delicate limit constructions. You can find more details in the survey [KT3] by Kashiwara and Tanisaki themselves.

2.8. Let me conclude this section with a character formula that follows from 2.6(2) and that is used when comparing character formulae for Kac-Moody algebras and quantum groups in Section 3.

Suppose that $\mathfrak{g}_{\mathrm{KM}}$ is an affine Lie algebra constructed as in 2.5 starting from a finite dimensional simple complex Lie algebra $\mathfrak{g}$. Then one can identify the root system R of $\mathfrak{g}$ with the subset $R_{\mathrm{KM}} \cap \sum_{i=1}^{r} \mathbf{C}\alpha_i$. Furthermore we can identify $\mathfrak{g}$ with the subalgebra of $\mathfrak{g}_{\mathrm{KM}}$ generated by all $\mathfrak{g}_{\mathrm{KM},\alpha}$ with $\alpha \in R$. Under this embedding of $\mathfrak{g}$ into $\mathfrak{g}_{\mathrm{KM}}$ each $\mathfrak{g}_\alpha$ with $\alpha \in R$ is identified with $\mathfrak{g}_{\mathrm{KM},\alpha}$ and $\mathfrak{h}$ is identified with $\sum_{i=1}^{r} \mathbf{C}h_i \subset \mathfrak{h}_{\mathrm{KM}}$; in fact each $h_i \in \mathfrak{h}$ is identified with $h_i \in \mathfrak{h}_{\mathrm{KM}}$.

We can find an element $d \in \mathfrak{h}_{\mathrm{KM}}$ such that $\alpha_0(d) = 1$ and $\alpha_i(d) = 0$ for all $i > 0$. We have then $\mathfrak{h}_{\mathrm{KM}} = \mathfrak{h} \oplus \mathbf{C}c \oplus \mathbf{C}d$ and we can characterise R as the set of all $\alpha \in R_{\mathrm{KM}}$ with $\alpha(d) = 0$. For all $i \in \mathbf{Z}$ set

$$\mathfrak{g}_{\mathrm{KM}}^{[i]} := \{\, x \in \mathfrak{g}_{\mathrm{KM}} \mid [d, x] = i\, x \,\}. \tag{1}$$

We get then $\mathfrak{g}_{\mathrm{KM}}^{[0]} = \mathfrak{g} \oplus \mathbf{C}c \oplus \mathbf{C}d$. Any $\mathfrak{g}_{\mathrm{KM}}^{[i]}$ with $i \neq 0$ is the direct sum of all root spaces $\mathfrak{g}_{\mathrm{KM},\alpha}$ with $\alpha(d) = i$. The explicit description of the root system R_{KM} implies that all $\mathfrak{g}_{\mathrm{KM}}^{[i]}$ are finite dimensional. We have $[\mathfrak{g}_{\mathrm{KM}}^{[i]}, \mathfrak{g}_{\mathrm{KM}}^{[j]}] \subset \mathfrak{g}_{\mathrm{KM}}^{[i+j]}$ for all i and j.

Set

$$\mathfrak{q}^+ := \bigoplus_{i>0} \mathfrak{g}_{\mathrm{KM}}^{[i]}, \qquad \mathfrak{q}^- := \bigoplus_{i<0} \mathfrak{g}_{\mathrm{KM}}^{[i]}, \qquad \mathfrak{p} := \bigoplus_{i\geq 0} \mathfrak{g}_{\mathrm{KM}}^{[i]} = \mathfrak{g}_{\mathrm{KM}}^{[0]} \oplus \mathfrak{q}^+. \tag{2}$$

These are subalgebras of $\mathfrak{g}_{\mathrm{KM}}$ with $\mathfrak{g}_{\mathrm{KM}} = \mathfrak{q}^- \oplus \mathfrak{p}$ as a vector space.

Set

$$Y := \{\, \mu \in \mathfrak{h}_{\mathrm{KM}}^* \mid \mu(h_i) \in \mathbf{N} \text{ for all } i \geq 1 \,\}. \tag{3}$$

Any $\mu \in Y$ defines a finite dimensional simple $\mathfrak{g}$–module $V(\mu_{|\mathfrak{h}})$ with highest weight $\mu_{|\mathfrak{h}}$. We extend $V(\mu_{|\mathfrak{h}})$ to a $\mathfrak{g}_{\mathrm{KM}}^{[0]}$ module $V_{\mathrm{KM}}(\mu)$ by letting c act as multiplication by $\mu(c)$ and d act as multiplication by $\mu(d)$. (Recall that c and d commute with $\mathfrak{g}$.) We extend $V_{\mathrm{KM}}(\mu)$ further to a $\mathfrak{p}$–module by letting $\mathfrak{q}^+$ act as 0; this works since $\mathfrak{q}^+$ is an ideal in $\mathfrak{p}$.

Now consider the induced $\mathfrak{g}_{\mathrm{KM}}$–module

$$M'_{\mathrm{KM}}(\mu) := U(\mathfrak{g}_{\mathrm{KM}}) \otimes_{U(\mathfrak{p})} V_{\mathrm{KM}}(\mu) \tag{4}$$

that we call a generalised Verma module. It is clearly a homomorphic image of the Verma module $M_{\mathrm{KM}}(\mu)$. The subspace $1 \otimes V_{\mathrm{KM}}(\mu)$ of $M'_{\mathrm{KM}}(\mu)$ is a finite dimensional $\mathfrak{g}$–submodule of $M'_{\mathrm{KM}}(\mu)$ and generates $M'_{\mathrm{KM}}(\mu)$ as a $\mathfrak{g}_{\mathrm{KM}}$–module. It follows that $M'_{\mathrm{KM}}(\mu)$ considered as a $\mathfrak{g}$–module is locally finite, i.e., a sum of finite dimensional $\mathfrak{g}$–submodules. (If V is a finite dimensional $\mathfrak{g}$–submodule in a $\mathfrak{g}_{KM}$–module M, then each $\mathfrak{g}_{KM}^{[i]} V$ is a finite dimensional $\mathfrak{g}$–submodule. Therefore the sum of all finite dimensional $\mathfrak{g}$–submodules in M is a $\mathfrak{g}_{KM}$–submodule of M.)

Now also all subquotients of the $\mathfrak{g}_{KM}$–module $M'_{KM}(\mu)$ are locally finite for $\mathfrak{g}$. This applies in particular to all composition factors. If $L_{KM}(\mu')$ is a composition factor of $M'_{KM}(\mu)$, then we get $\mu' \in Y$. Otherwise there exists $i \geq 1$ with $\mu(h_i) \notin \mathbf{N}$; then the $\mathfrak{g}$–module generated by a highest weight vector v in $L_{KM}(\mu')$ is infinite dimensional since it contains the infinitely many linearly independent vectors $y^m v$, $m \geq 0$, with $y \in \mathfrak{g}_{-\alpha_i}$, $y \neq 0$.

One gets now as in 2.2 that there exist integers $c_{\mu\nu}$ with $c_{\mu\mu} = 1$ such that

$$\operatorname{ch} L_{KM}(\mu) = \sum_{\nu \in Y, \nu \leq \mu} c_{\mu\nu} \operatorname{ch} M'_{KM}(\nu) \tag{5}$$

for any $\mu \in Y$.

One can identify the Weyl group W of R with the subgroup of W_{KM} generated by all s_i with $i \geq 1$. One has then $(w\lambda)_{|\mathfrak{h}} = w(\lambda_{|\mathfrak{h}})$ and $(w\bullet\lambda)_{|\mathfrak{h}} = w\bullet(\lambda_{|\mathfrak{h}})$ for all $w \in W$ and $\lambda \in \mathfrak{h}_{KM}^*$. For any $\mu \in Y$ we can write 1.7(3) in the form $\operatorname{ch} V(\mu_{|\mathfrak{h}}) = \operatorname{ch} U(\mathfrak{n}^-) \sum_{w \in W} \det(w) e(w\bullet\mu_{|\mathfrak{h}})$. This implies $\operatorname{ch} V_{KM}(\mu) = \operatorname{ch} U(\mathfrak{n}^-) \sum_{w \in W} \det(w) e(w\bullet\mu)$. (Use that $(w\mu)(c) = \mu(c)$ and $(w\mu)(d) = \mu(d)$ for $w \in W$.) One gets now

$$\operatorname{ch} M'_{KM}(\mu) = \operatorname{ch} U(\mathfrak{q}^-) \operatorname{ch} V_{KM}(\mu) = \sum_{w \in W} \det(w) \operatorname{ch} M_{KM}(w\bullet\mu) \tag{6}$$

for all $\mu \in Y$.

Suppose now that $\mu \in Y$ with $\mu(h_0) \in \mathbf{Z}$ and $(\mu + \rho_{KM}, \delta) < 0$. There exists $z \in W_{KM}$ such that $\lambda = z^{-1}\bullet\mu$ satisfies $(\lambda + \rho_{KM})(h_i) \leq 0$ for all i, $0 \leq i \leq r$. Choose z of minimal length with this property. Then one gets from the generalisation of 2.6(2) that

$$\operatorname{ch} L_{KM}(\mu) = \operatorname{ch} L_{KM}(z\bullet\lambda) = \sum_{x \in W_{KM}} \det(xz) P_{x,z}^{KM}(1) \operatorname{ch} M_{KM}(x\bullet\lambda).$$

Comparing with (5) and (6) we get for any $\nu \in Y$ that $c_{\mu\nu}$ is equal to the sum of all $\det(xz) P_{x,z}^{KM}(1)$ with $x \in W_{KM}$ and $x\bullet\lambda = \nu$. We get thus

$$\operatorname{ch} L_{KM}(z\bullet\lambda) = \sum_{x \in W_{KM}, x\bullet\lambda \in Y} \det(xz) P_{x,z}^{KM}(1) \operatorname{ch} M'_{KM}(x\bullet\lambda). \tag{7}$$

3 Quantum groups

The quantum groups that we consider here should more properly be called quantised enveloping algebras. One can associate such an algebra to any symmetrisable Kac–Moody algebra. However, we restrict ourselves here to quantum groups corresponding to finite dimensional semi-simple complex Lie algebras. For more information on quantum groups we refer to [CP], [Ja2], [Jo2], and [Lu4].

3.1. Let $\mathfrak{g}$ be a finite dimensional semi-simple complex Lie algebra as in 1.1. We keep the notations like $\mathfrak{h}$, R, R^+, X, X_+, W from 1.1. In particular $\alpha_1, \alpha_2, \ldots, \alpha_r$ will be a basis for the root system R. We suppose that the W–invariant form on $\mathfrak{h}^*$ is normalised such that $(\alpha, \alpha) = 2$ for all short roots in any irreducible component of R. This normalisation implies that $(\lambda, \alpha) \in \mathbf{Z}$ for all $\alpha \in R$ and $\lambda \in X$. Set $d_i = (\alpha_i, \alpha_i)/2$ for all i; we have then $d_i \in \{1, 2, 3\}$.

Fix a complex number q that is transcendental over the rational numbers. Set $q_i = q^{d_i}$ for all i. Now the *quantised enveloping algebra* $U_q(\mathfrak{g})$ is the associative algebra over $\mathbf{C}$ with $4r$ generators E_i, F_i, K_i, K_i^{-1} $(1 \le i \le r)$ and certain relations.

Among these relations there are the requirements that each K_i^{-1} is indeed the multiplicative inverse to K_i and that each K_i commutes with each K_j. So the subalgebra $U_q^0(\mathfrak{g})$ generated by all K_i and K_i^{-1} is commutative. One can show that all $K_1^{m_1} K_2^{m_2} \ldots K_r^{m_r}$ with
$$(m_1, m_2, \ldots, m_r) \in \mathbf{Z}^r$$
form a basis for $U_q^0(\mathfrak{g})$. Any $\lambda \in X$ defines an algebra homomorphism from $U_q^0(\mathfrak{g})$ to $\mathbf{C}$ that we again denote by λ:
$$\lambda\colon U_q^0(\mathfrak{g}) \longrightarrow \mathbf{C}, \quad K_1^{m_1} K_2^{m_2} \ldots K_r^{m_r} \mapsto q^{(\lambda, m_1\alpha_1 + m_2\alpha_2 + \cdots + m_r\alpha_r)}. \quad (1)$$

Now some of the other defining relations for $U_q(\mathfrak{g})$ can be formulated as $KE_iK^{-1} = \alpha_i(K)E_i$ and $KF_iK^{-1} = \alpha_i(K)^{-1}F_i = (-\alpha_i)(K)F_i$ for all i and all $K \in U_q^0(\mathfrak{g})$.

The remaining relations are the quantum Serre relations (see the references) and the condition that $E_iF_j - F_jE_i = \delta_{ij}(K_i - K_i^{-1})/(q_i - q_i^{-1})$ for all i and j.

Let $U_q^+(\mathfrak{g})$ be the subalgebra of $U_q(\mathfrak{g})$ generated by all E_i and $U_q^-(\mathfrak{g})$ the subalgebra generated by all F_i. One gets then a triangular decomposition of $U_q(\mathfrak{g})$: The multiplication map induces an isomorphism
$$U_q^-(\mathfrak{g}) \otimes U_q^0(\mathfrak{g}) \otimes U_q^+(\mathfrak{g}) \xrightarrow{\sim} U_q(\mathfrak{g}) \quad (2)$$
of vector spaces.

3.2. If V is a $U_q^0(\mathfrak{g})$–module, then one defines weight spaces

$$V_\lambda := \{\, v \in V \mid u\,v = \lambda(u)\,v \text{ for all } u \in U_q^0(\mathfrak{g}) \,\} \qquad (1)$$

for all $\lambda \in X$. The sum of these V_λ is direct. If all V_λ are finite dimensional and if V is equal to their sum, then one defines a formal character $\operatorname{ch} V$ as in 1.2.

For example, $U_q^+(\mathfrak{g})$ and $U_q^-(\mathfrak{g})$ are $U_q^0(\mathfrak{g})$–modules such that each K_i acts by conjugation: we have $K_i \cdot x = K_i\,x\,K_i^{-1}$ for any x in $U_q^+(\mathfrak{g})$ or $U_q^-(\mathfrak{g})$. For this action one gets

$$\operatorname{ch} U_q^+(\mathfrak{g}) = \sum_{\mu \in X} P(\mu)\,e(\mu) \quad \text{and} \quad \operatorname{ch} U_q^-(\mathfrak{g}) = \sum_{\mu \in X} P(-\mu)\,e(\mu) \qquad (2)$$

where P is Kostant's partition function from 1.6. In other words, we have $\operatorname{ch} U_q^+(\mathfrak{g}) = \operatorname{ch} U(\mathfrak{n}^+)$ and $\operatorname{ch} U_q^-(\mathfrak{g}) = \operatorname{ch} U(\mathfrak{n}^-)$ where we regard $U(\mathfrak{n}^+)$ and $U(\mathfrak{n}^-)$ as $\mathfrak{h}$–modules via the adjoint action.

One can show that there exists for each dominant weight $\lambda \in X_+$ a finite dimensional simple $U_q(\mathfrak{g})$–module $V_q(\lambda)$ generated by a highest weight vector v_λ of weight λ. This means that $E_i\,v_\lambda = 0$ for all i and that $K\,v_\lambda = \lambda(K)\,v_\lambda$ for all $K \in U_q^0(\mathfrak{g})$. Then $V_q(\lambda)$ is the direct sum of all $V_q(\lambda)_\mu$ with $\mu \in X$ and it turns out that the character of $V_q(\lambda)$ is given by Weyl's character formula: We have

$$\operatorname{ch} V_q(\lambda) = \operatorname{ch} V(\lambda) \qquad (3)$$

for all $\lambda \in X_+$.

The $V_q(\lambda)$ do not exhaust the isomorphism classes of finite dimensional simple $U_q(\mathfrak{g})$–modules. One gets for each $\tau = (\tau_1, \tau_2, \ldots, \tau_r) \in \{\pm 1\}^r$ a one dimensional $U_q(\mathfrak{g})$–module $\mathbf{C}_\tau$ where each K_i acts as multiplication by τ_i whereas all E_i and F_i act as 0.

An arbitrary finite dimensional simple $U_q(\mathfrak{g})$–module is isomorphic to a tensor product $\mathbf{C}_\tau \otimes V_q(\lambda)$ with $\lambda \in X_+$ and τ as above. For this tensor product to make sense we have to introduce a comultiplication $\Delta : U_q(\mathfrak{g}) \to U_q(\mathfrak{g}) \otimes U_q(\mathfrak{g})$ on $U_q(\mathfrak{g})$. It can be chosen such that $\Delta(E_i) = E_i \otimes 1 + K_i \otimes E_i$ and $\Delta(F_i) = F_i \otimes K_i^{-1} + 1 \otimes F_i$ and $\Delta(K_i) = K_i \otimes K_i$ for all i. There are also a counit and an antipode on $U_q(\mathfrak{g})$ giving it the structure of a Hopf algebra.

3.3. The main object of our interest in this section is not $U_q(\mathfrak{g})$ itself, but a modification where q is replaced by a root of unity. There are two quite different ways of constructing such a quantum group at a root of unity and the algebras one gets have quite different properties. We shall here look at Lusztig's version of the construction that involves divided powers. (For a survey on the other version, first investigated by De Concini and Kac, see [DCP].)

For any $n \in \mathbf{Z}$ set $[n]_i := (q_i^n - q_i^{-n})/(q_i - q_i^{-1})$ for $1 \leq i \leq r$; this is a non-zero element in $\mathbf{Z}[q_i, q_i^{-1}]$ except for $[0]_i = 0$. In case $n \geq 0$ set $[n]_i^! := [1]_i [2]_i \dots [n]_i$; this means in particular that $[0]_i^! = 1$. One then calls all $E_i^{(n)} := E_i^n / [n]_i^!$ and $F_i^{(n)} := F_i^n / [n]_i^!$ with $n \in \mathbf{N}$ and $1 \leq i \leq r$ divided powers.

Set $\mathcal{A} = \mathbf{Z}[q, q^{-1}]$. Let $U_q(\mathfrak{g})_{\mathcal{A}}$ denote the $\mathcal{A}$–subalgebra of $U_q(\mathfrak{g})$ generated by all $E_i^{(n)}$ and $F_i^{(n)}$ as above together with all $K_i^{\pm 1}$. One can show that $U_q(\mathfrak{g})_{\mathcal{A}}$ contains all elements of the form

$$\begin{bmatrix} K_i; a \\ n \end{bmatrix} := \prod_{j=1}^{n} \frac{K_i q_i^{a-j+1} - K_i^{-1} q_i^{-(a-j+1)}}{q_i^j - q_i^{-j}} \tag{1}$$

with $a \in \mathbf{Z}$ and $n \in \mathbf{N}$.

3.4. For any $\zeta \in \mathbf{C}$, $\zeta \neq 0$, set

$$U_\zeta(\mathfrak{g}) := U_q(\mathfrak{g})_{\mathcal{A}} \otimes_{\mathcal{A}} \mathbf{C} \tag{1}$$

where we regard $\mathbf{C}$ as an $\mathcal{A}$–algebra with the unique ring homomorphism $\mathcal{A} \to \mathbf{C}$ such that $q \mapsto \zeta$.

By abuse of notation we write $E_i^{(n)}$ for the element $E_i^{(n)} \otimes 1 \in U_\zeta(\mathfrak{g})$. We use a similar convention for $F_i^{(n)}$, K_i, and $\begin{bmatrix} K_i; a \\ n \end{bmatrix}$. Set $U_\zeta^+(\mathfrak{g})$ resp. $U_\zeta^-(\mathfrak{g})$ equal to the subalgebra of $U_\zeta(\mathfrak{g})$ generated over $\mathbf{C}$ by all $E_i^{(n)}$ resp. all $F_i^{(n)}$ with $n \in \mathbf{N}$. Set $U_\zeta^0(\mathfrak{g})$ equal to the subalgebra generated by all $K_i^{\pm 1}$ and all $\begin{bmatrix} K_i; a \\ n \end{bmatrix}$ as above. Then the multiplication map induces an isomorphism

$$U_\zeta^-(\mathfrak{g}) \otimes U_\zeta^0(\mathfrak{g}) \otimes U_\zeta^+(\mathfrak{g}) \xrightarrow{\sim} U_\zeta(\mathfrak{g}) \tag{2}$$

of vector spaces.

The Hopf algebra structure on $U_q(\mathfrak{g})$ induces first a Hopf algebra structure on $U_q(\mathfrak{g})_{\mathcal{A}}$ and then one on $U_\zeta(\mathfrak{g})$.

3.5. The algebra $U_\zeta^0(\mathfrak{g})$ is commutative. Any $\lambda \in X$ induces an algebra homomorphism $U_\zeta^0(\mathfrak{g}) \to \mathbf{C}$, again denoted by λ, such that

$$\lambda(K_i) = \zeta^{d_i \lambda(h_i)} \qquad \text{and} \qquad \lambda(\begin{bmatrix} K_i; a \\ n \end{bmatrix}) = \begin{bmatrix} \lambda(h_i) + a \\ n \end{bmatrix}_{i, \zeta}$$

where $\begin{bmatrix} m \\ n \end{bmatrix}_{i, \zeta}$ denotes the image of $\begin{bmatrix} m \\ n \end{bmatrix}_i \in \mathcal{A}$ under the homomorphism $\mathcal{A} \to \mathbf{C}$ with $q \mapsto \zeta$. (One shows first that the homomorphism from 3.1(1) maps $U_q(\mathfrak{g})_{\mathcal{A}} \cap U_q^0(\mathfrak{g})$ to $\mathcal{A}$. Then one observes that $U_\zeta^0(\mathfrak{g})$ can be identified with $(U_q(\mathfrak{g})_{\mathcal{A}} \cap U_q^0(\mathfrak{g})) \otimes_{\mathcal{A}} \mathbf{C}$.)

The induced homomorphism $U_\zeta^0(\mathfrak{g}) \to \mathbf{C}$ determines λ uniquely. We can thus identify X with a subset of $\mathrm{Hom}_{\mathbf{C}-\mathrm{alg}}(U_\zeta^0(\mathfrak{g}), \mathbf{C})$. We define for each $U_\zeta^0(\mathfrak{g})$–module V weight spaces V_λ in analogy to 3.2(1), similarly for ch V.

Given $\lambda \in X_+$ and a highest weight vector v_λ in $V_q(\lambda)$ as in 3.2 set

$$V_q(\lambda)_{\mathcal{A}} := U_q(\mathfrak{g})_{\mathcal{A}} v_\lambda \qquad \text{and} \qquad V_\zeta(\lambda) := V_q(\lambda)_{\mathcal{A}} \otimes_{\mathcal{A}} \mathbf{C}. \qquad (2)$$

Then $V_\zeta(\lambda)$ is a $U_\zeta(\mathfrak{g})$–module satisfying ch $V_\zeta(\lambda) = $ ch $V(\lambda)$. Each $V_\zeta(\lambda)$ has a unique simple factor module $L_\zeta(\lambda)$.

Each $\tau = (\tau_1, \tau_2, \ldots, \tau_r) \in \{\pm 1\}^r$ yields also for $U_\zeta(\mathfrak{g})$ a one dimensional module $\mathbf{C}_\tau$ where any K_i acts as τ_i and any $\left[\begin{smallmatrix} K_i ; a \\ n \end{smallmatrix} \right]$ as $\tau_i^n \left[\begin{smallmatrix} a \\ n \end{smallmatrix} \right]_{i,\zeta}$. An arbitrary finite dimensional simple $U_\zeta(\mathfrak{g})$–module has then the form $\mathbf{C}_\tau \otimes V_\zeta(\lambda)$ with $\lambda \in X_+$ and τ as above.

3.6. If ζ is not a root of unity, then all $V_\zeta(\lambda)$ are simple and the character of $L_\zeta(\lambda) = V_\zeta(\lambda)$ is given by Weyl's formula.

So suppose now that ζ is a root of unity. There is an integer $\ell \geq 1$ such that ζ is a primitive ℓ–th root of unity. Let us assume that $\ell > 1$ and that ℓ is prime to all non-zero entries of the Cartan matrix of $\mathfrak{g}$. This means that ℓ is odd and that it is prime to 3 in case $\mathfrak{g}$ has a component of type G_2.

In this case one can show for any $\lambda \in X_+$ that all composition factors of $V_\zeta(\lambda)$ have the form $L_\zeta(\mu)$ with $\mu \in X_+ \cap W_a \bullet_\ell \lambda$. Here W_a is the *affine Weyl group* of R (or rather of the dual root system), i.e., the group of affine transformations of $\mathfrak{h}^*$ generated by W and by all translations by elements in $\mathbf{Z}R$. Furthermore we define $w \bullet_\ell \nu$ for any $\nu \in \mathfrak{h}^*$ as follows: If $w \in W$, then $w \bullet_\ell \nu = w \bullet \nu$; if w is the translation by $\gamma \in \mathbf{Z}R$, then $w \bullet_\ell \nu = \nu + \ell \gamma$.

As in previous situations it now follows that there are integers $b_{\mu\nu}$ for any $\mu \in X_+$ with $b_{\mu\mu} = 1$ such that

$$\mathrm{ch}\, L_\zeta(\mu) = \sum_\nu b_{\mu\nu}\, \mathrm{ch}\, V_\zeta(\nu) = \sum_\nu b_{\mu\nu}\, \mathrm{ch}\, V(\nu) \qquad (1)$$

where ν runs over all $\nu \in X_+ \cap W_a \bullet_\ell \mu$ with $\nu \leq \mu$.

Note that $-\zeta$ is a primitive 2ℓ–th root of unity under our assumptions. Andersen has shown in [A3] that

$$\mathrm{ch}\, L_{-\zeta}(\mu) = \mathrm{ch}\, L_\zeta(\mu) \qquad (2)$$

for all $\mu \in X_+$.

3.7. Continue to assume that ζ is a primitive ℓ-th root of unity with $\ell > 1$ odd and prime to 3 in case G_2 is involved.

Set

$$C^- := \{\, \lambda \in X \mid -\ell \le (\lambda + \rho)\,(h_\alpha) \le 0 \text{ for all } \alpha \in R^+ \,\}. \qquad (1)$$

This is a fundamental domain for the $\bullet_\ell$-action of W_a on X. So any $\mu \in X$ can be written in the form $\mu = x \bullet_\ell \lambda$ with $\lambda \in C^-$ and $x \in W_a$. We shall below assume that we do so with x of minimal length.

Here the term *minimal length* refers to a certain generating system that turns W_a into a Coxeter group. Our choice here consists of those elements that act as reflections with respect to the "walls" of C^-. This means explicitly that we take all simple reflections s_i, $1 \le i \le r$, from W together with the affine reflections $s_{\beta,-1}$ where β runs over the dominant short roots of the irreducible components of the root system R. (By definition $s_{\beta,-1}$ is given by $s_{\beta,-1}(\nu) = s_\beta(\nu) - \beta = \nu - (\nu(h_\beta) + 1)\beta$, hence satisfies $s_{\beta,-1} \bullet_\ell \nu = \nu - ((\nu + \rho)(h_\beta) + \ell)\beta$.)

This choice of generating system determines also Kazhdan–Lusztig polynomials $P_{x,w}$ for all $x, w \in W_a$. In case $x, w \in W$ these polynomials constructed working with W_a coincide with those constructed working with W. This is why we do not introduce a new notation.

Now one can show (see 3.8) under mild restriction for ℓ: Let $\mu \in X_+$ and $\lambda \in C^-$ and $w \in W_a$ such that $\mu = w \bullet_\ell \lambda$. If w has minimal length for $\mu = w \bullet_\ell \lambda$, then

$$\operatorname{ch} L_\zeta(w \bullet_\ell \lambda) = \sum_{x \in W_a,\, x \bullet_\ell \lambda \in X_+} \det(wx)\, P_{x,w}(1)\, \operatorname{ch} V(x \bullet_\ell \lambda). \qquad (2)$$

Here I write $\det(y)$ for the determinant of the linear part of any $y \in W_a$: If $y = y_1 y_2$ with $y_1 \in W$ and y_2 a translation, then $\det(y) = \det(y_1)$.

This character formula was conjectured by Lusztig in [Lu2] in case all irreducible components of R have only one root length.

3.8. The character formula 3.7(2) follows from the character formula 2.8(7) [proved by Kashiwara and Tanisaki] thanks to work by Kazhdan and Lusztig in [KL3–6] and [Lu5]. Here I can only say very little about this work.

One reduces easily to the case where $\mathfrak{g}$ is simple. Assume for the moment in addition that all roots in R have the same length, i.e., that R is of type A, D, or E. Let β denote the unique dominant root in R. Consider the affine Kac–Moody algebra $\mathfrak{g}_{KM}$ constructed by the procedure from 2.5 working with this choice of β.

We identify $\mathfrak{g}$ as in 2.8 with a subalgebra of $\mathfrak{g}_{KM}$ and $\mathfrak{h}$ with a subspace of $\mathfrak{h}_{\mathrm{KM}}$. Furthermore W is identified with a subgroup of W_{KM}. Now one shows in the theory of affine Kac–Moody algebras that W_{KM} is isomorphic to the affine Weyl group W_a. An isomorphism takes any s_i with $i \geq 1$ considered as an element in W_{KM} to s_i considered as an element in W_a, and it takes s_0 to $s_{\beta,-1}$.

We use this isomorphism to identify W_a and W_{KM}. One checks that this is compatible with the definition of $\det(y)$ for $y \in W_a$ following 3.7(2). Furthermore, since the identification is compatible with the generating sets used to define the Kazhdan–Lusztig polynomials, we get also

$$P_{x,w} = P^{\mathrm{KM}}_{x,w} \qquad \text{for all } x, w \in W_a. \tag{1}$$

Set $\mathfrak{h}^*_{\mathrm{KM},-\ell} := \{ \mu \in \mathfrak{h}^*_{\mathrm{KM}} \mid (\mu + \rho_{\mathrm{KM}})(c) = -\ell \}$ with c as in 2.6. Then $\mathfrak{h}^*_{\mathrm{KM},-\ell}$ is stable under the dot action of W_{KM} since $\alpha_i(c) = 0$ for all i, $0 \leq i \leq r$. Now an elementary calculation shows that

$$(w \bullet \mu)_{|\mathfrak{h}} = w \bullet \ell(\mu_{|\mathfrak{h}}) \qquad \text{for all } \mu \in \mathfrak{h}^*_{\mathrm{KM},-\ell} \text{ and } w \in W_{\mathrm{KM}}. \tag{2}$$

(It suffices to take $w = s_i$ with $0 \leq i \leq r$.)

Consider now $\lambda \in C^- \subset \mathfrak{h}^*$. We can find an element $\widetilde{\lambda} \in \mathfrak{h}^*_{\mathrm{KM}}$ such that $(\widetilde{\lambda} + \rho_{\mathrm{KM}})(c) = -\ell$ and $\widetilde{\lambda}_{|\mathfrak{h}} = \lambda$. (Recall that $\mathfrak{h}_{\mathrm{KM}} = \mathfrak{h} \oplus \mathbf{C}c \oplus \mathbf{C}d$; so we choose $(\widetilde{\lambda} + \rho_{\mathrm{KM}})(d)$ arbitrarily.) We have under our assumptions

$$(\widetilde{\lambda} + \rho_{\mathrm{KM}})(h_0) = (\widetilde{\lambda} + \rho_{\mathrm{KM}})(c) - (\widetilde{\lambda} + \rho_{\mathrm{KM}})(h_\beta) = -\ell - (\lambda + \rho)(h_\beta) \leq 0,$$

hence $(\widetilde{\lambda} + \rho_{\mathrm{KM}})(h_i) \leq 0$ for all i, $0 \leq i \leq r$. Therefore 2.8(7) holds (with λ replaced by $\widetilde{\lambda}$) for all $z \in W_{\mathrm{KM}}$ with $z \bullet \widetilde{\lambda} \in Y$.

The definition of Y in 2.8(3) shows that $z \bullet \widetilde{\lambda} \in Y$ if and only if $(z \bullet \widetilde{\lambda})_{|\mathfrak{h}} \in X_+$. Set $\mathcal{C}$ equal to the category of all $\mathfrak{g}_{\mathrm{KM}}$–modules M of finite length such that all composition factors of M have the form $L_{\mathrm{KM}}(w \bullet \widetilde{\lambda})$ with $w \in W_{\mathrm{KM}}$ and $w \bullet \widetilde{\lambda} \in Y$. Suppose that we have an exact functor $\mathcal{F}$ from $\mathcal{C}$ to the category of finite dimensional $U_\zeta(\mathfrak{g})$–modules taking $M'_{\mathrm{KM}}(\mu)$ to $V_\zeta(\mu_{|\mathfrak{h}})$ and $L_{\mathrm{KM}}(\mu)$ to $L_\zeta(\mu_{|\mathfrak{h}})$ for each $\mu \in Y \cap W_{\mathrm{KM}} \bullet \widetilde{\lambda}$. Then $\mathcal{F}$ induces a homomorphism of Grothendieck groups, and 2.8(7) implies 3.7(2) using (1) and (2) since we can regard these character formulae as identities in the Grothendieck groups.

Kazhdan and Lusztig basically construct such a functor in [KL3–6]. However there are some differences. First of all, they do not work with $\mathfrak{g}_{\mathrm{KM}}$, but with a subalgebra $\mathfrak{g}'_{\mathrm{KM}}$ of codimension 1. This subalgebra is the direct sum of all $\mathfrak{g}_{\mathrm{KM},\alpha}$ with $\alpha \in R_{\mathrm{KM}}$ and of $\mathfrak{h}'_{\mathrm{KM}} := \mathfrak{h} \oplus \mathbf{C}c$. So they drop the element $d \in \mathfrak{h}_{\mathrm{KM}}$. One now has first to check that the $\mathfrak{g}_{\mathrm{KM}}$–modules $L_{\mathrm{KM}}(\mu)$ still are irreducible when restricted to $\mathfrak{g}'_{\mathrm{KM}}$ and

that 2.8(7) is also a character formula for simple $\mathfrak{g}'_{KM}$–modules. For this
see Prop. 8.1 in [So4]; the result goes back to Polo.

Then Thm. 38.1 in [KL6] says that there is an equivalence of categories
between suitable categories of $\mathfrak{g}'_{KM}$–modules and of $U_{-\zeta}(\mathfrak{g})$–modules. (In
type E the proof in [KL6] requires that ℓ is not too small; it suffices to
assume $\ell > 31$, see [KL6], Cor. 2 to Lemma 31.5.) According to 9.1
in [Lu5] this equivalence takes $M'_{KM}(\mu)$ to $V_{-\zeta}(\mu_{|\mathfrak{h}})$ for each $\mu \in Y$ with
$(\mu + \rho_{KM})(c) = -\ell$. It then has to take the simple head $L_{KM}(\mu)$ to
$L_{-\zeta}(\mu_{|\mathfrak{h}})$. One gets now a character formula for simple $U_{-\zeta}(\mathfrak{g})$–modules.
To conclude the argument one applies 3.6(2).

In case R has two root lengths, this procedure has to be modified, see
[Lu5], 9.2. In the end one has to use a generalisation of 2.6(2) where the
weights no longer belong to X^{KM}.

4 Prime characteristic

Finally we turn to the analogues in characteristic $p > 0$ of the complex
semi-simple groups and their Lie algebras. Their representations have
been investigated for a longer period that those of Kac–Moody algebras
or quantum groups. However, our present knowledge in prime characteristic is weaker than that in the other areas, and some of the strongest
results in this area have been proved by comparing prime characteristic
and quantum groups at a root of unity. — For more background on
algebraic groups let me refer you as in 1.3 to [Bo1], [Hu2], and [Sp3];
for more information on representations of algebraic groups in prime
characteristic one may consult [Ja3].

4.1. Let k be an algebraically closed field of characteristic $p > 0$. We
consider a connected and simply connected algebraic group G_k over k;
denote its Lie algebra by $\mathfrak{g}_k$ and choose a maximal torus T_k in G_k.
We assume that G_k has the same type as our complex Lie algebra $\mathfrak{g}$.
This means that the root system of G_k with respect to T_k (the non-zero
weights of T_k on $\mathfrak{g}_k$ for the adjoint representation) can be identified with
the root system R of $\mathfrak{g}$. The assumption that G_k is simply connected implies that we can identify the group of characters of T_k with the group X
of integral weights from 1.1.

One can regard G_k as a universal Chevalley group as described (e.g.)
in [Hu1], 27.4. In particular one can identify $\mathfrak{g}_k$ with $\mathfrak{g}_\mathbf{Z} \otimes_\mathbf{Z} k$ where $\mathfrak{g}_\mathbf{Z}$
is the $\mathbf{Z}$–span of a Chevalley basis for $\mathfrak{g}$ as in [Hu1], 25.2.

4.2. A G_k–module is a vector space V over k with a homomorphism

$G_k \to GL(V)$ of algebraic groups. Then V is the direct sum of its weight spaces $V_\lambda = \{ v \in V \mid t v = \lambda(t) v$ for all $t \in T_k \}$. If V is finite dimensional, then we can define a formal character $\mathrm{ch}\, V \in \mathbf{Z}[X]$ as before. This character is W–invariant.

One can now show: Any simple G_k–module is finite dimensional. For any dominant $\mu \in X_+$ there exists a simple G_k–module $L_k(\mu)$ such that $\dim L_k(\mu)_\mu = 1$ and such that $L_k(\mu)_\nu \neq 0$ implies $\nu \leq \mu$. These two properties determine $L_k(\mu)$ uniquely up to isomorphism. We call $L_k(\mu)$ the *simple G_k–module with highest weight μ*. Every simple G_k–module is isomorphic to $L_k(\mu)$ for exactly one $\mu \in X_+$.

One can construct these simple modules using reduction modulo p: Inside the simple $\mathfrak{g}$–module $V(\mu)$ with highest weight μ one chooses a minimal admissible lattice $V_\mathbf{Z}(\mu)$ as in [Hu1], 27.3. Then $V_k(\mu) := V_\mathbf{Z}(\mu) \otimes_\mathbf{Z} k$ can be regarded as a G_k–module via the identification of G_k with a Chevalley group. Now $L_k(\mu)$ is the unique simple quotient of $V_k(\mu)$. The character of $V_k(\mu)$ is given by Weyl's character formula since the construction yields

$$\mathrm{ch}\, V_k(\mu) \;=\; \mathrm{ch}\, V(\mu) \qquad \text{for all } \mu \in X_+. \tag{1}$$

One usually calls $V_k(\mu)$ the *Weyl module with highest weight μ*.

4.3.   For any $\mu \in X_+$ the composition factors of the Weyl module $V_k(\mu)$ are $L_k(\mu)$ with multiplicity 1 and certain $L_k(\nu)$ with $\nu \in X_+$ and $\nu < \mu$. This implies that $\mathrm{ch}\, L_k(\mu)$ is a $\mathbf{Z}$–linear combination of the $\mathrm{ch}\, V_k(\nu) = \mathrm{ch}\, V(\nu)$ with $\nu \leq \mu$.

One can show more precisely that all composition factors of $V_k(\mu)$ have the form $L_k(\nu)$ with $\nu \in X_+ \cap W_a \bullet_p \mu$. Here W_a is the affine Weyl group of R as in 3.6 and the $\bullet_p$–action is defined as the $\bullet_\ell$–action from 3.6 taking $\ell = p$ there, i.e., we have $w \bullet_p \nu = w \bullet \nu$ for $w \in W$ whereas $w \bullet_p \nu = \nu + p\gamma$ in case w is the translation by $\gamma \in \mathbf{Z}R$. So we get as in 3.6: There are integers $b_{\mu\nu}$ for any $\mu \in X_+$ with $b_{\mu\mu} = 1$ such that

$$\mathrm{ch}\, L_k(\mu) \;=\; \sum_\nu b_{\mu\nu}\, \mathrm{ch}\, V(\nu) \tag{1}$$

where ν runs over all $\nu \in X_+ \cap W_a \bullet_p \mu$ with $\nu \leq \mu$.

In 1979 Lusztig stated in [Lu1] a conjecture for the coefficients $b_{\mu\nu}$ provided that μ is not "too large" and belongs to a set $\mathcal{M}_p$ that we now define. (The need for such a restriction on μ will be explained in 4.4.) Let h denote the maximum of all $\rho(h_\alpha) + 1$ with $\alpha \in R^+$; this is the *Coxeter number* of the root system R. For example, if $G_k = SL_n(k)$, then $h = n$. Set now

$$\mathcal{M}_p := \{ \mu \in X_+ \mid (\mu + \rho)(h_\alpha) \leq p(p - h + 2) \text{ for all } \alpha \in R^+ \}. \quad (2)$$

Note that $(\mu+\rho)(h_\alpha) > 0$ for all $\alpha \in R^+$; so we have $\mathcal{M}_p = \emptyset$ if $p \leq h-2$. As in 3.7 the set

$$C^- := \{ \lambda \in X \mid -p \leq (\lambda + \rho)(h_\alpha) \leq 0 \text{ for all } \alpha \in R^+ \}$$

is a fundamental domain for the $\bullet_p$–action of W_a on X. Lusztig's conjecture says now: Write $\mu \in X_+$ in the form $\mu = w\bullet_p\lambda$ with $\lambda \in C^-$ and $w \in W_a$ such that w has minimal length for $\mu = w\bullet_p\lambda$. If $w\bullet_p\lambda \in \mathcal{M}_p$, then

$$\operatorname{ch} L_k(w\bullet_p\lambda) = \sum_{x \in W_a,\, x\bullet_p\lambda \in X_+} \det(wx)\, P_{x,w}(1)\, \operatorname{ch} V(x\bullet_p\lambda). \quad (3)$$

Here as in 3.7 "minimal length" refers to the generating system of W_a given by the reflections with respect to the walls of C^-; similarly the Kazhdan–Lusztig polynomials are taken with respect to this system.

Now a comparison with 3.7(2) shows that we can reformulate Lusztig's conjecture as follows: If ζ is a primitive p–th root of unity, then

$$\operatorname{ch} L_k(\mu) = \operatorname{ch} L_\zeta(\mu) \qquad \text{for all } \mu \in \mathcal{M}_p. \quad (4)$$

Of course, when Lusztig made his conjecture, quantum groups were still unknown. At that time it was known that (3) holds for groups of rank at most 2 as well as for $G_k = SL_4(k)$.

In [AJS] it was proved that (4) holds if p is large enough. More precisely, given a root system R there exists an (unknown) bound $m(R)$ such that (4) holds whenever $p > m(R)$. We got this type of result by showing that there are systems of linear equations with coefficients in $\mathbf{Z}$ so that (4) holds if and only if the dimensions of the solution spaces do not change when one reduces the coefficients modulo p.

4.4. Recall that $\alpha_1, \alpha_2, \ldots, \alpha_r$ is our base for R. Set

$$X_p := \{ \mu \in X_+ \mid \mu(h_{\alpha_i}) < p \text{ for all } i,\, 1 \leq i \leq r \}. \quad (1)$$

Each $\mu \in X_+$ has a unique decomposition $\mu = p\mu_1+\mu_0$ with $\mu_0 \in X_p$ and $\mu_1 \in X_+$. In this situation *Steinberg's tensor product theorem* implies that

$$L_k(\mu) \simeq L_k(p\mu_1) \otimes L_k(\mu_0). \quad (2)$$

For any $\chi = \sum_{\nu \in X} a_\nu e(\nu) \in \mathbf{Z}[X]$ write $\chi^{(p)} = \sum_{\nu \in X} a_\nu e(p\nu) \in \mathbf{Z}[X]$.

(Here we have $a_\nu \in \mathbf{Z}$ for all ν.) Another consequence of Steinberg's tensor product theorem is

$$\operatorname{ch} L_k(p\mu) = \left(\operatorname{ch} L_k(\mu)\right)^{(p)} \qquad \text{for all } \mu \in X_+. \tag{3}$$

So we get for $\mu = p\mu_1 + \mu_0$ as above

$$\operatorname{ch} L_k(p\mu_1 + \mu_0) = \left(\operatorname{ch} L_k(\mu_1)\right)^{(p)} \operatorname{ch} L_k(\mu_0). \tag{4}$$

This implies: If we know all $\operatorname{ch} L_k(\mu)$ with $\mu \in X_p$, then we can compute inductively all $\operatorname{ch} L_k(\mu)$ with $\mu \in X_+$.

An elementary calculation shows: If $p \geq 2h - 3$, then X_p is contained in the set $\mathcal{M}_p$ from 4.3(2). So Lusztig's conjecture 4.3(3) leads to (conjectural) character formulae for all $\operatorname{ch} L_k(\mu)$ with $\mu \in X_+$ as long as $p \geq 2h - 3$.

Lusztig has proved an analogue to Steinberg's tensor product theorem for quantum groups. If ζ is a primitive p–th root of unity and if $\mu = p\mu_1 + \mu_0$ as above, then Lusztig's result implies

$$\operatorname{ch} L_\zeta(p\mu_1 + \mu_0) = \left(\operatorname{ch} V(\mu_1)\right)^{(p)} \operatorname{ch} L_\zeta(\mu_0). \tag{5}$$

A comparison of (5) and (4) indicates the need for a restriction on $w \bullet_p \lambda$ in 4.3(3): If we assume to start with that 4.3(4) holds for all $\mu_0 \in X_p$, i.e., that $\operatorname{ch} L_k(\mu_0) = \operatorname{ch} L_\zeta(\mu_0)$ for all these μ_0, then it will hold for $\mu = p\mu_1 + \mu_0$ as above if and only if $\operatorname{ch} L_k(\mu_1) = \operatorname{ch} V(\mu_1)$. Now the definition of $\mathcal{M}_p$ was done in such a way that the equality $\operatorname{ch} L_k(\mu_1) = \operatorname{ch} V(\mu_1)$ holds for all $\mu \in \mathcal{M}_p$. (Of course $\mathcal{M}_p$ was defined before quantum groups were introduced; the original justification for this bound takes longer to explain.)

One could now hope that 4.3(4) might hold for all $\mu \in X_p$, and for $p > h$ this seems to be a realistic conjecture. However, for smaller primes things go wrong: Andersen and I found an example for $G = SL_{p+3}$ where 4.3(4) fails for some $\mu \in X_p$, see [A2], 7.9. Note that in this example $h = p+3 > p$. At this point there is no conjecture that predicts character formulae for all $L_k(\mu)$ with $\mu \in X_p$ when p is small.

4.5. The set X_p has another important property: The $L_k(\mu)$ with $\mu \in X_p$ are also simple as modules for the Lie algebra $\mathfrak{g}_k$ of G_k. In fact, a theorem due to Curtis says that the $L_k(\mu)$ with $\mu \in X_p$ are a system of representatives for the isomorphism classes of simple *restricted* $\mathfrak{g}_k$–modules.

In order to explain the last statement we have to recall that the Lie algebra of an algebraic group in characteristic p has an extra structure: It comes with a p–th power map $x \mapsto x^{[p]}$ that makes the Lie algebra into a Lie p–algebra. If one regards $x \in \mathfrak{g}_k$ as a derivation (with certain invariance properties) on the algebra $k[G_k]$ of regular functions on G_k, then the p–th power of x is again a derivation (with the same invariance properties), hence an element of $\mathfrak{g}_k$ that we denote by $x^{[p]}$.

We call a representation $\varphi : \mathfrak{g}_k \to \mathrm{End}_k(V)$ and the corresponding $\mathfrak{g}_k$–module V restricted if φ satisfies $\varphi(x^{[p]}) = \varphi(x)^p$ for all $x \in \mathfrak{g}_k$. If M is a G_k–module, then the induced $\mathfrak{g}_k$–module is automatically restricted.

Regarding all $L_k(\mu)$ with $\mu \in X_p$ as simple $\mathfrak{g}_k$–modules is an essential ingredient in the proof of 4.3(4) for large p in [AJS]. Lusztig's conjecture is equivalent to a conjecture on the characters of the projective covers of these $L_k(\mu)$ in the category of all restricted $\mathfrak{g}_k$–modules, and these projective covers are the main objects in [AJS]. For a more detailed survey of that proof see [So2].

4.6. The approach to Lusztig's conjecture 4.3(3) described above, via quantum groups and Kac–Moody algebras to the geometry of flag varieties, goes back to an idea of Lusztig formulated shortly after the discovery of quantum groups. Of course, this is a rather indirect approach and one may ask whether there is not a more direct way. Now Bezrukavnikov has recently announced that one can avoid the tour via quantum groups, see [Be], Remark 3.13.

Let J_0 denote the ideal in $U(\mathfrak{g}_k)$ generated by all $x^p - x^{[p]}$ with $x \in \mathfrak{g}_k$. So a $\mathfrak{g}_k$–module is restricted if and only if it is annihilated by J_0. Set I_{0k} equal to the annihilator of the trivial one dimensional $\mathfrak{g}_k$–module k in the algebra $U(\mathfrak{g}_k)^{G_k}$ of those elements in the enveloping algebra that are invariant under the adjoint action of G_k. Over $\mathbf{C}$ the analogous algebra $U(\mathfrak{g})^G$ coincides with the centre of $U(\mathfrak{g})$. But in prime characteristic $U(\mathfrak{g}_k)^{G_k}$ is a proper subalgebra of the centre of $U(\mathfrak{g}_k)$; the centre of $U(\mathfrak{g}_k)$ contains in addition all $x^p - x^{[p]}$ with $x \in \mathfrak{g}_k$.

Let $\mathcal{C}_k$ denote the category of all finite dimensional $\mathfrak{g}_k$–modules that are annihilated by I_{0k} and by a power of J_0. A special case of Theorem 5.3.1 in [BMR] says: If p is greater than the Coxeter number h of R, then the bounded derived category $D^b(\mathcal{C}_k)$ is naturally equivalent with the bounded derived category $D^b(\mathcal{C}\mathrm{oh}(\mathcal{B}_k^{(1)}))$ of coherent sheaves over $\mathcal{O}_{\mathcal{B}_k^{(1)}}$. Here $\mathcal{B}_k$ is the flag variety of G_k and $\mathcal{B}_k^{(1)}$ indicates a Frobenius twist. So $\mathcal{B}_k^{(1)}$ coincides with $\mathcal{B}_k$ as a topological space with a sheaf of rings; but we change the k–algebra structure on these rings: If U is an open subset, the $\mathcal{O}_{\mathcal{B}_k^{(1)}}(U) = \mathcal{O}_{\mathcal{B}_k}(U)$, but any $a \in k$ acts on $\mathcal{O}_{\mathcal{B}_k^{(1)}}(U)$ as $\sqrt[p]{a}$ does on $\mathcal{O}_{\mathcal{B}_k}(U)$.

The equivalence of derived categories yields isomorphisms of Grothendieck groups $K(\mathcal{C}_k) \simeq K(\mathcal{C}\mathrm{oh}(\mathcal{B}_k^{(1)})) \simeq K(\mathcal{C}\mathrm{oh}(\mathcal{B}_k))$ (Note that one can identify the categories $\mathcal{C}\mathrm{oh}(\mathcal{B}_k^{(1)})$ and $\mathcal{C}\mathrm{oh}(\mathcal{B}_k)$.) The Grothendieck group $K(\mathcal{C}_k)$ is (for $p > h$) a free $\mathbf{Z}$–module of rank $|W|$; a basis are the classes of the simple modules $L_k(w \cdot 0 + p\rho_w)$ with $w \in W$ where $\rho_w \in X$ is chosen such that $w \cdot 0 + p\rho_w \in X_p$.

According to [Be], Cor. 3.12 the images of these simple modules in $K(\mathcal{C}\mathrm{oh}(\mathcal{B}_k))$ are independent of p in case $p \gg 0$. For this to make sense, one identifies the complexification $K(\mathcal{C}\mathrm{oh}(\mathcal{B}_k)) \otimes_{\mathbf{Z}} \mathbf{C}$ with the Borel–Moore homology $H^{\mathrm{BM}}_{\bullet}(\mathcal{B})$ of the flag variety $\mathcal{B}$ of G over $\mathbf{C}$. More precisely the claim in [Be] is that the classes of the simple modules are mapped to a "canonical" basis for $H^{\mathrm{BM}}_{\bullet}(\mathcal{B})$ constructed by Lusztig.

To get from this result to Lusztig's conjecture 4.3(3) requires some work not explained in [Be]. For example, one has to identify the images of the projective covers of the simple modules and one has to work with an enriched category where objects are graded by the weight lattice X.

Conclusion

The search for an algebraic proof of Weyl's character formula — the main topic of the first section of this survey — has changed our way of looking at those characters. It is now most natural to think of the character of a finite dimensional simple $\mathfrak{g}$–module as a linear combination of "simpler" characters, namely of the characters of Verma modules.

This point of view is then repeated in the generalisations we consider. We always try to express the characters of the simple modules in terms of known characters of "standard" modules. So far one has not found more "direct" formulae. Also inductive formulae similar to Freudenthal's are missing.

In all cases values of Kazhdan–Lusztig polynomials occur as coefficients (up to sign) when we write the character of a simple module as a linear combination of characters of standard modules. These polynomials are defined as coefficients expressing one basis of a suitable Hecke algebra in terms of another basis. They can then be computed inductively.

The known proofs relate these polynomials rather indirectly to the character formulae. They rely on the fact that the Kazhdan–Lusztig polynomials describe certain geometric data on flag varieties. On the other hand, the truth of the (Kazhdan–)Lusztig conjectures has usually strong applications to the structure of the standard modules. See (e.g.) [Ja3], Section II.C, for the prime characteristic case.

There is an additional case where Kazhdan–Lusztig polynomials (or rather some "periodic" generalisations by Lusztig) play a role. This is for certain non-restricted representations of $\mathfrak{g}_k$. In this case we have so far only experimental evidence from some examples and no general results. But the work in [BMR] and [Be] looks like steps in the right direction.

Bibliography

[A1] H. H. Andersen: A new proof of old character formulas, pp. 193–207 in R. Fossum et al. (eds.), *Invariant Theory*, Proc. Denton, TX, 1986 (Contemp. Math. **88**), Providence, RI, 1989 (Amer. Math. Soc.)

[A2] H. H. Andersen: Finite-dimensional representations of quantum groups, pp. 1–18 in W. F. Haboush, B. J. Parshall (eds.), *Algebraic Groups and their Generalizations: Quantum and Infinite-dimensional Methods*, Proc. University Park, Penn. 1991 (Proc. Sympos. Pure Math. **56:2**), Providence, R. I. 1994 (Amer. Math. Soc.)

[A3] H. H. Andersen: Quantum groups at roots of ± 1, *Commun. Algebra* **24** (1996), 3269–3282

[AJS] H. H. Andersen, J. C. Jantzen, W. Soergel: Representations of quantum groups at a pth root of unity and of semisimple groups in characteristic p: independence of p, *Astérisque* **220** (1994)

[BB] A. Beilinson, J. Bernstein: Localisation de $\mathfrak{g}$-modules, *C. R. Acad. Sci. Paris* (I) **292** (1981), 15–18

[BG] J. N. Bernstein, S. I. Gel'fand: Tensor products of finite- and infinite-dimensional representations of semisimple Lie algebras, *Compositio Math.* **41** (1980), 245–285

[BGG1] I. N. Bernshtein, I. M. Gel'fand, S. I. Gel'fand: Structure of representations generated by vectors of highest weight, *Funct. Anal. Appl.* **5** (1971), 1–8, translated from: Структура представлений, порожденных векторами старшего веса, Функц. анализ и его прил. **5:1** (1971)

[BGG2] I. N. Bernstein, I. M. Gelfand, S. I. Gelfand: Differential operators on the base affine space and a study of $\mathfrak{g}$-modules, pp. 21–64 in: I. M. Gelfand (ed.), *Lie Groups and their Representations* (Proc. Budapest 1971), London 1975 (A. Hilger)

[Be] R. Bezrukavnikov: Noncommutative counterparts of the Springer resolution, pp. 1119–1144 in: M. Sanz-Solé et al. (eds.), *Proceedings of the International Congress of Mathematicians — Madrid 2006*, vol. **II**, Zürich 2006 (European Math. Soc.)

[BMR] R. Bezrukavnikov, I. Mirković, D. Rumynin: Localization of modules for a semisimple Lie algebra in prime characteristic, *Annals of Math.* (to appear) [preprint: `math.RT/0205144`]

[Bo1] A. Borel: *Linear Algebraic Groups*, 2nd ed. (Graduate Texts in Math. **126**), New York etc. 1991 (Springer)

[Bo2] A. Borel: Hermann Weyl and Lie groups, pp. 53–82 in: K. Chandrasekaran (ed.), *Hermann Weyl 1885 – 1985*, Heidelberg etc. 1986 (Springer) = pp. 136–165 in Oeuvres · Collected Papers, vol. IV

[Br1] R. Brauer: *Über die Darstellung der Drehungsgruppe durch Gruppen linearer Substitutionen*, Diss., Berlin (1926) = pp. 335–403 in Collected Papers, vol. III

[Br2] R. Brauer: Sur la multiplication des caractéristiques des groupes continus et semi-simples, *C. R. Acad. Sci. Paris* **204** (1937), 1784–1786 = pp. 472–474 in Collected Papers, vol. III

[BtD] Th. Bröcker, T. tom Dieck: *Representations of Compact Lie Groups* (Graduate Texts in Math. **98**), New York etc. 1985 (Springer)

[BK] J-L. Brylinski, M. Kashiwara: Kazhdan-Lusztig conjecture and holonomic systems, *Invent. math.* **64** (1981), 387–410

[Ca] R. W. Carter: *Lie Algebras of Finite and Affine Type* (Cambridge Studies in Advanced Math. **96**), Cambridge 2005 (Cambridge Univ.)

[Ca] P. Cartier: On H. Weyl's character formula, *Bull. Amer. Math. Soc.* **67** (1961), 228–230

[Cs] L. Casian: Proof of the Kazhdan-Lusztig conjecture for Kac-Moody algebras (the characters ch $L_{\omega\rho-\rho}$), *Adv. Math.* **119** (1996), 207–281

[CP] V. Chari, A. Pressley: *A Guide to Quantum Groups*, Cambridge 1994 (Cambridge Univ.)

[Ch] C. Chevalley: Sur la classification des algèbres de Lie simples et de leurs représentations. *C. R. Acad. Sci. Paris* **227** (1948), 1136–1138

[DCP] C. De Concini, C. Procesi: Quantum groups, pp. 31–140 in: G. Zampieri, A. D'Agnolo (eds.), *D-modules, Representation Theory, and Quantum Groups*, Proc. Venezia 1992 (Lecture Notes in Math. **1565**), Berlin etc. 1993 (Springer)

[De] M. Demazure: Sur la formule des caractères de H. Weyl, *Invent. math.* **9** (1969/1970), 249–252

[DGK] V. V. Deodhar, O. Gabber, V. Kac: Structure of some categories of representations of infinite-dimensional Lie algebras, *Adv. in Math.* **45** (1982), 92–116

[Do] S. Donkin: *Rational Representations of Algebraic Groups* (Lecture Notes in Math. **1140**), Berlin etc. 1985 (Springer)

[E] T. J. Enright: *Lectures on Representations of Complex Semisimple Lie Groups*, Lecture notes by Vyjayanthi Sundar (TIFR Lectures on Math. and Physics **66**), Berlin etc. 1981 (Springer)

[Fi] P. Fiebig: The combinatorics of category $\mathcal{O}$ over symmetrizable Kac-Moody algebras, *Transform. Groups* **11** (2006), 29–49

[F] H. Freudenthal: Zur Berechnung der Charaktere der halbeinfachen Lieschen Gruppen I, *Indag. Math.* **16** (1954), 369–376

[HC] Harish-Chandra: On some applications of the universal enveloping algebra of a semisimple Lie algebra. *Trans. Amer. Math. Soc.* **70** (1951), 28–96

[Hu1] J. E. Humphreys: *Introduction to Lie Algebras and Representation Theory* (Graduate Texts in Math. **9**), New York etc. 1972 (Springer)

[Hu2] J. E. Humphreys: *Linear Algebraic Groups* (Graduate Texts in Math. **21**), New York etc. 1975 (Springer)

[Ja1] J. C. Jantzen: Zur Charakterformel gewisser Darstellungen halbeinfacher Gruppen und Lie-Algebren, *Math. Z.* **140** (1974), 127–149

[Ja2] J. C. Jantzen: *Lectures on Quantum Groups* (Graduate Studies in Math. ̃**6**), Providence, R. I., 1996 (Amer. Math. Soc.)

[Ja3] J. C. Jantzen: *Representations of Algebraic Groups*, 2nd ed. (Math. Surveys and Monographs **107**), Providence, R. I., 2003 (Amer. Math. Soc.)

[Jo1] A. Joseph: Dixmier's problem for Verma and principal series submodules, *J. London Math. Soc. (2)* **20** (1979), 193–204

[Jo2] A. Joseph: *Quantum Groups and Their Primitive Ideals* (Ergebnisse der Math. (3) **29**), Berlin etc. 1995 (Springer)

[Kc1] V. G. Kac: Infinite-dimensioned Lie algebras and Dedekind's η-function, *Funct. Anal. Appl.* **8** (1974), 68–70, translated from: Бесконечномерные алгебры Ли и η-функция Дедекинда, Функц. анализ и его прил. **8**:**1** (1971), 77–78

[Kc2] V. G. Kac: *Infinite Dimensional Lie Algebras*, 3rd ed., Cambridge 1990 (Cambridge Univ.)

[KK] V. G. Kac, D. A. Kazhdan: Structure of representations with highest weight of infinite-dimensional Lie algebras, *Adv. in Math.* **34** (1979), 97–108

[Ks] M. Kashiwara: Kazhdan-Lusztig conjecture for a symmetrizable Kac-Moody Lie algebra pp. 407–433 in: P. Cartier et al. (eds.), *The Grothendieck Festschrift II* (Progress in Math. **87**), Boston, MA, 1990 (Birkhäuser)

[KT1] M. Kashiwara, T. Tanisaki: Kazhdan-Lusztig conjecture for symmetrizable Kac-Moody Lie algebra II. Intersection cohomologies of Schubert varieties, pp. 159–195 in: A. Connes et al. (eds.), *Operator Algebras, Unitary Representations, Enveloping Algebras, and Invariant Theory*, Proc. Paris 1989, (Progress in Math. **92**), Boston, MA, 1990 (Birkhäuser)

[KT2] M. Kashiwara, T. Tanisaki: Kazhdan-Lusztig conjecture for affine Lie algebras with negative level, *Duke Math. J.* **77** (1995), 21–62

[KT3] M. Kashiwara, T. Tanisaki: On Kazhdan-Lusztig conjectures, *Sugaku Expositions* **11** (1998), 177–195, transl. from: *Sūgaku* **47** (1995), 268–285

[KT4] M. Kashiwara, T. Tanisaki: Kazhdan-Lusztig conjecture for affine Lie algebras with negative level II. Non-integral case, *Duke Math. J.* **84** (1996), 771–813

[KT5] M. Kashiwara, T. Tanisaki: Kazhdan-Lusztig conjecture for symmetrizable Kac-Moody Lie algebras III. Positive rational case, *Asian J. Math.* **2** (1998), 779–832

[KT6] M. Kashiwara, T. Tanisaki: Characters of irreducible modules with non-critical highest weights for affine Lie algebras, pp. 275–296 in: J. Wang, Z. Lin (eds.), *Representations and Quantizations*, Proc. Shanghai 1998, Beijing 2000 (China High. Educ. Press)

[KL1] D. Kazhdan, G. Lusztig: Representations of Coxeter groups and Hecke algebras, *Invent. math.* **53** (1979), 165–184

[KL2] D. Kazhdan, G. Lusztig: Schubert varieties and Poincaré duality, pp. 185–203 in: R. Osserman, A. Weinstein (eds.), *Geometry of the Laplace Operator*, Proc. Honolulu 1979 (Proc. Symp. Pure Math. **36**), Providence, R. I. 1980 (Amer. Math. Soc.)

[KL3] D. Kazhdan, G. Lusztig: Tensor structures arising from affine Lie algebras I, *J. Amer. Math. Soc.* **6** (1993), 905–947

[KL4] D. Kazhdan, G. Lusztig: Tensor structures arising from affine Lie algebras II, *J. Amer. Math. Soc.* **6** (1993), 949–1011

[KL5] D. Kazhdan, G. Lusztig: Tensor structures arising from affine Lie algebras III, *J. Amer. Math. Soc.* **7** (1994), 335–381

[KL6] D. Kazhdan, G. Lusztig: Tensor structures arising from affine Lie algebras IV, *J. Amer. Math. Soc.* **7** (1994), 383–453

[Ko] B. Kostant: A formula for the multiplicity of a weight, *Trans. Amer. Math. Soc.* **93** (1959), 53–73

[Ku1] S. Kumar: Extension of the category $\mathcal{O}^g$ and a vanishing theorem for the Ext functor for Kac–Moody algebras, *J. Algebra* **108** (1987), 472–491

[Ku2] S. Kumar: Demazure character formula in arbitrary Kac-Moody setting, *Invent. math.* **89** (1987), 395–423

[Ku3] S. Kumar: Toward proof of Lusztig's conjecture concerning negative level representations of affine Lie algebras, *J. Algebra* **164** (1994), 515–527

[Li1] P. Littelmann: A Littlewood-Richardson rule for symmetrizable Kac-Moody algebras, *Invent. math.* **116** (1994), 329–346

[Li2] P. Littelmann: Paths and root operators in representation theory, *Ann. of Math. (2)* **142** (1995), 499–525

[Lu1] G. Lusztig: Some problems in the representation theory of finite Chevalley groups, pp. 313–317 in B. Cooperstein, G. Mason (eds.), *The Santa Cruz Conference on Finite Groups*, Proc. 1979 (Proc. Symp. Pure Math. **37**), Providence, R. I. 1980 (Amer. Math. Soc.)

[Lu2] G. Lusztig: Modular representations and quantum groups, pp. 59–77 in A. J. Hahn et al. (eds.), *Classical Groups and Related Topics*, Proc. Beijing 1987 (Contemp. Math. **82**), Providence, R. I. 1989 (Amer. Math. Soc.)

[Lu3] G. Lusztig: On quantum groups, *J. Algebra* **131** (1990), 466–475

[Lu4] G. Lusztig: *Introduction to Quantum Groups* (Progress in Math. **110**), Boston, MA, 1993 (Birkhäuser)

[Lu5] G. Lusztig: Monodromic systems on affine flag manifolds, *Proc. Roy. Soc. London* (A) **445** (1994), 231–246 and **450** (1995), 731–732

[M] O. Mathieu: Formules de caractères pour les algèbres de Kac–Moody générales, *Astérisque* **159-160** (1988)

[MP] R. V. Moody, A. Pianzola: *Lie Algebras with Triangular Decompositions*, New York, 1995 (Wiley)

[S1] I. Schur: *Über eine Klasse von Matrizen, die sich einer gegebenen Matrix zuordnen lassen*, Diss., Berlin 1901 = pp. 1–72 in Ges. Abh. Band I

[S2] I. Schur: Neue Anwendung der Integralrechnung auf Probleme der Invariantentheorie II. Über die Darstellung der Drehungsgruppe durch lineare homogene Substitutionen, *Sitzungsber. Preuss. Akad. Wiss.* **1924**, 297–321 = pp. 460–484 in Ges. Abh. Band II

[Sl] P. Slodowy: The early development of the representation theory of semisimple Lie groups: A. Hurwitz, I. Schur, H. Weyl, *Jber. Deutsche Math.-Verein.* **101** (1999), 97–115

[So1] W. Soergel: Kategorie $\mathcal{O}$, perverse Garben und Moduln über den Koinvarianten zur Weylgruppe, *J. Amer. Math. Soc.* **3** (1990), 421–445

[So2] W. Soergel: Roots of unity and positive characteristic, pp. 315–338 in: B. Allison, G. Cliff (eds.), *Representations of Groups*, Proc. Banff 1994 (CMS Conf. Proc. **16**), Providence, R. I. 1995 (Amer. Math. Soc.)

[So3] W. Soergel: Kazhdan-Lusztig polynomials and a combinatoric[s] for tilting modules, *Represent. Theory* **1** (1997), 83–114

[So4] W. Soergel: Character formulas for tilting modules over Kac–Moody algebras, *Represent. Theory* **2** (1998), 432–448

[Sp1] T. A. Springer: Weyl's character formula for algebraic groups, *Invent. math.* **5** (1968), 85–105

[Sp2] T. A. Springer: Quelques applications de la cohomologie d'intersection, pp. 249–273 [exp. 589] in: *séminaire bourbaki 1981/82* (Astérisque **92-93**), Paris 1982 (Soc. Math. France)

[Sp3] T. A. Springer: *Linear Algebraic Groups* (Progress in Math. **9**), 2nd ed., Boston etc. 1998 (Birkhäuser)

[St] R. Steinberg: A general Clebsch-Gordan theorem, *Bull. Amer. Math. Soc.* **67** (1961), 406–407

[Wa] Z. X. Wan: *Introduction to Kac-Moody Algebra*, Teaneck, NJ, 1991 (World Scientific)

[W1] H. Weyl: Theorie der Darstellung kontinuierlicher halb-einfacher Gruppen durch lineare Transformationen I, *Math. Z.* **23** (1925), 271–309 = pp. 543–579 in Ges. Abh. Band II

[W2] H. Weyl: Theorie der Darstellung kontinuierlicher halb-einfacher Gruppen durch lineare Transformationen II, *Math. Z.* **24** (1926), 328–376 = pp. 580–628 in Ges. Abh. Band II

[W3] H. Weyl: Theorie der Darstellung kontinuierlicher halb-einfacher Gruppen durch lineare Transformationen III, *Math. Z.* **24** (1926), 377–395 and 789–791 = pp. 629–647 in Ges. Abh. Band II

[W4] H. Weyl: *The Classical Groups, Their Invariants and Representations* (Princeton Math. Series **1**), 2nd ed., Princeton 1946 (Princeton Univ.)

12

The Classification of Affine Buildings

Richard M. Weiss

Department of Mathematics
Tufts University
rweiss@tufts.edu

1 Introduction

The theory of affine buildings reveals fascinating links between group theory, Euclidean geometry and number theory. In particular, reflections, the Weyl chambers of root systems and valuations of fields all play a central role in their classification.

The study of affine buildings was begun by Bruhat and Tits in [4] and the classification of affine buildings of rank at least four was completed by Tits in [11]. When combined with the classification of Moufang polygons carried out in [12], the Bruhat-Tits classification covers also affine buildings of rank three under the assumption in this case that the building at infinity is Moufang.

Our goal here is to give a very brief overview of this work. All the details can be found in the forthcoming book [14] (as well, of course, as in [4] and [11]).

In this article we regard buildings exclusively as certain edge-colored graphs. For different points of view, see [1]. Other excellent sources of results about affine buildings are [3], [5] and [8].

2 Buildings

Let Σ be an edge-colored graph and let I denote the set of colors appearing on the edges of Σ. We call $|I|$ the *rank* of Σ. For each subset J of I let Σ_J be the graph obtained from Σ by deleting all the edges whose color is *not* in J (but without deleting any vertices). A *J-residue* of Σ for some subset J of I is a connected component of the graph Σ_J. Thus two distinct J-residues (for a fixed subset J of I) are always disjoint.

A *panel* is a residue of rank one. In other words, a panel is a maximal connected monochromatic subgraph.

A *chamber system* is a connected edge-colored graph Σ such that all panels are complete graphs having at least two vertices. Thus, in particular, for every vertex x of a chamber system and every color $i \in I$, there exists a vertex y such that $\{x, y\}$ is an edge of color i. According to convention, the vertices of a chamber system are called *chambers*. A chamber system is called *thin* (resp. *thick*) if all its panels are of size exactly two (resp. at least three).

A chamber system of rank one is just a complete graph. Suppose that Σ is a chamber system of rank two. Let Γ be the graph whose vertices are the panels of Σ, where two panels are joined by an edge whenever their intersection contains a chamber. Two panels of the same color are always disjoint. Thus Γ is bipartite. Since Σ is connected, so is Γ. Since every panel of Σ contains at least two chambers, every vertex of Γ has at least two neighbors. Suppose, conversely, that Γ is a connected bipartite graph such that every vertex has at least two neighbors. Then there is a unique decomposition of the vertex set of Γ into "black" and "white" vertices, so that every edge joins two vertices of different colors. Let Σ be the edge-colored graph with color set $I = \{\text{black, white}\}$ whose vertices are the edges of Γ, where two edges of Γ are joined by a black (resp. white) edge of Σ whenever their intersection contains a black (resp. white) vertex of Γ. Then Σ is a chamber system of rank two. These two constructions are inverses of each other. Thus a chamber system of rank two is essentially the same thing as a connected bipartite graph.

Let Π be a *Coxeter graph* with *vertex* set I. This means that Π is an edge-colored graph whose edge "colors" are elements in the set

$$\{n \in \mathbb{Z} \mid n \geq 3\} \cup \{\infty\}.$$

The "color" of an edge of a Coxeter graph is usually referred to as its "label." We denote the label on an edge e by m_e. Associated with Π is the corresponding *Coxeter group* W. This is the group generated by the vertex set I of Π that is defined by the relations $i^2 = 1$ for all $i \in I$, $(ij)^{m_e} = 1$ for each edge e such that $m_e \neq \infty$, where i and j are the elements of I joined by e, and $(ij)^2 = 1$ for all pairs of distinct elements i and j of I that are not adjacent in Π. Let Σ_Π be the edge-colored graph with vertex set W and color set I, where two elements x and y of W are joined by an edge of color i whenever $y = xi$. Then Σ_Π is a chamber system of rank $|I|$ with panels of size two (i.e. Σ_Π is thin).

Notice that I is now simultaneously the vertex set of Π, a distinguished generating set of W and the set of colors on the edges of Σ_Π.

If, for example, Π has just two vertices and n is the label on its one edge, then W is the dihedral group D_{2n} of order $2n$ and Σ_Π is a circuit having $2n$ vertices whose edges are colored alternately black and white.

We can now give the definition of a building. Let Π be a Coxeter diagram with vertex set I. A *building of type* Π is a chamber system Δ with color set I containing a distinguished set of subgraphs called *apartments* all isomorphic (as edge-colored graphs) to the chamber system Σ_Π such that the following hold:

(i) For every pair of chambers of Δ, there exists an apartment containing them both.

(ii) If A_1 and A_2 are apartments both containing chambers x and y, then there exists a (color-preserving) isomorphism from A_1 to A_2 fixing x and y.

(iii) If A_1 and A_2 are apartments both containing a chamber x and both containing chambers in a panel P, then there exists a (color-preserving) isomorphism from A_1 to A_2 fixing x and mapping $A_1 \cap P$ to $A_2 \cap P$.

We observe explicitly that nothing is said in the definition of a building about the automorphism group of the building; in particular, the isomorphisms in conditions (ii) and (iii) are not necessarily induced by automorphisms of Δ.

A building of type Π is called *irreducible* if the Coxeter diagram Π is connected. Buildings which are not irreducible can be decomposed as a kind of direct product and studied one "component" at a time. From now on, we will simply assume that all buildings are irreducible, usually without saying this explicitly.

Notice that Σ_Π is itself a building of type Π. This is the unique *thin* building of type Π. From now on, we will assume that all buildings are thick, usually without saying this explicitly.

Before going on to the next section, we introduce the notion of a *root*: A *reflection* of the Coxeter chamber system Σ_Π is a color-preserving automorphism of Σ_Π fixing an edge. If r is a reflection of Σ_Π, we denote by M_r the set of edges fixed by r. A *wall* is a set of edges of Σ_Π of the form M_r for some reflection r. Every edge of Σ_Π lies in a unique wall. Let $M = M_r$ be a wall and let Σ_Π^M denote the graph obtained from Σ_Π by deleting all the edges in M but without deleting any chambers. Then the graph Σ_Π^M has exactly two connected components, and they are interchanged by the reflection r. A *root* of Σ_Π is the set of chambers in a connected component of Σ_Π^M for some wall M.

If Δ is a building of type Π, then each apartment is isomorphic to Σ_Π. It thus makes sense to talk about roots of an apartment of Δ.

3 Spherical Buildings

A Coxeter diagram Π is called *spherical* if the corresponding Coxeter group W (i.e. the chamber set of the chamber system Σ_Π) is finite. A building is called *spherical* if it is of spherical type. Equivalently (since the apartments of a building of type Π are all isomorphic to Σ_Π), a building is spherical if its apartments are finite.

Suppose that Π is connected and has only two vertices. Then Π is spherical if and only if the label n on its unique edge is finite. Suppose that, in fact, n is finite, and let Δ be a building of type Π. Then Δ is a chamber system of rank two. Let Γ be the corresponding bipartite graph as described in §1. Then every vertex of Γ has at least three neighbors (i.e. Γ is "thick"), the diameter of Γ (with respect to the metric on the vertex set of Γ which counts the minimal number of edges needed to go from one vertex to another) is n and the length of its minimal circuits (i.e. its *girth*) is $2n$. Moreover, all thick bipartite graphs of diameter n and girth $2n$ arise in this way.

A thick bipartite graph of diameter n and girth $2n$ is called a *generalized polygon* (or, more specifically, a *generalized n-gon*). Thus spherical buildings of rank two whose Coxeter diagram has label n and generalized n-gons are essentially the same thing.

Let Γ be a *generalized triangle* (i.e. a generalized 3-gon). Then Γ is, in particular, bipartite, so there is a unique decomposition of the vertex set of Γ into "white" and "black" vertices so that every edge joins vertices of different color (as was already observed in §1). We now call the white vertices "points" and the black vertices "lines" and we declare that a "point" is incident with a "line" if they are connected by an edge of Γ. In this fashion we obtain a geometry of points and lines in which any two "points" lie on a unique "line" and any two "lines" contain a unique "point."[1] Thus Γ is the "incidence graph" of a projective plane.

1 Here is the proof: Let x and y be two points. (We omit the quotation marks.) Every path in Γ beginning and ending at points must pass alternately through points and lines and hence has even length. Thus the distance from x to y is even. Since the diameter of Γ is three, the distance from x to y must be two. In other words, there is a line z adjacent to both x and y. If z' were a second such line, then (x, z, y, z', x) would be a circuit of length four. Since the girth of Γ is six, it

Conversely, the incidence graph of an arbitrary projective plane is a generalized triangle. Thus generalized triangles and projective planes are essentially the same thing.

The most typical projective planes arise as the set of one-dimensional subspaces (the "points") and two-dimensional subspaces (the "lines") of a right vector space over a field K (skew or commutative), where incidence is given by containment. Choose such a vector space V and let Γ be the corresponding generalized triangle. If $\{v_1, v_2, v_3\}$ is a basis of V over K, then the subgraph of Γ spanned by the three "points" $\langle v_1 \rangle$, $\langle v_2 \rangle$ and $\langle v_3 \rangle$ and the three "lines" $\langle v_1, v_2 \rangle$, $\langle v_2, v_3 \rangle$ and $\langle v_3, v_1 \rangle$ is an apartment Σ of Γ (i.e. a circuit of length six), and every apartment of Γ arises in this way. The vector space V is, in some sense, constructed from a combinatorial object, namely the basis $\{v_1, v_2, v_3\}$, and an algebraic object, namely the field K. Thus also the generalized triangle Γ can be viewed as a geometrical object constructed from a combinatorial object, namely the apartment Σ, and an algebraic object, namely the field, which serves to transform the thin apartment Σ into the thick building Δ.

In [9], Tits classified all spherical buildings of rank ℓ at least three. In [12], this classification was extended to the case $\ell = 2$ under the assumption that the spherical building is *Moufang*.[2]

The classification of Moufang spherical buildings of rank at least two shows that these buildings are all like the projective planes described above in that they are uniquely determined by the structure of a fixed apartment Σ (i.e. by the Coxeter diagram Π) and an algebraic structure Λ (a kind of "thickening agent"). The algebraic structure Λ depends on the family of buildings under consideration. For some families it is a field but for others, Λ is a skew-field, an anisotropic quadratic space, a Jordan division algebra, a composition algebra or one of an assortment of more exotic structures. In each case, the algebraic structure Λ parametrizes the root groups associated with all the roots of a fixed apartment Σ of Δ, and Δ is, in turn, uniquely determined by this "root datum."

---

follows that z is unique. The proof that every two lines intersect in a unique point is obtained by simply interchanging the words "point" and "line."

2 In fact, *all* irreducible thick spherical buildings of rank least three as well as all the residues of rank two of such a building are Moufang. This is a consequence of the main theorem 4.1.2 in [9]; see 11.6 in [13]. Here is the definition of the Moufang condition: Let a be a root of an apartment Σ (as defined at the end of §2) in a spherical building Δ. The *root group* U_a is the group consisting of all automorphisms of Δ which acts trivially on every panel containing two chambers in a. The Moufang property says that for every root a, the root group U_a acts transitively on the set of all apartments of Δ containing a (among which Σ is only one).

276   *Richard M. Weiss*

In each case, the algebraic structure Λ is defined over a field which we denote by K.[3] We say, in fact, that the building Δ itself is *defined over* K. In some cases, K equals Λ but in some cases it does not. For example, when Λ is an anisotropic quadratic space or a Jordan division algebra, K is the field over which the underlying vector space J of Λ is defined. In these two cases, some of the root groups are isomorphic to the additive group of K and the others to the additive group J.

Let Ξ denote the set of roots of our fixed apartment Σ. There is always a canonical subset Ξ_0 of Ξ such that for each $a \in \Xi_0$, U_a is isomorphic to the additive group of K.[4] The subset Ξ_0 of Ξ will be important in §5.

4 Affine Buildings

Suppose that Π is the Coxeter diagram with two vertices whose unique edge has the label ∞. Then the bipartite graphs corresponding to buildings of type Π are trees such that every vertex has at least two neighbors, and the apartments of these buildings are the subtrees A such that each vertex of A has exactly two neighbors in A. Thus we can think of the apartments as one-dimensional affine spaces partitioned by the integers. This is the simplest example of an *affine* building.

An *affine Coxeter diagram* is a diagram Π each connected component of which is one of the irreducible diagrams in Figure 12.1 and an *affine building* is a building whose Coxeter diagram is affine.

Each diagram in Figure 12.1 has a name of the form $\tilde{X}_\ell$, where X_ℓ is the name of one of the spherical diagrams in Figure 12.2. Here ℓ is the number of vertices of X_ℓ and also the number of vertices of $\tilde{X}_\ell$ minus one. In each case the diagram X_ℓ can be obtained from $\tilde{X}_\ell$ by deleting a single vertex x and all the edges containing x. (The vertices of the diagram $\tilde{X}_\ell$ with this property are called *special*. Thus, for example, every vertex of $\tilde{A}_\ell$ for $\ell \geq 2$ is special, the special vertices of $\tilde{B}_\ell$ are the two at the left and the special vertices of $\tilde{C}_\ell$ are the vertices at the two extremes.)

We call the spherical diagram X_ℓ the *derived diagram* of $\tilde{X}_\ell$. For

3 We must, in fact, allow K to be a skew-field or an octonion division ring in some cases. If Δ is the spherical building associated with the F-rational points of an absolutely simple algebraic group G of F-rank at least two, then Δ is Moufang and F is either the center of K or the intersection of the center of K with the fixed point set of a certain involution of K.

4 This statement must be slightly modified for the Moufang quadrangles of indifferent and exceptional type, but we omit the details in this survey. If Π is simply laced, then $\Xi_0 = \Xi$.

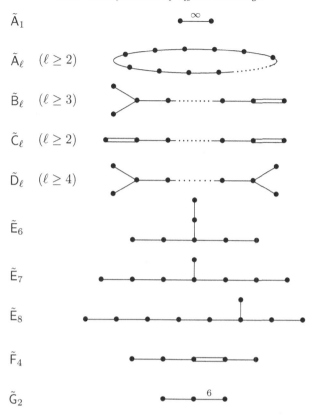

Fig. 12.1. The Irreducible Affine Coxeter Diagrams

each $\ell \geq 3$, the Coxeter diagrams $\tilde{B}_\ell$ and $\tilde{C}_\ell$ are distinct but have the same derived diagram. Every other diagram in Figure 12.1 is uniquely determined by its derived diagram.

We now fix a connected affine diagram $\Pi = \tilde{X}_\ell$ and let Σ_Π be the corresponding chamber system. Then Σ_Π has a natural representation in terms of "alcoves" in a Euclidean space V of dimension ℓ which we now describe.

Let Φ be the roots system called X_ℓ, let V denote the ambient Euclidean space, let

$$B := \{\alpha_1, \alpha_2, \ldots, \alpha_\ell\}$$

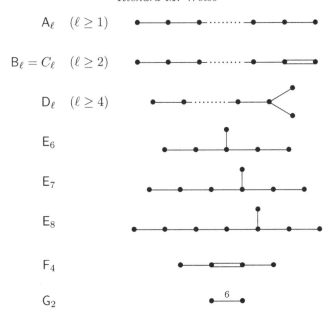

$\mathsf{A}_\ell \quad (\ell \geq 1)$

$\mathsf{B}_\ell = C_\ell \quad (\ell \geq 2)$

$\mathsf{D}_\ell \quad (\ell \geq 4)$

E_6

E_7

E_8

F_4

G_2

Fig. 12.2. Some Spherical Coxeter Diagrams

be a basis of Φ, let

$$s_\alpha(v) = v - 2\frac{v \cdot \alpha}{\alpha \cdot \alpha}\alpha$$

for all $\alpha \in \Phi$ and all $v \in V$, let

$$S = \{s_{\alpha_i} \mid i \in [1, \ell]\}$$

and let $W = \langle S \rangle$. The group W is the *Weyl group* of Φ. Up to isomorphism (W, S) is the Coxeter system corresponding to the derived diagram X_ℓ of $\tilde{\mathsf{X}}_\ell$.

For each pair $(\alpha, k) \in \Phi \times \mathbb{Z}$, let

$$s_{\alpha,k}(v) = s_\alpha(v) + 2k\alpha/(\alpha \cdot \alpha)$$

for all $v \in V$. The elements $s_{\alpha,k}$ are called *affine reflections* of V. let $\tilde{\alpha}$ be the highest root of Φ with respect to the basis B as defined in Proposition 25 in Chapter VI, Section 1.8, of [2], let

$$\tilde{S} = \{s_{\alpha_i,0} \mid i \in [1, \ell]\} \cup \{s_{\tilde{\alpha},1}\}$$

and let $\tilde{W}$ be the group generated by $\tilde{S}$. Then $(\tilde{W}, \tilde{S})$ is isomorphic to the Coxeter system corresponding to the diagram $\tilde{\mathsf{X}}_\ell$.

For each pair $(\alpha, k) \in \Phi \times \mathbb{Z}$, we denote by $H_{\alpha,k}$ the *affine hyperplane*

$$\{v \in V \mid \alpha \cdot v = k\}.$$

Let

$$X = \bigcup_{(\alpha,k)\in\Phi\times\mathbb{Z}} H_{\alpha,k}.$$

An *alcove* of Φ is a connected component of $V \backslash X$. One alcove, for example, is

$$\{v \in V \mid v \cdot \alpha_i > 0 \text{ for all } i \in [1, \ell] \text{ and } v \cdot \tilde{\alpha} < 1\}.$$

The Weyl group $\tilde{W}$ acts sharply transitively on the set of alcoves. Each alcove is bounded by exactly $\ell + 1$ of the hyperplanes in X. Let Γ be the graph whose vertices are the alcoves, where two distinct alcoves are adjacent in Γ whenever there is a hyperplane in X bounding both of them. Then the edges of Γ can be colored (with color set the vertex set of the Coxeter diagram $\tilde{X}_\ell$) to make Γ isomorphic to Σ_Π.

Here, for example, is how the alcoves look if $\Pi = \tilde{C}_2$:

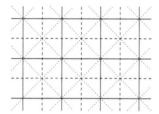

The affine hyperplanes in the set X are the straight lines in this picture and the alcoves in this picture are the open triangles. Two alcoves are adjacent when they are separated by a segment in one of the affine hyperplanes, and if two alcoves are separated by a segment, then the color of the corresponding edge of Σ_Π is the "color" of the segment, either "dotted," "dashed" or "solid."

A point v in the ambient space V of Φ is called *special* if for each $\alpha \in \Phi$, there exists $k \in \mathbb{Z}$ such that $v \in H_{\alpha,k}$. A *Weyl chamber* of the root system Φ is a set of the form

$$\bigcap_{1 \le i \le \ell} \{v \in V \mid v \cdot \beta_i > 0\}$$

for some basis $\{\beta_1, \beta_2, \ldots, \beta_\ell\}$ of Φ. A *sector* of Σ_Π is the set of alcoves contained in a fixed translate of a Weyl chamber whose apex is a special point of V. If $\Pi = \tilde{C}_2$, for example, then the special points are those at the intersection of four hyperplanes of the form $H_{\alpha,k}$, and the sectors are the set of alcoves contained in the wedge bounded by two hyperplanes of the form $H_{\alpha,k}$ making an angle of 45 degrees.

Two sectors of Σ_Π are called *parallel* of their intersection is also a sector. We denote the parallel class of a sector S by S^∞ and we denote the set of all parallel classes of sectors in Σ_Π by Σ_Π^∞. Then Σ_Π^∞ has, canonically, the structure of an apartment in a building of type X_ℓ. If $\Pi = \tilde{\mathsf{C}}_2$, for example, it is easy to see that there are eight parallel classes of sectors and they are arranged naturally in the form of a circuit.

For each root $\alpha \in \Phi$ and each integer k, let $K_{\alpha,k}$ be the set of alcoves in the half-space

$$\{v \in V \mid \alpha \cdot v > -k\}$$

(i.e. on one side of the hyperplane $H_{\alpha,k}$). Then $K_{\alpha,k}$ is a root of Σ_Π and every root is of this form. Let $K_{\alpha,k}^\infty$ be the set of parallel classes of the sectors contained in the root $K_{\alpha,k}$. Then $K_{\alpha,k}^\infty = K_{\alpha,0}^\infty$ for all k, and the map $\alpha \mapsto K_{\alpha,0}^\infty$ is a bijection from the root system Φ to the set of roots in of Σ_Π^∞ (as defined at the end of §1).

Now suppose that Δ is a building of type $\Pi = \tilde{\mathsf{X}}_\ell$. Then each apartment of Δ is isomorphic to the chamber system Σ_Π, which we can think of now as described in terms of alcoves. In particular, each apartment contains sectors. A *sector* of Δ is a sector in one of its apartments. We declare that two sectors of Δ (not necessarily in the same apartment) are *parallel* if their intersection is a sector and we denote by S^∞ the parallel class containing a sector S. Then the set Δ^∞ of sector equivalence classes has, canonically, the structure of a building of type X_ℓ called the *building at infinity* of Δ. Since X_ℓ is a spherical diagram, Δ^∞ is, in fact, a spherical building. For each apartment A of Δ, the set A^∞ of parallel classes S^∞ for all sectors S contained in A is an apartment of Δ^∞, and the map $A \mapsto A^\infty$ is a bijection from the set of all apartments of Δ to the set of all apartments of Δ^∞.

5 Bruhat-Tits Buildings

We define a *Bruhat-Tits* building to be an (irreducible thick) affine building of rank at least three whose building at infinity is Moufang (as defined in §3). We suppose from now on that Δ is a Bruhat-Tits building of type $\Pi = \tilde{\mathsf{X}}_\ell$ and that A is an apartment of Δ. Thus $\ell \geq 2$ and by the classification of Moufang spherical buildings summarized in §3, the building at infinity Δ^∞ is uniquely determined by the root groups associated with all the roots of the fixed apartment $\Sigma := A^\infty$ (i.e. by the *root datum of* Δ^∞ *based at* Σ), and this root datum is, in turn, uniquely determined by a suitable algebraic structure Λ. (Thus Λ is a field or a skew-field or an anisotropic quadratic space a Jordan division algebra, etc., according to the family of Moufang spherical buildings to which Δ^∞ belongs.)

The elements in the root groups of Δ^∞ all have unique extensions to automorphisms of Δ.[5] As indicated at the end of §4, we can identify the apartment A of Δ with the chamber system Σ_Π described in terms of a root system and alcoves in §4. Thus, in particular, the roots of A are of the form $K_{\alpha,k}$ for $(\alpha, k) \in \Phi \times \mathbb{Z}$ (where $K_{\alpha,k}$ is as defined at the end of §4) and the map $\alpha \to K^\infty_{\alpha,0}$ is a bijection from Φ to the set of roots of Σ. Let $a = K^\infty_{\alpha,0}$ be a root of Σ. Then for each $g \in U_a^*$, there exists a unique $k \in \mathbb{Z}$ such that the fixed point set of g in A is the root $K_{\alpha,k}$. Let φ_a denote the map $g \mapsto k$ from U_a^* to $\mathbb{Z}$. We extend φ_a to U_a by setting $\varphi_a(1) = \infty$.

We use additive notation for the group U_a even though U_a is, in some cases, non-abelian. If $g, h \in U_a$, then $g + h$ acts trivially on the set of alcoves in the half-space

$$\{v \in V \mid \alpha \cdot v \geq -m\},$$

where $m = \min\{\varphi_a(g), \varphi_a(h)\}$. Thus

$$\varphi_a(g + h) \geq \min\{\varphi_a(g), \varphi_b(h)\} \tag{5.1}$$

for all roots a of Σ and all $g, h \in U_a$. Since inverses have the same fixed point sets, we also have $\varphi_a(-g) = \varphi_a(g)$ for each $g \in U_a^*$. It follows that the map ∂_a from $U_a \times U_a$ to $\mathbb{R}$ given by

$$\partial_a(g, h) = 2^{-\varphi_a(g-h)} \tag{5.2}$$

for all $(g, h) \in U_a \times U_a$ is a metric on U_a. The group U_a must, in fact, be complete with respect to this metric for all roots a of Σ.[6]

Let

$$\varphi = \{\varphi_a \mid a \in \Xi\},$$

where Ξ denotes the set of roots of the apartment Σ. The inequality (5.1) is called condition (V1) in [8] (on page 139). There are two more

---

5 The proof of this (in [11]) and the proof (in [4]) of the existence of the building at infinity Δ^∞ are two of the high points in the classification.

6 This is a consequence of the fact that, for the sake of simplicity, we are implicitly working with the set of *all* apartments of Δ in this article rather than a *system of apartments*; see page 129 in [8].

conditions on the set φ called (V2) and (V3) on page 139 of [8] which can be derived similarly.

Now let Ξ_0 and K be as described at the end of §3 and choose $a \in \Xi_0$. Thus the root group U_a is isomorphic to the additive group of K. As the inequality (5.1) suggests, φ_a is determined by a *discrete valuation* ν of K.[7] Since U_a must be complete with respect to the metric ∂_a defined in equation (5.2), the field K must be complete with respect to ν. The existence of such a valuation ν is thus a necessary condition for Δ to exist (given the spherical building Δ^∞).

Now let Δ_0 be an *arbitrary* spherical building of rank at least two satisfying the Moufang condition, let Σ be an apartment of Δ_0 and let Ξ be the set of roots of Σ. A *valuation* of the root datum of Δ_0 based at Σ is a collection of surjective maps φ_a from U_a^* to $\mathbb{Z}$, one for each root $a \in \Xi$, satisfying the three conditions (V1), (V2) and (V3). Note that we are now using the word "valuation" in two different ways, to refer to a discrete valuation of a field and to refer to a valuation of a root datum.

We can now formulate a brief version of the classification of Bruhat-Tits buildings:

Theorem. *Let Δ_0 be a Moufang spherical building of type Π_0 defined over K (as defined at the end of §3) and suppose that K is complete with respect to a discrete valuation ν. Let Σ be an apartment of Δ_0, let Ξ be the set of roots of Σ and let Ξ_0 be as described at the end of §3.[8] Then the following hold:*

(i) *There is a unique valuation φ of the root datum of Δ_0 based at Σ such that φ_a is determined by ν for each $a \in \Xi_0$.*

(ii) *There is a unique Moufang spherical building $\hat{\Delta}_0$ of type Π_0 containing Δ_0 as a subbuilding and a unique valuation*

$$\hat{\varphi} = \{\hat{\varphi}_a \mid a \in \Xi\}$$

of the root datum of $\hat{\Delta}_0$ based at Σ such that for each root $a \in \Xi$, the root group $\hat{U}_a$ of $\hat{\Delta}_0$ contains the root group U_a as a subgroup, the map φ_a is the restriction of $\hat{\varphi}_a$ to U_a and the root group $\hat{U}_a$ is the completion of U_a with respect to the metric given in equation (5.2).

---

7 A discrete valuation of a field K is a surjective homomorphism ν from the multiplicative group K^* to the additive group $\mathbb{Z}$ such that $\nu(u + v) \geq \min\{\nu(u), \nu(v)\}$ for all $u, v \in K$ such that $u \neq -v$.

8 For the sake of simplicity, we are ignoring the Moufang quadrangles of indifferent and exceptional type (as indicated in the footnote at the end of §3) in the formulation of this result. In fact, similar assertions hold in these cases as well.

(iii) There is a unique affine Coxeter diagram Π *with derived diagram* Π_0 *and a unique Bruhat-Tits building* Δ *of type* Π *such that* $\Delta^\infty \cong \hat{\Delta}_0$ *and the valuation* $\hat{\varphi}$ *arises from* Δ *as described above.*

Now suppose that Δ_0 is an arbitrary Moufang spherical building with defining field K. If there exists an affine building Δ such that $\Delta^\infty \cong \Delta_0$, then, as we observed in §5, K must be complete with respect to a discrete valuation (so the Theorem applies) and, in fact, all the root groups U_a must be complete with respect to the metric ∂_a defined in equation (5.2), so $\hat{\Delta}_0 = \Delta_0$.

If Δ_0 is the spherical building associated with the F-rational points of an absolutely simple algebraic group G of F-rank at least two (in which case $F \subset K$ and K is finite-dimensional over F) and K is complete with respect to a discrete valuation, then it is always true that $\hat{\Delta}_0 = \Delta_0$. Furthermore, F is complete with respect to a discrete valuation if and only if K is, and if either F or K is complete with respect to a discrete valuation, then the valuation is unique. We conclude that Δ_0 is the building at infinity of a Bruhat-Tits building Δ if and only if F is complete with respect to a discrete valuation, and if F is complete with respect to a discrete valuation, then the Bruhat-Tits building Δ and its type Π are uniquely determined by F and G.[9]

The notion of a valuation of a root datum φ makes perfectly good sense if the requirement that "each φ_a maps U_a^* surjectively to $\mathbb{Z}$" is replaced by "each φ_a maps U_a^* to $\mathbb{R}$." If these images are non-discrete subgroups of $\mathbb{R}$, then φ no longer corresponds to an affine building. In [4] and [11], however, Bruhat and Tits showed that there is a class of geometric structures which are classified by these "non-discrete" valuations of a root datum. These structures structures are now usually referred to as *non-discrete buildings* although they are not really buildings. See also [6] and [7] for more details about them

Bibliography

[1] P. Abramenko and K. S. Brown, *Approaches to Buildings*, Springer, Berlin, Heidelberg, New York, 2007.

[2] N. Bourbaki, *Elements of Mathematics: Lie Groups and Lie Algebras, Chapters 4–6*, Springer, Berlin, Heidelberg, New York, 1968.

[3] K. S. Brown, *Buildings*, Springer, Berlin, Heidelberg, New York, 1989.

---

9 The standard reference for the connection between algebraic groups over a local field and affine building is [10].

[4] F. Bruhat and J. Tits, Groupes réductifs sur un corps local, I. Données radicielles valuées, *Publ. Math. I.H.E.S.* **41** (1972), 5-252.

[5] P. Garrett, *Buildings and classical groups*, Chapman & Hall, London, 1997.

[6] B. Kleiner and Leeb, B., Rigidity of quasi-isometries for symmetric spaces and Euclidean buildings, *Inst. Hautes Études Sci. Publ. Math.* **86** (1997), 115-197.

[7] A. Parreau, Immeubles affines: construction par les normes et étude des isométries, in: *Crystallographic Groups and Their Generalizations (Kortrijk, 1999)*, Contemp. Math. **262**, Amer. Math. Soc., Providence, 2000, pp. 263-302.

[8] M. A. Ronan, *Lectures on Buildings*, Academic Press, London, New York, San Diego, 1989.

[9] J. Tits, *Buildings of Spherical Type and finite BN-pairs*, Lecture Notes in Mathematics, vol. 386, Springer, Berlin, Heidelberg, New York, 1974.

[10] J. Tits, Reductive groups over local fields, in *Proc. Symp. Pure Math.* **33**, Part 1 (*Automorphic Forms, Representations and L-Functions*, Corvallis 1977), pp. 29-69, *Amer. Math. Soc.*, 1979.

[11] Tits, J., Immeubles de type affine, in *Buildings and the Geometry of Diagrams* (Como 1984), pp. 159-190, Lecture Notes in Mathematics **1181**, Springer, New York, Heidelberg, Berlin, 1986.

[12] J. Tits and R. M. Weiss, *Moufang Polygons*, Springer Monographs in Mathematics, Springer, Berlin, Heidelberg, New York, 2002.

[13] R. M. Weiss, *The Structure of Spherical Buildings*, Princeton University Press, 2006.

[14] R. M. Weiss, *The Structure of Affine Buildings*, Annals of Mathematics Studies, Princeton University Press, to appear.

13
Emmy Noether and Hermann Weyl.[1]

Peter Roquette

Mathematisches Institut
Universität Heidelberg
roquette@uni-hd.de

Contents

1	**Preface**	**285**
2	**Introduction**	**287**
3	**The first period: until 1915**	**288**
	3.1 Their mathematical background	289
	3.2 Meeting in Göttingen 1913	291
4	**The second period: 1915-1920**	**293**
5	**The third period: 1920-1932**	**295**
	5.1 Innsbruck 1924 and the method of abstraction	296
	5.2 Representations: 1926/27	299
	5.3 A letter from N to W: 1927	302
	5.4 Weyl in Göttingen: 1930-1933	304
6	**Göttingen exodus: 1933**	**307**
7	**Bryn Mawr: 1933-1935**	**312**
8	**The Weyl-Einstein letter to the NYT**	**314**
9	**Appendix: documents**	**318**
	9.1 Weyl's testimony	318
	9.2 Weyl's funeral speech	319
	9.3 Letter to the New York Times	321
	9.4 Letter of Dr. Stauffer-McKee	323

1 Preface

We are here for a conference in honor of Hermann Weyl and so I may be allowed, before touching the main topic of my talk, to speak about my personal reminiscences of him.

It was in the year 1952. I was 24 and had my first academic job

---

[1] This is the somewhat extended manuscript of a talk presented at the Hermann Weyl conference in Bielefeld, September 10, 2006.

at München when I received an invitation from van der Waerden to give a colloquium talk at Zürich University. In the audience of my talk I noted an elder gentleman, apparently quite interested in the topic. Afterwards – it turned out to be Hermann Weyl – he approached me and proposed to meet him next day at a specific point in town. There he told me that he wished to know more about my doctoral thesis, which I had completed two years ago already but which had not yet appeared in print. Weyl invited me to join him on a tour on the hills around Zürich. On this tour, which turned out to last for several hours, I had to explain to him the content of my thesis which contained a proof of the Riemann hypothesis for function fields over finite base fields. He was never satisfied with sketchy explanations, his questions were always to the point and he demanded every detail. He seemed to be well informed about recent developments.

This task was not easy for me, without paper and pencil, nor black-board and chalk. So I had a hard time. Moreover the pace set by Weyl was not slow and it was not quite easy to keep up with him, in walking as well as in talking.

Much later only I became aware of the fact that this tour was a kind of examination, Weyl wishing to find out more about that young man who was myself. It seems that I did not too bad in this examination, for some time later he sent me an application form for a grant-in-aid from the Institute for Advanced Study in Princeton for the academic year 1954/55. In those years Weyl was commuting between Zürich and Princeton on a half-year basis. In Princeton he had found, he wrote to me, that there was a group of people who were working in a similar direction.

Hence I owe to Hermann Weyl the opportunity to study in Princeton. The two academic years which I could work and learn there turned out to be important for my later mathematical life. Let me express, posthumously, my deep gratitude and appreciation for his help and concern in this matter.

The above story shows that Weyl, up to his last years, continued to be active helping young people find their way into mathematics. He really cared. I did not meet him again in Princeton; he died in 1955.

Let us now turn to the main topic of this talk as announced in the title.

2 Introduction

Both Hermann Weyl and Emmy Noether belonged to the leading group of mathematicians in the first half of 20th century, who shaped the image of mathematics as we see it today.

Emmy Noether was born in 1882 in the university town of Erlangen, as the daughter of the renowned mathematician Max Noether. We refer to the literature for information on her life and work, foremost to the empathetic biography by Auguste Dick [8] which has appeared in 1970, the 35th year after Noether's tragic death. It was translated into English in 1981. For more detailed information see, e.g., the very carefully documented report by Cordula Tollmien [47]. See also Kimberling's publications on Emmy Noether, e.g., his article in [5].

When the Nazis had come to power in Germany in 1933, Emmy Noether was dismissed from the University of Göttingen and she emigrated to the United States. She was invited by Bryn Mawr College as a visiting professor where, however, she stayed and worked for 18 months only, when she died on April 14, 1935 from complications following a tumor operation.[2]

Quite recently we have found the text, hitherto unknown, of the speech which Hermann Weyl delivered at the funeral ceremony for Emmy Noether on April 17, 1935.[3] That moving text puts into evidence that there had evolved a close emotional friendship between the two. There was more than a feeling of togetherness between immigrants in a new and somewhat unfamiliar environment. And there was more than high esteem for this women colleague who, as Weyl has expressed it[4], was "superior to him in many respects". This motivated us to try to find out more about their mutual relation, as it had developed through the years.

We would like to state here already that we have not found many documents for this. We have not found letters which they may have exchanged.[5] Neither did Emmy Noether cite Hermann Weyl in her papers nor vice versa[6]. After all, their mathematical activities were going into somewhat different directions. Emmy Noether's creative power was directed quite generally towards the clarification of mathematical struc-

---

2 See footnote 49.
3 See [37]. We have included in the appendix an English translation of Weyl's text; see section 9.2.
4 See [53].
5 With one exception; see section 5.3.
6 There are exceptions; see section 4.

tures and concepts through abstraction, which means leaving all unnecessary entities and properties aside and concentrating on the essentials. Her basic work in this direction can be subsumed under algebra, but her methods eventually penetrated all mathematical fields, including number theory and topology.

On the other side, Hermann Weyl's mathematical horizon was widespread, from complex and real analysis to algebra and number theory, mathematical physics and logic, also continuous groups, integral equations and much more. He was a mathematical generalist in a broad sense, touching also philosophy of science. His mathematical writings have a definite flair of art and poetry, with his book on symmetry as a culmination point [54].

We see that the mathematical style as well as the extent of Weyl's research work was quite different from that of Noether. And from all we know the same can be said about their way of living. So, how did it come about that there developed a closer friendly relationship between them? Although we cannot offer a clear cut answer to this question, I hope that the reader may find something of interest in the following lines.

3 The first period: until 1915

In the mathematical life of Emmy Noether we can distinguish four periods.[7] In her first period she was residing in Erlangen, getting her mathematical education and working her way into abstract algebra guided by Ernst Fischer, and only occasionally visiting Göttingen. The second period starts in the summer of 1915 when she came to Göttingen for good, in order to work with Klein and Hilbert. This period is counted until about 1920. Thereafter there begins her third period, when her famous paper "*Idealtheorie in Ringbereichen*" (Ideal theory in rings) appeared, with which she "*embarked on her own completely original mathematical path*" – to cite a passage from Alexandroff's memorial address [2]. The fourth period starts from 1933 when she was forced to emigrate and went to Bryn Mawr.

---

7 Weyl [53] distinguishes three epochs but they represent different time intervals than our periods.

3.1 Their mathematical background

Hermann Weyl, born in 1885, was about three years younger than Emmy Noether. In 1905, when he was 19, he entered Göttingen University (after one semester in München). On May 8, 1908 he obtained his doctorate with a thesis on integral equations, supervised by Hilbert.

At about the same time (more precisely: on December 13, 1907) Emmy Noether obtained her doctorate from the University of Erlangen, with a thesis on invariants supervised by Gordan. Since she was older than Weyl we see that her way to Ph.D. was longer than his. This reflects the fact that higher education, at that time, was not as open to females as it is today; if a girl wished to study at university and get a Ph.D. then she had to overcome quite a number of difficulties arising from tradition, prejudice and bureaucracy. Noether's situation is well described in Tollmien's article [47].[8]

But there was another difference between the status of Emmy Noether and Hermann Weyl at the time of their getting the doctorate.

On the one side, Weyl was living and working in the unique Göttingen mathematical environment of those years. Weyl's thesis belongs to the theory of integral equations, the topic which stood in the center of Hilbert's work at the time, and which would become one of the sources of the notion of "Hilbert space". And Weyl's mathematical curiosity was not restricted to integral equations. In his own words, he was captivated by all of Hilbert's mathematics. Later he wrote:[9]

I resolved to study whatever this man [Hilbert] had written. At the end of my first year I went home with the "Zahlbericht" under my arm, and during the summer vacation I worked my way through it - without any previous knowledge of elementary number theory or Galois theory. These were the happiest months of my life, whose shine, across years burdened with our common share of doubt and failure, still comforts my soul.

We see that Weyl in Göttingen was exposed to and responded to the new and exciting ideas which were sprouting in the mathematical world at the time. His mathematical education was strongly influenced by his advisor Hilbert.

On the other side, Noether lived in the small and quiet mathematical world of Erlangen. Her thesis, supervised by Paul Gordan, belongs to classical invariant theory, in the framework of so-called symbolic computations. Certainly this did no longer belong to the main problems which

---

8 For additional material see also Tollmien's web page: www.tollmien.com.
9 Cited from the Weyl article in "MacTutor History of Mathematics Archive".

dominated mathematical research in the beginning of the 20th century. It is a well-known story that after Hilbert in 1888 had proved the finiteness theorem of invariant theory which Gordan had unsuccessfully tried for a long time, then Gordan did not accept Hilbert's existence proof since that was not constructive in his (Gordan's) sense. He declared that Hilbert's proof was "theology, not mathematics". Emmy Noether's work was fully integrated into Gordan's formalism and so, in this way, she was not coming near to the new mathematical ideas of the time.[10] In later years she described the work of her thesis as rubbish (*"Mist"* in German[11]). In a letter of April 14, 1932 to Hasse she wrote:

Ich habe das symbolische Rechnen mit Stumpf und Stil verlernt.
I have completely forgotten the symbolic calculus.

We do not know when Noether had first felt the desire to update her mathematical background. Maybe the discussions with her father helped to find her way; he corresponded with Felix Klein in Göttingen and so was well informed about the mathematical news from there. She herself reports that it was mainly Ernst Fischer who introduced her to what was then considered "modern" mathematics. Fischer came to Erlangen in 1911, as the successor of the retired Gordan.[12] In her curriculum vitae which she submitted in 1919 to the Göttingen Faculty on the occasion of her *Habilitation*, Noether wrote:

Wissenschaftliche Anregung verdanke ich wesentlich dem persönlichen mathematischen Verkehr in Erlangen und in Göttingen. Vor allem bin ich Herrn E. Fischer zu Dank verpflichtet, der mir den entscheidenden Anstoß zu der Beschäftigung mit abstrakter Algebra in arithmetischer Auffassung gab, was für all meine späteren Arbeiten bestimmend blieb.
I obtained scientific guidance and stimulation mainly through personal mathematical contacts in Erlangen and in Göttingen. Above all I am indebted to Mr. E. Fischer from whom I received the decisive impulse to study abstract algebra from an arithmetical viewpoint, and this remained the governing idea for all my later work.

Thus it was Fischer under whose direction Emmy Noether's mathe-

---

10 Well, Noether had studied one semester in Göttingen, winter 1903/04. But she fell ill during that time and had to return to her home in Erlangen, as Tollmien [47] reports. We did not find any indication that this particular semester has had a decisive influence on her mathematical education.

11 Cited from Auguste Dick's Noether biography [8].

12 More precisely: Gordan retired in 1910 and was followed by Erhard Schmidt who, however, left Erlangen one year later already and was followed in turn by Ernst Fischer. – The name of Fischer is known from the *Fischer-Riesz theorem* in functional analysis.

matical outlook underwent the *"transition from Gordan's formal stand-point to the Hilbert method of approach"*, as Weyl stated in [53].

We may assume that Emmy Noether studied, like Weyl, all of Hilbert's papers, at least those which were concerned with algebra or arithmetic. In particular she would have read the paper [14] where Hilbert proved that every ideal in a polynomial ring is finitely generated; in her famous later paper [29] she considered arbitrary rings with this property, which today are called "Noetherian rings". We may also assume that Hilbert's *Zahlbericht* too was among the papers which Emmy Noether studied; it was the standard text which every young mathematician of that time read if he/she wished to learn algebraic number theory. We know from a later statement that she was well acquainted with it – although at that later time she rated it rather critically[13], in contrast to Weyl who, as we have seen above, was enthusiastic about it. But not only Hilbert's papers were on her agenda; certainly she read Steinitz' great paper *"Al-gebraische Theorie der Körper"* [44] which marks the start of abstract field theory. This paper is often mentioned in her later publications, as the basis for her abstract viewpoint of algebra.

3.2 Meeting in Göttingen 1913

Hermann Weyl says in [53], referring to the year 1913:

. . . She must have been to Göttingen about that time, too, but I suppose only on a visit with her brother Fritz. At least I remember him much better than her from my time as a Göttinger Privatdozent, 1910-1913.

We may conclude that he had met Emmy Noether in Göttingen about 1913, but also that she did not leave a lasting impression on him on that occasion.

As Tollmien [47] reports, it was indeed 1913 when Emmy Noether visited Göttingen for a longer time (together with her father Max Noether). Although we have no direct confirmation we may well assume that she met Weyl during this time. In the summer semester 1913 Weyl gave two talks in the *Göttinger Mathematische Gesellschaft*. In one session he reported on his new book *"Die Idee der Riemannschen Fläche"* (The idea of the Riemann surface) [51], and in another he presented his proof on the equidistribution of point sequences modulo 1 in arbitrary dimensions

---

13 In a letter of November 17, 1926 to Hasse; see [22]. Olga Taussky-Todd [45] reports from later time in Bryn Mawr, that once *"Emmy burst out against the* Zahlbericht, *quoting also Artin as having said that it delayed the development of algebraic number theory by decades"*.

[52] – both pieces of work have received the status of a "classic" by now. Certainly, Max Noether as a friend of Klein will have been invited to the sessions of the *Mathematische Gesellschaft*, and his daughter Emmy with him. Before and after the session people would gather for discussion, and from all we know about Emmy Noether she would not have hesitated to participate in the discussions. From what we have said in the foregoing section we can conclude that her mathematical status was up-to-date and well comparable to his, at least with respect to algebra and number theory.

Unfortunately we do not know anything about the possible subjects of the discussions of Emmy Noether with Weyl. It is intriguing to think that they could have talked about Weyl's new book "*The Idea of the Riemann surface*". Weyl in his book defines a Riemann surface axiomatically by structural properties, namely as a connected manifold X with a complex 1-dimensional structure. This was a completely new approach, a structural viewpoint. Noether in her later period used to emphasize on every occasion the structural viewpoint. The structure in Weyl's book is an *analytic* one, and he constructs an algebraic structure from this, namely the field of meromorphic funtions, using the so-called Dirichlet principle – whereas Emmy Noether in her later papers always starts from the function field as an *algebraic* structure. See, e.g., her report [27]. There she did not cite Weyl's book but, of course, this does not mean that she did not know it.

We observe that the starting idea in Weyl's book was the definition and use of an axiomatically defined topological space[14]. We wonder whether this book was the first instance where Emmy Noether was confronted with the axioms of what later was called a topological space. It is not without reason to speculate that her interest in topology was inspired by Weyl's book. In any case, from her later cooperation with Paul Alexandroff we know that she was acquainted with problems of topology; her contribution to algebraic topology was the notion of "Betti group" instead of the "Betti number" which was used before. Let us cite Alexandroff in his autobiography [1]:

In the middle of December Emmy Noether came to spend a month in Blaricum. This was a brilliant addition to the group of mathematicians around Brouwer. I remember a dinner at Brouwer's in her honour during which she explained the definition of the Betti groups of complexes, which spread around quickly and completely transformed the whole of topology.

---

14 The Hausdorff axiom was not present in the first edition. This gap was filled in later editions.

This refers to December 1925. Blaricum was the place where L. E. J. Brouwer lived.

We have mentioned this contact of Emmy Noether to the group around Brouwer since Weyl too did have mathematical contact with Brouwer. In fact, in his book *"The idea of the Riemann surface"* Weyl mentioned Brouwer as a source of inspiration. He writes:

In viel höherem Maße, als aus den Zitaten hervorgeht, bin ich dabei durch die in den letzten Jahren erschienenen grundlegenden topologischen Untersuchungen Brouwers, dessen gedankliche Schärfe und Konzentration man bewundern muss, gefördert worden ...

I have been stimulated – much more than the citations indicate – by the recent basic topological investigations of Brouwer, whose ideas have to be admired in their sharpness and concentration.

Brouwer's biographer van Dalen reports that Weyl and Brouwer met several times in the early 1920s [49]. By the way, Emmy Noether, Hermann Weyl and L. E. J. Brouwer met in September 1920 in Bad Nauheim, at the meeting of the DMV.[15]

Returning to the year 1913 in Göttingen: In the session of July 30, 1913 of the *Göttinger Mathematische Gesellschaft*, Th. v. Kármán reported on problems connected with a recent paper on turbulence by Emmy Noether's brother Fritz. Perhaps Fritz too was present in Göttingen on this occasion, and maybe this was the incident why Weyl had remembered not only Emmy but also Fritz? That he remembered Fritz "much better" may be explained by the topic of Fritz' paper; questions of turbulence lead to problems about partial differential equations, which was at that time more close to Weyl's interests than were algebraic problems which Emmy pursued.

4 The second period: 1915-1920

In these years Emmy Noether completed several papers which are of algebraic nature, mostly about invariants, inspired by the Göttingen mathematical atmosphere dominated by Hilbert. She also wrote a report in the *Jahresbericht der DMV* on algebraic function fields, in which Noether compares the various viewpoints of the theory: analytic, geometric and algebraic (which she called "arithmetic") and she points out the analogies to the theory of number fields. That was quite well known to the people working with algebraic functions, but perhaps not written up systematically as Emmy Noether did. Generally, these papers of

---

15 DMV = *Deutsche Mathematiker Vereinigung* = German Mathematical Society.

hers can be rated as good work, considering the state of mathematics of the time, but not as outstanding. It is unlikely that Hermann Weyl was particularly interested in these papers; perhaps he didn't even know about them.[16]

But this would change completely with the appearance of Noether's paper on invariant variation problems [26] (*"Invariante Variationsprobleme"*). The main result of this paper is of fundamental importance in many branches of theoretical physics even today. It shows a connection between conservation laws in physics and the symmetries of the theory. It is probably the most cited paper of Emmy Noether up to the present day. In 1971 an English translation appeared [46], and in 2004 a French translation with many comments [21].

In 1918, when the paper appeared, its main importance was seen in its applicability in the framework of Einstein's relativity theory. Einstein wrote to Hilbert in a letter of May 24, 1918:

Gestern erhielt ich von Frl. Noether eine sehr interessante Arbeit über Invarianten-bildung. Es imponiert mir, dass man diese Dinge von so allgemeinem Standpunkt übersehen kann ... Sie scheint ihr Handwerk zu verstehen.

Yesterday I received from Miss Noether a very interesting paper on the formation of invariants. I am impressed that one can handle those things from such a general viewpoint ... She seems to understand her job.

Einstein had probably met Emmy Noether already in 1915 during his visit to Göttingen.

Emmy Noether's result was the fruit of a close cooperation with Hilbert and with Klein in Göttingen during the past years. As Weyl [53] reports, *"Hilbert at that time was over head and ears in the general theory of relativity, and for Klein, too, the theory of relativity brought the last flareup of his mathematical interest and production"*. Emmy Noether, although she was doubtless influenced, not only assisted them but her work was a genuine production of her own. In particular, the connection of invariants with the symmetry groups, with its obvious reference to Klein's Erlanger program, caught the attention of the world of mathematicians and theoretical physicists.[17] Noether's work in this direction has been described in detail in, e.g., [39], [21], [57].

It is inconceivable that Hermann Weyl did not take notice of this important work of Emmy Noether. At that time Weyl, who was in cor-

---

16 In those years Weyl was no more in Göttingen but held a professorship at ETH in Zürich.

17 In [39] it is said that nevertheless "few mathematicians and even fewer physisists ever read Noether's original article ...".

respondence with Hilbert and Einstein, was also actively interested in the theory of relativity; his famous book *"Raum, Zeit, Materie"* (Space, Time, Matter) had just appeared. Emmy Noether had cited Weyl's book[18], and almost certainly she had sent him a reprint of her paper. Thus, through the medium of relativity theory there arose mathematical contact between them.[19] Although we do not know, it is well conceivable that there was an exchange of letters concerning the mathematical theory of relativity. From now on Weyl would never remember her brother Fritz better than Emmy.

In 1919 Emmy Noether finally got her *Habilitation*. Already in 1915 Hilbert and Klein, convinced of her outstanding qualification, had recommended her to apply for *Habilitation*. She did so, but it is a sad story that it was unsuccessful because of her gender although her scientific standing was considered sufficient. The incident is told in detail in Tollmien's paper [47]. Thus her *Habilitation* was delayed until 1919 after the political and social conditions had changed.

We see again the difference between the scientific careers of Weyl and of Emmy Noether. Weyl had his *Habilitation* already in 1910, and since 1913 he held a professorship in Zürich. Emmy Noether's *Habilitation* was possible only nine years later than Weyl's. As is well known, she never in her life got a permanent position; although in the course of time she rose to become one of the leading mathematicians in the world.

5 The third period: 1920-1932

The third period of Noether's mathematical life starts with the great paper *"Idealtheorie in Ringbereichen"* (Ideal theory in rings) [29].[20] After Hilbert had shown in 1890 that in a polynomial ring (over a field as base) every ideal is finitely generated, Noether now takes this property as an axiom and investigates the primary decomposition of ideals in arbitrary rings satisfying this axiom. And she reformulates this axiom as an "ascending chain condition" for ideals. Nowadays such rings are called "Noetherian". The paper appeared in 1921.

We note that she was nearly 40 years old at that time. The mathemat-

---

18 The citation is somewhat indirect. Noether referred to the literature cited in a paper by Felix Klein [20], and there we find Weyl's book meontioned. In a second paper of Noether [30] Weyl's book is cited directly.

19 Added in proof: We read in [39] that there is a reference to Noether in Weyl's book, tucked away in a footnote.

20 Sometimes the earlier investigation jointly with Schmeidler [28] is also counted as belonging to this period.

ical life of Emmy Noether is one of the counterexamples to the dictum that mathematics is a science for the young and the most creative work is done before 40. Emmy Noether would not have been a candidate for the Fields medal if it had already existed at that time.

5.1 Innsbruck 1924 and the method of abstraction

We do not know whether and how Weyl took notice of the above-mentioned paper of Noether [29]. But her next great result, namely the follow-up paper [32] on the ideal theory of what are now called Dedekind rings, was duly appreciated by Weyl. At the annual DMV meeting in 1924 in Innsbruck Noether reported about it [31]. And Weyl was chairing that session; so we know that he was informed first hand about her fundamental results.

In her talk, Emmy Noether defined Dedekind rings by axioms and showed that every ring satisfying those axioms admits a unique factorization of ideals into prime ideals. Well, Noether did not use the terminology "Dedekind ring"; this name was coined later. Instead, she used the name "5-axioms-ring" since in her enumeration there were 5 axioms. Then she proved that the ring of integers in a number field satisfies those axioms, and similarly in the funtion field case. This is a good example of Noether's "method of abstraction". By working solely with those axioms she first *generalized* the problem, and it turned out that by working in this generalization the proof of prime decomposition is *simplified* if compared with the former proofs (two of which had been given by Hilbert [15]).

How did Weyl react to Noether's method of abstraction? At that time, this method met sometimes with skepticism and even rejection by mathematicians. But Hilbert in various situations had already taken first steps in this direction and so Weyl, having been Hilbert's doctorand, was not against Noether's method. After all, in his book "Space, Time, Matter" Weyl had introduced vector spaces by axioms, not as n-tuples[21].

Weyl's reaction can be extracted implicitly from an exchange of letters with Hasse which happened seven years later. The letter of Weyl is dated December 8, 1931. At that time Weyl held a professorship in Göttingen (since 1930) as the successor of Hilbert. Thus Emmy Noether was now his colleague in Göttingen. Hasse at that time held a professorship in Marburg (also since 1930) as the successor of Hensel. The occasion of

---

21 This has been expressly remarked by MacLane [23].

Weyl's letter was the theorem that every simple algebra over a number field is cyclic; this had been established some weeks ago by Brauer, Hasse and Noether, and the latter had informed Weyl about it. So Weyl congratulated Hasse for this splendid achievement. And he recalled the meeting in Innsbruck 1924 when he first had met Hasse.

For us, Hasse's reply to Weyl's letter is of interest.[22] Hasse answered on December 15, 1931. First he thanked Weyl for his congratulations, but at the same time pointed out that the success was very essentially due also to the elegant theory of Emmy Noether, as well as the p-adic theory of Hensel. He also mentioned Minkowski in whose work the idea of the Local-Global principle was brought to light very clearly. And then Hasse continued, recalling Innsbruck:

Auch ich erinnere mich sehr gut an Ihre ersten Worte zu mir anläßlich meines Vortrages über die erste explizite Reziprozitätsformel für höheren Exponenten in Innsbruck. Sie zweifelten damals ein wenig an der inneren Berechtigung solcher Untersuchungen, indem Sie ins Feld führten, es sei doch gerade Hilberts Verdienst, die Theorie des Reziprozitätsgesetzes von den expliziten Rechnungen früherer Forscher, insbesondere Kummers, befreit zu haben.

I too remember very well your first words to me on the occasion of my talk in Innsbruck, about the first explicit reciprocity formula for higher exponent. You somewhat doubted the inner justification of such investigations, by pointing out that Hilbert had freed the theory of the reciprocity law from the explicit computations of former mathematicians, in particular Kummer's.

We conclude: Hasse in Innsbruck had talked on explicit formulas and Weyl had critized this, pointing out that Hilbert had embedded the reciprocity laws into more structural results. Probably Weyl had in mind the product formula for the so-called Hilbert symbol which, in a sense, comprises all explicit reciprocity formulas.[23] For many, like Weyl, this product formula was the final word on reciprocity while for Hasse this was the starting point for deriving explicit, constructive reciprocity formulas, using heavily the p-adic methods of Hensel.

We can fairly well reconstruct the situation in Innsbruck: Emmy Noether's talk had been very abstract, and Hasse's achievement was in some sense the opposite since he was bent on explicit formulas, and quite involved ones too.[24] Weyl had been impressed by Emmy Noether's

---

22 We have found Hasse's letter in the Weyl legacy in the archive of the ETH in Zürich.

23 But Hilbert was not yet able to establish his product formula in full generality. We refer to the beautiful and complete treatment in Hasse's class field report, Part 2 [10] which also contains the most significant historic references.

24 Hasse's Innsbruck talk is published in [9]. The details are found in volume 154 of Crelle's Journal where Hasse had published 5 papers on explicit reciprocity laws.

achievements which he considered as continuing along the lines set by Hilbert's early papers on number theory. In contrast, he considered Hasse's work as pointing not to the future but to the mathematical past.

We have mentioned here these letters Weyl-Hasse in order to put into evidence that already in 1924, Weyl must have had a very positive opinion on Emmy Noether's methods, even to the point of preferring it to explicit formulas.

But as it turned out, Hasse too had been impressed by Noether's lecture. In the course of the years after 1924, as witnessed by the Hasse-Noether correspondence [22], Hasse became more and more convinced about Noether's abstract methods which, in his opinion, served to clarify the situation; he used the word "*durchsichtig*" (lucid). Hasse's address at the DMV meeting in Prague 1929 [11] expresses his views very clearly. Hensel's p-adic methods could also be put on an abstract base, due to the advances in the theory of valuations.[25] But on the other hand, Hasse was never satisfied with abstract theorems only. In his cited letter to Weyl 1931 he referred to his (Hasse's) class field report Part II [10] which had appeared just one year earlier. There, he had put Artin's general reciprocity law[26] as the base, and from this structural theorem he was able to derive all the known reciprocity formulas. Hasse closed his letter with the following:

... Ich kann aber natürlich gut verstehen, daß Dinge wie diese expliziten Reziprozitätsformeln einem Manne Ihrer hohen Geistes- und Geschmacksrichtung weniger zusagen, als mir, der ich durch die abstrakte Mathematik Dedekind-E. Noetherscher Art nie restlos befriedigt bin, ehe ich nicht zum mindesten *auch* eine explizite, formelmäßige konstruktive Behandlung daneben halten kann. Erst von der letzteren können sich die eleganten Methoden und schönen Ideen der ersteren wirklich vorteilhaft abheben.

... *But of course I well realize that those explicit reciprocity formulas may be less attractive to a man like you with your high mental powers and taste, as to myself. I am never fully satisfied by the abstract mathematics of Dedekind-E. Noether type before I can* also *supplement it by at least one explicit, computational and constructive treatment. It is only in comparison with the latter that the elegant methods and beautiful ideas of the former can be appreciated advantageously.*

Here, Hasse touches a problem which always comes up when, as Emmy

25 For this see [36].
26 By the way, this was the first treatment of Artin's reciprocity law in book form after Artin's original paper 1927. We refer to our forthcoming book on the Artin-Hasse correspondence.

Noether propagated, the abstract methods are put into the foreground. Namely, abstraction and axiomatization is not to be considered as an end in itself; it is a method to deal with concrete problems of substance. But Hasse was wrong when he supposed that Weyl did not see that problem. Even in 1931, the same year as the above cited letters, Weyl gave a talk on abstract algebra and topology as two ways of mathematical comprehension [55]. In this talk Weyl stressed the fact that axiomatization is not only a way of securing the logical truth of mathematical results, but that it had become a powerful tool of concrete mathematical research itself, in particular under the influence of Emmy Noether. But he also said that abstraction and generalization do not make sense without mathematical substance behind it. This is close to Hasse's opinion as expressed in his letter above.[27] The mathematical work of both Weyl and Hasse puts their opinions into evidence.

At the same conference [55] Weyl also said that the *"fertility of these abstracting methods is approaching exhaustion"*. This, however, met with sharp protests by Emmy Noether, as Weyl reports in [53]. In fact, today most of us would agree with Noether. The method of abstracting and axiomatizing has become a natural and powerful tool for the mathematician, with striking successes until today. In Weyl's letter to Hasse (which we have not cited fully) there are passages which seem to indicate that in principle he (Weyl) too would agree with Emmy Noether. For, he encourages Hasse to continue his work in the same fashion, and there is no mention of an impending "exhaustion". He closes his letter with the following sentence which, in our opinion, shows his (Weyl's) opinion of how to work in mathematics:

Es freut mich besonders, daß bei Ihnen die in Einzelleistungen sich bewährende wissenschaftliche Durchschlagskraft sich mit geistigem Weitblick paart, der über das eigene Fach hinausgeht.

In particular I am glad that your scientific power, tested in various special accomplishments, goes along with a broad view stretching beyond your own special field.

5.2 Representations: 1926/27

We have made a great leap from 1924 to 1931. Now let us return and proceed along the course of time. In the winter semester 1926/27 Her-

---

27 Even more clearly Hasse has expressed his view in the foreword to his beautiful and significant book on abelian fields [13].

mann Weyl stayed in Göttingen as a visiting professor, and he lectured on representations of continuous groups. In [53] he reports:

I have a vivid recollection of her [Emmy Noether]*...She was in the audience; for just at that time the hypercomplex number systems and their representations had caught her interest and I remember many discussions when I walked home after the lectures, with her and von Neumann, who was in Göttingen as a Rockefeller Fellow, through the cold, dirty, rain-wet streets of Göttingen.*

This gives us information not only about the weather conditions in Göttingen in winter time but also that a lively discussion between Weyl and Emmy Noether had developed.

We do not know precisely when Emmy Noether first had become interested in the representation theory of groups and algebras, or "hypercomplex systems" in her terminology. In any case, during the winter semester 1924/25 in Göttingen she had given a course on the subject. And in September 1925 she had talked at the annual meeting of the DMV in Danzig on "*Group characters and ideal theory*". There she advocated that the whole representation theory of groups should be subsumed under the theory of algebras and their ideals. She showed how the Wedderburn theorems for algebras are to be interpreted in representation theory, and that the whole theory of Frobenius on group characters is subsumed in this way. Although she announced a more detailed presentation in the *Mathematische Annalen*, the mathematical public had to wait until 1929 for the actual publication [33][28]. Noether was not a quick writer; she developed her ideas again and again in discussions, mostly on her walks with students and colleagues into the woods around Göttingen, and in her lectures.

The text of her paper [33] consists essentially of the notes taken by van der Waerden at her lecture in the winter semester 1927/28. Although the main motivation of Noether was the treatment of Frobenius' theory of representations of finite groups, it turned out that finite groups are treated on the last two pages only – out of a total of 52 pages. The main part of the paper is devoted to introducing and investigating general abstract notions, capable of dealing not only with the classical theory of finite group representations but with much more. Again we see the power of Noether's abstracting methods. The paper has been said to constitute "*one of the pillars of modern linear algebra*".[29]

---

28 This appeared in the "Mathematische Zeitschrift" and not in the "Annalen" as announced by Noether in Danzig.
29 Cited from [7] who in turn refers to Bourbaki.

We can imagine Emmy Noether in her discussions with Weyl on the cold, wet streets in Göttingen 1926/27, explaining to him the essential ideas which were to become the foundation of her results in her forthcoming paper [33]. We do not know to which extent these ideas entered Weyl's book [56] on classical groups. After all, the classical groups which are treated in Weyl's book are *infinite* while Noether's theory aimed at the representation of *finite* groups. Accordingly, in Noether's work there appeared a *finiteness condition* for the algebras considered, namely the descending chain condition for (right) ideals. If one wishes to use Noether's results for infinite groups one first has to generalize her theory such as to remain valid in more general cases too. Such a generalization did not appear until 1945; it was authored by Nathan Jacobson [16]. He generalized Noether's theory to simple algebras containing at least one irreducible right ideal.

At this point let me tell a story which I witnessed in 1947. I was a young student in Hamburg then. In one of the colloquium talks the speaker was F. K. Schmidt who recently had returned from a visit to the US, and he reported on a new paper by Jacobson which he had discovered there.[30] This was the above mentioned paper [16]. F. K. Schmidt was a brilliant lecturer and the audience was duly impressed. In the ensuing discussion Ernst Witt, who was in the audience, commented that all this had essentially been known to Emmy Noether already.

Witt did not elaborate on his comment. But he had been one of the "Noether boys" in 1932/33, and so he had frequently met her. There is no reason to doubt his statement. It may well have been that she had told him, and perhaps others too, that her theory could be generalized in the sense which later had been found by Jacobson. Maybe she had just given a hint in this direction, without details, as was her usual custom. In fact, reading Noether's paper [33] the generalization is obvious to any reader who is looking for it.[31] It is fascinating to think that the idea for such a generalization arose from her discussions with Weyl in Göttingen in 1927, when infinite groups were discussed and the need to generalize her theory became apparent.

By the way, Jacobson and Emmy Noether met in 1934 in Princeton, when she was running a weekly seminar. We cannot exclude the possi-

---

30 In those post-war years it was not easy to get hold of books or journals from foreign countries, and so the 1945 volume of the "Transactions" was not available at the Hamburg library.

31 A particularly short and beautiful presentation is to be found in Artin's article [4] where he refers to Tate.

bility that she had given a hint to him too, either in her seminar or in
personal discussions. After all, this was her usual style, as reported by
van der Waerden [48].

5.3 A letter from N to W: 1927

As stated in the introduction we have not found letters from Emmy
Noether to Weyl, with one exception. That exception is kept in the
archive of the ETH in Zürich. It is written by Emmy Noether and dated
March 12, 1927. This is shortly after the end of the winter semester
1926/27 when Weyl had been in Göttingen as reported in the previous
section. Now Weyl was back in Zürich and they had to write letters
instead of just talking.

The letter concerns Paul Alexandroff and Heinz Hopf and their plan
to visit Princeton in the academic year 1927/28.

We have already mentioned Alexandroff in section 3.2 in connection
with Noether's contributions to topology. From 1924 to 1932 he spent
every summer in Göttingen, and there developed a kind of friendly relationship between him and Emmy Noether. The relation of Noether to
her "Noether boys" has been described by André Weil as like a mother
hen to her fledglings [50]. Thus Paul Alexandroff was accepted by Emmy
Noether as one of her fledglings. In the summer semester 1926 Heinz
Hopf arrived in Göttingen as a postdoc from Berlin and he too was accepted as a fledgling. Both Alexandroff and Hopf became close friends
and they decided to try to go to Princeton University in the academic
year 1927/28.

Perhaps Emmy Noether had suggested this; in any case she helped
them to obtain a Rockefeller grant for this purpose. It seems that Weyl
also had lent a helping hand, for in her letter to him she wrote:

... Jedenfalls danke ich Ihnen sehr für Ihre Bemühungen; auch Alexandroff und
Hopf werden Ihnen sehr dankbar sein und es scheint mir sicher, dass wenn die
formalen Schwierigkeiten erst einmal überwunden sind, Ihr Brief dann von we-
sentlichem Einfluss sein wird.

*... In any case I would like to thank you for your help; Alexandroff and Hopf
too will be very grateful to you. And I am sure that if the formal obstacles will
be overcome then your letter will be of essential influence.*

The "formal obstacles" which Noether mentioned were, firstly, the fact
that originally the applicants (Alexandroff and Hopf) wished to stay
for a period less than an academic year in Princeton (which later they
extended to a full academic year), and secondly, that Hopf's knowledge

of the English language seemed not to be sufficient in the eyes of the Rockefeller Foundation (but Noether assured them that Hopf *wanted* to learn English).[32] But she mentioned there had been letters sent to Lefschetz and Birkhoff and that at least the latter had promised to approach the Rockefeller Foundation to make an exception.

Alexandroff was in Moscow and Hopf in Berlin at the time, and so the mother hen acted as representative of her two chickens.[33]

About Alexandroff's and Hopf's year in Princeton we read in the Alexandroff article of MacTutor's History of Mathematics archive:

Aleksandrov and Hopf spent the academic year 1927-28 at Princeton in the United States. This was an important year in the development of topology with Aleksandrov and Hopf in Princeton and able to collaborate with Lefschetz, Veblen and Alexander.

The letter from Noether to Weyl shows that both N. and W. were instrumental in arranging this important Princeton year for Alexandroff and Hopf. Both were always ready to help young mathematicians to find their way.

REMARK: Later in 1931, when Weyl had left Zürich for Göttingen, it was Heinz Hopf who succeeded Weyl in the ETH Zürich. At those times it was not uncommon that the leaving professor would be asked for nominations if the faculty wished to continue his line. We can well imagine that Weyl, who originally would have preferred Artin[34], finally nominated Heinz Hopf for this position. If so then he would have discussed it with Emmy Noether since she knew Hopf quite well. It may even have been that she had taken the initiative and proposed to Weyl the nomination of Hopf. In fact, in the case of Alexandroff she did so in a letter to Hasse dated October 7, 1929 when it was clear that Hasse would change from Halle to Marburg. There she asked Hasse whether he would propose the name of Alexandroff as a candidate in Halle.[35] It seems realistic to assume that in the case of Heinz Hopf she acted similarly.

---

32 It seems that his knowledge of English had improved in the course of time since Heinz Hopf had been elected president of the IMU (International Mathematical Union) in 1955 till 1958.

33 Probably the letters from Noether and Weyl to the Rockefeller Foundation, i.e., to Trowbridge, on this matter are preserved in the Rockefeller archives in New York but we have not checked this.

34 In fact, in 1930 Artin received an offer from the ETH Zürich which, however, he finally rejected.

35 This however, did not work out. The successor of Hasse in Halle was Heinrich Brandt, known for the introduction of "Brandt's gruppoid" for divisor classes in simple algebras over number fields.

REMARK 2: The above mentioned letter of Noether to Weyl contains
a postscript which gives us a glimpse of the mathematical discussion
between the two (and it is the only written document for this). It reads:

Die Mertens-Arbeit, von der ich Ihnen sprach, steht Monatshefte, Bd. 4. Ich
dachte an den Schluss, Seite 329. Es handelt sich hier aber doch nur um De-
terminanten-Relationen, sodass es für Sie wohl kaum in Betracht kommt.

*The Mertens paper which I mentioned to you is contained in volume 4 of
the* Monatshefte. *I had in mind the end of the paper, page 329. But this is
concerned with determinant relations only, hence it will perhaps not be relevant
to your purpose.*

The Mertens paper is [25]. We have checked the cited page but did not
find any hint which would connect to Weyl's work. Perhaps someone
else will be able to interpret Noether's remark.

5.4 Weyl in Göttingen: 1930-1933

Weyl in [53] reports:

*When I was called permanently to Göttingen in 1930, I earnestly tried to obtain
from the Ministerium a better position for her* [Emmy Noether], *because I
was ashamed to occupy such a preferred position beside her whom I knew my
superior as a mathematician in many respects.*

We see that by now, Weyl was completely convinced about the mathematical stature of Emmy Noether. After all, Emmy Noether in 1930 was
the world-wide acknowledged leader of abstract algebra, and her presence in Göttingen was the main attraction for young mathematicians
from all over the world to visit the Mathematical Institute and study
with her.

It would be interesting to try to find out which "better position" Weyl
had in mind in his negotiations with the *Ministerium* in Berlin. Maybe
he wished tenure for her, and an increase of her salary. The archives in
Berlin will perhaps have the papers and reports of Weyl's negotiations.
From those papers one may be able to extract the reasons for the rejection. But the opposition against Noether's promotion did not only come
from the Ministerium in Berlin. It seems that a strong opposition came
also from among the mathematician colleagues in Göttingen, for Weyl
continues with his report as follows:

... nor did an attempt [succeed] *to push through her election as a member of
the Göttinger Gesellschaft der Wissenschaften. Tradition, prejudice, external
considerations, weighted the balance against her scientific merits and scientific
greatness, by that time denied by no one.*

I do not know whether there exist minutes of the meetings of the *Göttinger Mathematische Gesellschaft* in 1930. If so then it would be interesting to know the traditional, biased and external arguments which Weyl said were put forward against Emmy Noether from the members of the *Mathematische Gesellschaft*. Was it still mainly her gender? Or was it the opposition to her "abstract" mathematical methods? In any case, the decision not to admit Emmy Noether as a member of the *Göttinger Mathematische Gesellschaft* is to be regarded as an injustice to her and a lack of understanding of the development of modern mathematics. After all, the Emmy Noether of 1930 was quite different from the Emmy Noether of 1915. Now in 1930, she had already gone a long way *"on her own completely original mathematical path"*, and her *"working and conceptual methods had spread everywhere"*. She could muster high-ranking colleagues and students who were fascinated by her way of mathematical thinking.

Nevertheless it seems that there was some opposition against her abstract methods, also in Göttingen among the mathematicians. Olga Taussky-Todd recalls in [45] her impression of the Göttingen mathematical scene:

...not everybody liked her [Emmy Noether]*, and not everybody trusted that her achievements were what they later were accepted to be.*

One day Olga Taussky had been present when one of the senior professors talked very roughly to Emmy Noether. (Later he apologized to her for this insult.) When Emmy Noether had her 50th birthday in 1932 then, as Olga Taussky recalls, nobody at Göttingen had taken notice of it, although at that time all birthdays were published in the *Jahresbericht of the DMV*.[36] Reading all this, I can understand Emmy Noether when later in 1935 she said to Veblen about her time in the USA:

The last year and a half had been the very happiest in her whole life, for she was appreciated in Bryn Mawr and Princeton as she had never been appreciated in her own country.[37]

Thus it seems that Weyl's statement that *"her scientific merits and scientific greatness by that time was denied by no one"* did not describe the situation exactly. Perhaps, since Weyl was the "premier professor" of mathematics at Göttingen, and since he was known to respect and

---

36 But Hasse in Marburg had sent her a birthday cake, together with a paper which he had dedicated to her. The paper was [12]. See [22].

37 Cited from a letter of Abraham Flexner to President Park of Bryn Mawr; see [37].

acknowledge Noether's merits and scientific greatness, nobody dared to tell him if he disagreed. Olga Taussky-Todd remembers that

"outside of Göttingen, Emmy was greatly appreciated in her country."

We may add that this was not only so in her country but also world-wide. And of course also in Göttingen there was an ever-growing fraction of mathematicians, including Weyl, who held Noether in high esteem. As to Hermann Weyl, let us cite MacLane who was a student at Göttingen in the period 1931-33. We read in [24]:

When I first came to Göttingen I spoke to Professor Weyl and expressed my interest in logic and algebra. He immediately remarked that in algebra Göttingen was excellently represented by Professor Noether; he recommended that I attend her courses and seminars ... By the time of my arrival she was Ausserordentlicher Professor. However, it was clear that in the view of Weyl, Hilbert, and the others, she was right on the level of any of the full professors. Her work was much admired and her influence was widespread.

MacLane sometimes joined the hiking parties (*Ausflug*) of Emmy Noether and her class to the hills around Göttingen. Noether used these hiking parties to discuss *"algebra, other mathematical topics and Russia"*.[38] It seems that Weyl too joined those excursions occasionally. There is a nice photo of Noether with Weyl and family, together with a group of mathematicians posing in front of the *"Gasthof Vollbrecht"*. The photo is published in [5] and dated 1932. Since Artin is seen as a member of the hiking party, it seems very probable that the photo was taken on the occasion of Artin's famous Göttingen lectures on class field theory which took place from February 29 to March 2, 1932.[39] This was a big affair and a number of people came from various places in order to listen to Artin lecturing on the new face of class field theory. The lectures were organized by Emmy Noether. Since she was not a full professor and, accordingly, had no personal funds to organize such meetings we suppose that one of her colleagues, probably Weyl, had made available the necessary financial means for this occasion. In any case we see that by now she was able to get support for her activities in Göttingen, not only for the Artin lectures but also for other speakers.

The International Congress of Mathematicians took place in September 1932 in Zürich. Emmy Noether was invited to deliver one of the

38 1928/29 Emmy Noether had been in Moscow as a visiting professor, on the invitation of Alexandroff whom she knew from Göttingen.
39 The photo is also contained in the Oberwolfach photo collection online. Probably the photo was taken by Natascha Artin, the wife of Emil Artin.

main lectures there. Usually, proposals for invited speakers at the IMU conferences were submitted by the presidents of the national mathematical organizations which were members of the IMU. In 1931/32 Hermann Weyl was president ("*Vorsitzender*") of the DMV. So it appears that Weyl had his hand in the affair when it came to proposing Emmy Noether as a speaker from Germany. The proposal had to be accepted by the executive committee. The nomination of Emmy Noether was accepted and this shows the great respect and admiration which Emmy Noether enjoyed on the international scale.

Emmy Noether's Zürich lecture can be considered as the high point in her mathematical career.

6 Göttingen exodus: 1933

The year 1933 brought about the almost complete destruction of the unique mathematical scene in Göttingen. In consequence of the anti-semitic political line of the Nazi government many scientists of Jewish origin had to leave the university, as well as those who were known to be critical towards the new government. The Göttingen situation in 1933 has often been described, and so we can refer to the literature, e.g., [40], [42].

Emmy Noether was of Jewish origin and so she too was a victim of the new government policy. On May 5, 1933 Emmy Noether obtained the message that she was put "temporarily on leave" from lecturing at the university. When Hasse heard this, he wrote a letter to her; we do not know the text of his letter but from her reply we may conclude that he asked whether he could be of help. Emmy Noether replied on May 10, 1933:

Lieber Herr Hasse!
Vielen herzlichen Dank für Ihren guten freundschaftlichen Brief! Die Sache ist aber doch für mich sehr viel weniger schlimm als für sehr viele andere: rein äußerlich habe ich ein kleines Vermögen (ich hatte ja nie Pensionsberechtigung), sodaß ich erst einmal in Ruhe abwarten kann; im Augenblick, bis zur definitiven Entscheidung oder etwas länger, geht auch das Gehalt noch weiter. Dann wird wohl jetzt auch einiges von der Fakultät versucht, die Beurlaubung nicht definitiv zu machen; der Erfolg ist natürlich im Moment recht fraglich. Schließlich sagte Weyl mir, daß er schon vor ein paar Wochen, wo alles noch schwebte, nach Princeton geschrieben habe wo er immer noch Beziehungen hat. Die haben zwar wegen der Dollarkrise jetzt auch keine Entschlußkraft; aber Weyl meinte doch daß mit der Zeit sich etwas ergeben könne, zumal Veblen im vorigen Jahr viel daran lag, mich mit Flexner, dem Organisator des neuen Instituts, bekannt zu machen. Vielleicht

kommt einmal eine sich eventuell wiederholende Gastvorlesung heraus, und im übrigen wieder Deutschland, das wäre mir natürlich das liebste. Und vielleicht kann ich Ihnen sogar auch einmal so ein Jahr Flexner-Institut verschaffen - das ist zwar Zukunftsphantasie - wir sprachen doch im Winter davon . . .

Dear Mr. Hasse!
Thanks very much for your good, friendly letter! But for myself, the situation is much less dire than for many others: in fact I have a small fortune (after all I was never entitled to pension) and hence for the time being I can quietly wait and see. Also, the salary payments continue until the final decision or even somewhat longer. Moreover the Faculty tries to avert my suspension to become final; at the moment, however, there is little hope for success. Finally, Weyl told me that some weeks ago already when things were still open, he had written to Princeton where he still has contacts. At the moment, however, because of the dollar crisis they don't have much freedom there for their decisions; but Weyl believes that in the course of time there may arise something, in particular since Veblen last year was eager to introduce me to Flexner, the organizer of the new Institute. Perhaps there will emerge a visiting professorship which may be iterated, and in the meantime Germany again, this would be the best solution for me, naturally. And maybe I will be able to manage for you too a year in the Flexner Institute – but this is my fantasy for the future – we have talked last winter about this. . .

The first impression while reading this letter is her complete selflessness, which is well-known from other reports on her life and which is manifest here again. She does not complain about her own situation but only points out that for other people things may be worse. Reading further, we see that the Faculty in Göttingen tries to keep her; this shows that she was respected there as a scientist and teacher although she still did not have a tenured position. Hermann Weyl was a full professor and hence a member of the Faculty committee; we can surely assume that he was one of the driving forces in trying to save Emmy Noether for a position in Göttingen. In fact, in his memorial speech [53] Weyl said:

It was attempted, of course, to influence the Ministerium and other responsible and irresponsible bodies so that her position might be saved. I suppose there could hardly have been in any other case such a pile of enthusiastic testimonials filed with the Ministerium as was sent in on her behalf. At that time we really fought; there was still hope left that the worst could be warded off . . .

And finally, in the above Noether letter we read that, independent of these attempts, Weyl had written to Princeton on her behalf. We do not know whom in Princeton Weyl had adressed. Since Noether mentions in her letter Veblen and Flexner, it seems probable that Weyl had written to one or both of them. Abraham Flexner was the spiritual founder and the first director of the newly-founded Institute for Advanced Study in

Princeton. Oswald Veblen was the first permanent mathematics professor of the IAS. Certain indications suggest that Weyl had written to Lefschetz too; see next section. Solomon Lefschetz had the position of full professor at Princeton University.

One year earlier, in the late summer of 1932, Weyl had rejected an offer to join the IAS as a permanent member. But now, since the political situation had deteriorated, he inquired whether it was possible to reverse his decision. (It was.) From Noether's letter we infer that Weyl did not only write on his own behalf but also on Noether's. This fact alone demonstrates the very high esteem in which he held Noether as a mathematician and as a personality.[40]

But of course, the best solution would be that Noether could stay in Göttingen. This was what Weyl wished to achieve foremost, as we cited above. (It was in vain.) Weyl reports in [53]:

I have a particularly vivid recollection of these months. Emmy Noether, her courage, her frankness, her unconcern about her own fate, her conciliatory spirit, were, in the middle of all the hatred and meanness, despair and sorrow surrounding us, a moral solace.

That stormy time of struggle in the summer of 1933 in Göttingen drew them closer together. This is also evident from the words Weyl used two years later in his speech at her funeral:[41]

You did not believe in evil, indeed it never occurred to you that it could play a role in the affairs of man. This was never brought home to me more clearly than in the last summer we spent together in Göttingen, the stormy summer of 1933. In the midst of the terrible struggle, destruction and upheaval that was going on around us in all factions, in a sea of hate and violence, of fear and desperation and dejection - you went your own way, pondering the challenges of mathematics with the same industriousness as before. When you were not allowed to use the institute's lecture halls you gathered your students in your own home. Even those in their brown shirts were welcome; never for a second did you doubt their integrity. Without regard for your own fate, openhearted and without fear, always conciliatory, you went your own way. Many of us believed that an enmity had been unleashed in which there could be no pardon; but you remained untouched by it all.

Parallel to the attempts of the Faculty to keep Noether in Göttingen, Hasse took the initiative and collected testimonials[42] which would put

---

40 In the course of time, Weyl used his influence in American academic circles to help many other mathematicians as well.

41 See section 9.2.

42 The German word is "*Gutachten*". I am not sure whether the translation into "*testimonial*" is adequate. My dictionary offers also "*opinion*" or "*expertise*" or

into evidence that Emmy Noether was a scientist of first rank and hence it would be advantageous for the scientific environment of Göttingen if she did not leave. Hasse collected 14 such testimonials. Together they were sent to the *Kurator* of the university who was to forward them to the *Ministerium* in Berlin. Recently we have found the text of those testimonials which are kept in the Prussian State archives in Berlin; we plan to publish them separately. The names of the authors are:

H. Bohr, Kopenhagen
Ph. Furtwängler, Wien
G.H. Hardy, Cambridge
H. Hasse, Marburg
O. Perron, München
T. Rella, Wien
J.A. Schouten, Delft
B. Segre, Bologna
K. Shoda, Osaka
C. Siegel, Frankfurt
A. Speiser, Zürich
T. Takagi, Tokyo
B.L. van der Waerden, Leipzig
H. Weyl, Göttingen

We see that also Hermann Weyl wrote a testimonial. We have included it in the appendix, translated into English; see section 9.1. Note that Weyl compared Emmy Noether to Lise Meitner, the nuclear physicist. In the present situation this comparison may have been done since Meitner, also of Jewish origin, was allowed to stay in Berlin continuing her research with Otto Hahn in their common laboratory. After all, the initiatives of Hasse and of Weyl were to obtain a similar status for Emmy Noether in Göttingen.

As is well-known, this was in vain. Perhaps those testimonials were never read after the *Kurator* of Göttingen University wrote to the *Ministerium* that Emmy Noether's political opinions were based on "Marxism".[43]

Let us close this section with some lines from a letter of Weyl to Heinrich Brandt in Halle. The letter is dated December 15, 1933; at that

---

"*letter of recommendation*". I have chosen "*testimonial*" since Weyl uses this terminology.

43 See [47]. – By the way, there was another such initiative started, namely in favor of Courant who also had been "*beurlaubt*" from Göttingen University. That was signed by 28 scientists including Hermann Weyl and Helmut Hasse. Again this was not successful, although this time the *Kurator's* statement was not as negative as in Noether's case. (We have got this information from Constance Reid's book on Courant [34].)

time Weyl and Noether were already in the USA. Brandt was known to be quite sceptical towards abstract methods in mathematics; he did not even like Artin's beautiful presentation of his own (Brandt's) discovery, namely that the ideals and ideal classes of maximal orders in a simple algebra over a number field form a groupoid under multiplication.[44] (The notion of "groupoid" is Brandt's invention.) Weyl's letter is a reply to one from Brandt which, however, is not known to us. Apparently Brandt had uttered some words against Noether's abstracting method, and Weyl replied explaining his own viewpoint:[45]

... So wenig mir persönlich die "abstrakte" Algebra liegt, so schätze ich doch ihre Leistungen und ihre Bedeutung offenbar wesentlich höher ein, als Sie das tun. Es imponiert mir gerade an Emmy Noether, daß ihre Probleme immer konkreter und tiefer geworden sind.

Personally, the "abstract" algebra doesn't suit me well, but apparently I do estimate its achievements and importance much higher than you are doing. I am particularly impressed that Emmy Noether's problems have become more and more concrete and deep.

Weyl continues as follows. It is not known whether Brandt had written some comments on Noether's Jewish origin and connected this with her abstract way of thinking, or perhaps Weyl's letter was triggered by the general situation in Germany and especially in Göttingen:

Warum soll ihr, der Hebräerin, nicht zustehen, was in den Händen des "Ariers" Dedekind zu großen Ergebnissen geführt hat? Ich überlasse es gern Herrn Spengler und Bieberbach, die mathematische Denkweise nach Völkern und Rassen zu zerteilen. Daß Göttingen den Anspruch verloren hat, mathematischer Vorort zu sein, gebe ich Ihnen gerne zu – was ist denn überhaupt von Göttingen übrig geblieben? Ich hoffe und wünsche, daß es eine seiner alten Tradition würdige Fortsetzung durch neue Männer finden möge; aber ich bin froh, daß ich es nicht mehr gegen einen Strom von Unsinn und Fanatismus zu stützen brauche!

Why should she, as of Hebrew descent, not be entitled to do what had led to such great results in the hands of Dedekind, the "Arian"? I leave it to Mr. Spengler and Mr. Bieberbach to divide the mathematical way of thinking according to nations and races. I concede that Göttingen has lost its role as a high-ranking mathematical place – what is actually left of Göttingen? I hope and wish that Göttingen would find a continuation by new men, worthy of its long tradition; but I am glad that I do not have to support it against a torrent of nonsense and fanaticism.

---

44 Artin's paper is [3].
45 I would like to thank M. Göbel for sending me copies of this letter from the Brandt archive in Halle. The letter is published in [17], together with other letters Brandt-Weyl and Brandt-Noether.

7 Bryn Mawr: 1933-1935

As we have seen in the foregoing section, Weyl had written to Princeton
on behalf of Emmy Noether, and this was in March or April 1933 already.
Since he was going to join the Institute for Advanced Study in Princeton,
one would assume that he had recommended accepting Emmy Noether
as a visiting scientist of the Institute. We know that some people at
the Institute were interested in getting Noether to Princeton, for at
the International Zürich Congress Oswald Veblen had been eager to
introduce Emmy Noether to the Institute's director, Abraham Flexner.
(See Noether's letter to Hasse, cited in the foregoing section.)

But as it turned out, Emmy Noether did not receive an invitation as a
visitor to the Institute. We do not know the reason for this; perhaps the
impending dollar crisis, mentioned in Noether's letter to Hasse, forced
the Institute to reduce its available funds. Or, may there have been
other reasons as well? On the other hand, from the documents which
we found in the archive of Bryn Mawr College it can be seen that the
Institute for Advanced Study contributed a substantial amount towards
the salary of Emmy Noether in Bryn Mawr.

We do not know who was the first to suggest that Bryn Mawr College
could be a suitable place for Emmy Noether. Some evidence points to the
conclusion that it was Solomon Lefschetz. In fact, we have found a letter,
dated June 12, 1933 already, adressed to the *"Emergency Committee in
Aid of Displaced German Scholars"*, where he discusses future aspects
for Emmy Noether and proposes Bryn Mawr.[46] Lefschetz had visited
Göttingen two years ago and so he knew Emmy Noether personally.
Lefschetz' letter is quite remarkable since, firstly, he clearly expresses
that Emmy Noether, in his opinion, was a leading figure in contemporary
mathematics; secondly we see that he had taken already practical steps
to provide Bryn Mawr with at least part of the necessary financial means
in order to offer Emmy Noether a stipend. Let us cite the relevant
portions of that letter:

*Dear Dr. Duggan: I am endeavoring to make connections with some wealthy
people in Pittsburgh, one of them a former Bryn Mawr student, with a view
of raising a fund to provide a research associateship at Bryn Mawr for Miss
Emmy Noether. As you may know, she is one of the most distinguished victims
of the Hitler cold pogrom and she is victimized doubly; first for racial reasons
and second, owing to her sex. It occured to me that it would be a fine thing
to have her attached to Bryn Mawr in a position which would compete with*

46 We have found this letter in the archives of the New York Public Library.

no one and would be created ad hoc; the most distinguished feminine math-
ematician connected with the most distinguished feminine university. I have
communicated with Mrs. Wheeler, the Head of the Department at Bryn Mawr,
and she is not only sympathetic but thoroughly enthusiastic for this plan,

So far as I know, your organization is the only one which is endeavoring
to do anything systematic to relieve the situation of the stranded German sci-
entists. As I do not think random efforts are advisable, I wish first of all to
inform you of my plan. Moreover, if I were to succeed only partially, would it
be possible to get any aid from your organization? I would greatly appreciate
your informing me on this point at your earliest convenience.

In the preliminary communication with my intended victims I mentioned
the following proposal: to contribute enough annually to provide Miss Noether
with a very modest salary, say $ 2000, and a retiring allowance of $ 1200.

Yours very sincerely, S. Lefschetz.

Already one month later the committee granted the sum of $ 2000 to
Bryn Mawr for Emmy Noether.

There arises the question from whom Lefschetz had got the information, at that early moment already, that Emmy Noether had been
suspended.[47] We are inclined to believe that it was Hermann Weyl. I
do not know whether the correspondence of Lefschetz of those years has
been preserved in some archive, and where. Perhaps it will be possible
to find those letters and check.

Emmy Noether arrived in Bryn Mawr in early November 1933. Her
first letter from Bryn Mawr to Hasse is dated March 6, 1934. She reported, among other things, that since February she gave a lecture once
a week at the Institute for Advanced Study in Princeton. In this lecture
she had started with representation modules and groups with operators.
She mentions that Weyl too is lecturing on representation theory, and
that he will switch to continuous groups later. It appears that the Göttingen situation of 1926/27 was repeating. And we imagine Hermann
Weyl and Emmy Noether walking after her lectures around the Campus
of Princeton University[48] instead of Göttingen's narrow streets, vividly
discussing new aspects of representation theory.

In the book [34] on Courant we read:

Weyl sent happy letters from Princeton. In Fine Hall, where Flexner's group
was temporarily housed, German was spoken as much as English. He frequently
saw Emmy Noether ...

47 Emmy Noether had been "beurlaubt", i.e., temporarily suspended from her du-
   ties, in May 1933. Observing that mail from Europe to USA used about 2-3
   weeks at that time, we conclude that Lefschetz must have started working on his
   Noether-Bryn Mawr idea immediately after receiving the news about her suspension. Noether was finally dismissed from university on September 9, 1933.
48 The Institute's Fuld Hall had not yet been built and the School of Mathematics
   of the Institute was temporarily housed in Fine Hall on the University Campus.

Perhaps in the Courant legacy we can find more about Weyl and Noether in Princeton, but we have not been able yet to check those sources.

Every week Emmy Noether visited the Brauers in Princeton; Richard Brauer was assistant to Weyl in that year and perhaps sometimes Hermann Weyl also joined their company. The name of Hermann Weyl appears several times in her letters to Hasse from Bryn Mawr. In November 1934 she reports that she had studied Weyl's recent publication on Riemann matrices in the Annals of Mathematics.

Emmy Noether died on April 14, 1935. One day later Hermann Weyl cabled to Hasse:

hasse mathematical institute gottingen – emmy noether died yesterday – by sudden collapse after successful – operation of tumor[49] *few days ago – burial wednesday bryn mawr – weyl*

At the burial ceremony on Wednesday Weyl spoke on behalf of her German friends and colleagues. We have included an English translation of this moving text in the appendix; see section 9.2. One week later he delivered his memorial lecture in the large auditorium of Bryn Mawr College. That text is published and well known [53].

8 The Weyl-Einstein letter to the NYT

On Sunday May 5, 1935 the New York Times published a "Letter to the Editor", signed by Albert Einstein and headed by the following title:

Professor Einstein Writes in Appreciation of a Fellow-Mathematician.

We have included the text of this letter in our appendix; see section 9.3.

Reading this letter one is struck by the almost poetic style which elevates the text to one of the pearls in the literature on mathematics. The text is often cited, the last citation which I found is in the "*Mitteilungen*" of the DMV, 2007 where Jochen Brüning tries to connect mathematics with poetry [6]. But because of this character of style it has been doubted whether the text really was composed by Einstein himself. If not then this would not have been the first and not the last incident where Einstein had put his name under a text which was not conceived by himself – provided that in his opinion the subject was

---

49 President Park of Bryn Mawr had sent a detailed report, dated May 16, 1935, to Otto Nöther in Mannheim, a cousin of Emmy Noether. A copy of that letter is preserved. There it is stated that according to the the medical diagnosis of the doctors who operated her, Emmy Noether suffered from a "pelvis tumor".

worth-while to support. Since Weyl's poetic style was known it was not considered impossible that the text was composed by Hermann Weyl.

Some time ago I have come across a letter signed by Dr. Ruth Stauffer-McKee. I include a copy of that letter in the appendix; see section 9.4. In particular I refer to the last paragraph of the letter. Based on the information provided by Stauffer I came to the conclusion that, indeed, the text was essentially written by Weyl. I have expressed this opinion in my talk in Bielefeld and also in a "Letter to the Editor" of the *"Mitteilungen der Deutschen Mathematiker-Vereinigung"* [38].

However, recently I have been informed that Einstein's draft of this letter in his own handwriting has been found by Siegmund-Schultze[50] in the Einstein archive in Jerusalem. The article is to appear in the next issue of the *Mitteilungen der DMV* [43]. This then settles the question of authorship in favor of Einstein. But what had induced Ruth Stauffer to claim that Weyl had "inspired" Einstein's letter?

In order to understand Stauffer's letter let us explain its background.

In 1972 there appeared a paper on Emmy Noether in the American Mathematical Monthly, authored by Clark Kimberling [18]. Among other information the paper contains the text of Einstein's letter to the New York Times. Kimberling had obtained the text from an article in the Bryn Mawr Alumnae Bulletin where it had been reprinted in 1935. Together with that text, we find in [18] the following:

A note in the files of the Bryn Mawr Alumnae Bulletin reads, "The above was inspired, if not written, by Dr. Hermann Weyl, eminent German mathematician. Mr. Einstein had never met Miss Noether."

(Here, by "above" was meant the text of the Einstein letter to the New York Times.)

While the first sentence of that "note" can be considered as an affirmation of the guess that Weyl had conceived the text of Einstein's letter, the second sentence is hard to believe. Emmy Noether often visited the Institute for Advanced Study in Princeton, the same place where Einstein was, and it seems improbable that they did not meet there. After all, Einstein was already in May 1918 well informed about Noether's achievements, when he wrote to Hilbert praising her work [26]. And in December that year, after receiving the printed version of this work, he wrote to Felix Klein and recommended her *Habilitation*. In the 1920s, Einstein had a correspondence with Emmy Noether who acted as referee

---

50 I would like to thank R. Siegmund-Schultze for a number of interesting comments and corrections to this article.

for papers which were submitted to the *Mathematische Annalen*. It is hard to believe that in Princeton he would have avoided meeting Emmy Noether, whom he esteemed so highly. Moreover, we have already mentioned in section 4 that Einstein probably had met Noether in 1915 in Göttingen. Also, on the DMV-meetings 1909 in Salzburg and 1913 in Wien both Einstein and Emmy Noether presented talks and there was ample opportunity for them to meet.

Thus it seemed that the "note" which Kimberling mentioned had been written by someone who was not well informed about the situation in the early thirties. Actually, that "note" was not printed in the Bryn Mawr Alumnae Bulletin but it was added later by typewriter, maybe only on the copy which was sent to Kimberling. It is not known who had been the author of that "note".

In the same volume of the American Mathematical Monthly where his article [18] had appeared, Kimberling published an *Addendum* saying that Einstein's former secretary, Miss Dukas, had objected to the statement that the letter written by Einstein was "*inspired, if not written by Dr. Hermann Weyl*". She insisted that the letter was written by Einstein himself at the request of Weyl.

This, however, induced Ruth Stauffer to write the above mentioned letter to the editor of the American Mathematical Monthly, which we are citing in section 9.4. Ruth Stauffer had been a Ph.D. student of Emmy Noether in Bryn Mawr and in her letter she recalls vividly the mathematical atmosphere in Princeton at that time.

On this evidence we were led to believe that the statement of Einstein's secretary Dukas may be due to a mix-up on her part. For, only shortly before Noether's death Einstein had written another letter in which he recommends that Emmy Noether's situation in Bryn Mawr College should be improved and put on a more solid base. At that time President Park of Bryn Mawr had tried to obtain testimonies on Emmy Noether, which could be used in order to get funds for a more permanent position.[51] Einstein's testimony is dated January 8, 1935 and is written in German; we have found it in the archives of the Institute for Advanced Study in Princeton. Its full text reads:

Fräulein Dr. Emmy Noether besitzt unzweifelhaft erhebliches schöpferisches Talent, was jeweilen von nicht sehr vielen Mathematikern einer Generation gesagt werden kann. Ihr die Fortsetzung der wissenschaftlichen Arbeit zu ermöglichen, be-

---

51 This was successful, but Emmy Noether died before she got to know about it. – Other testimonials, by Solomon Lefschetz, Norbert Wiener and George D. Birkhoff are published in Kimberling's article [19].

deutet nach meiner Ansicht die Erfüllung einer Ehrenpflicht und wirkliche Förderung wissenschaftlicher Forschung.

Without doubt Miss Dr. Emmy Noether commands significant and creative talent; this cannot be said of many mathematicians of one generation. In my opinion it is an obligation of honor to provide her with the means to continue her scientific work, and indeed this will be a proper support of scientific research.

It is apparent that the style of this is quite different from the style of the letter to the New York Times.

Although we now know that Miss Dukas was right and Einstein had composed his NYT-letter with his own hand, there remains the question as to the basis of Stauffer's contentions.

Stauffer was a young student and what she reports is partly based on what she heard from Mrs. Wheeler. But the latter, who was head of the mathematics department of Bryn Mawr College at the time, had studied in Göttingen with Hilbert in the same years as Hermann Weyl had; so they were old acquaintances and it seems probable that Weyl himself had told her the story as it had happened. Thus it may well have been that first Weyl had sent his obituary on Emmy Noether to the New York Times, and that this was returned with the suggestion that Einstein should write an obituary – as Ruth Stauffer narrates. And then Einstein wrote his letter "at the request of Weyl", as Miss Dukas has claimed. Whether there was any cooperation between Einstein and Weyl while drafting the letter is not known. But we can safely assume that both had talked if not about the text of the letter but certainly about Emmy Noether's personality, her work and her influence on mathematics at large. In this way Stauffer's claim may be justified that Weyl had "inspired" Einstein in writing his letter.

REMARK: It has been pointed out to me by several people that the very last sentence in the English version of Einstein's letter deviates in its meaning from the original German text wheras otherwise the translation seems to be excellent.[52] In the English version it is said that Noether's last years in Bryn Mawr were made the *"happiest and perhaps most fruitful years of her entire career"*, but the German text does not refer to her entire career and only pointed out that death came to her *"mitten in froher und fruchtbarer Arbeit"*. I do not know who had translated the German text into English. There is a letter of Abraham Flexner, the director of the Institute for Advanced Study in Princeton, addressed to Einstein and dated April 30, 1935, in which Flexner thanks Einstein for

---

52 The German text is published in my "Letter to the Editor" [38].

the "*beautiful tribute to Miss Noether*" and continues: "*I shall translate it into English and send it to the New York Times, through which it will reach, I think, many of those who should know of her career.*" But it does not seem justified, I believe, to conclude that Flexner personally did the translation job. He was quite busy with all kinds of responsibilities and certainly he had contacts to experts who would have been willing and competent to do it.[53]

FINAL REMARK: Weyl's solidarity with Emmy Noether extended to her brother and family. Emmy's brother Fritz had emigrated to Russia where he got a position at the university in Tomsk. In 1937 he was arrested and sentenced to 25 years in prison because of alleged espionage for Germany. In the Einstein archive in Jerusalem we have found a letter, dated April 1938 and signed by Einstein, addressed to the Russian minister of foreign affairs Litvinov. In this letter Einstein appeals to the minister in favor of Fritz Noether, whom he (Einstein) is sure to be innocent. In the Einstein archive, right after this letter, is preserved a curriculum vitae of Fritz Noether in Weyl's handwriting. Thus again it appears that Weyl has "inspired" Einstein to write such a letter.[54]

Among Weyl's papers I found a number of letters from 1938 and the following years, which show that he cared for the two sons of Fritz Noether, Hermann and Gottfried, who had to leave the Soviet Union after their father had been sentenced. Weyl saw to it that they obtained immigrant visa to the United States, and that they got sufficient means to finance their university education. Both became respected members of the scientific community.

9 Appendix: documents

9.1 Weyl's testimony

The following text[55] is from the testimonial, signed by Hermann Weyl on July 12, 1933 and sent by Hasse to the Ministerium in Berlin together with 13 other testimonials. We have found these testimonials in the Prussian state archive Berlin.

---

53 Siegmund-Schultze [43] advocates reasons to assume that indeed, Flexner himself did the translation job.

54 The appeal of Einstein was in vain. In 1941, when German troops were approaching the town of Orjol where Fritz was kept in prison, he was sentenced to death and immediately executed. See, e.g., [41].

55 Translated from German by Ian Beaumont.

Emmy Noether has attained a prominent position in current mathematical research – by virtue of her unusual deep-rooted prolific power, and of the central importance of the problems she is working on together with their interrelationships. Her research and the promising nature of the material she teaches enabled her in Göttingen to attract the largest group of students. When I compare her with the two woman mathematicians whose names have gone down in history, *Sophie Germain* and *Sonja Kowalewska*, she towers over them due to the originality and intensity of her scientific achievements. The name Emmy Noether is as important and respected in the field of mathematics as *Lise Meitner* is in physics.

She represents above all "Abstract Algebra". The word "abstract" in this context in no way implies that this branch of mathematics is of no practical use. The prevailing tendency is to solve problems using suitable visualizations, i.e. appropriate formation of concepts, rather than blind calculations. Fräulein Noether is in this respect the legitimate successor of the great German number theorist R. Dedekind. In addition, Quantum Theory has made Abstract Algebra the area of mathematics most closely related to physics.

In this field, in which mathematics is currently experiencing its most active progress, *Emmy Noether* is the recognised leader, both nationally and internationally.

Hermann Weyl

9.2 Weyl's funeral speech

The following text[56] *was spoken by Hermann Weyl on Emmy Noether's funeral on April 18, 1935. We have found this text in the legacy of Grete Hermann, which is preserved in the "Archiv der sozialen Demokratie" in Bonn.*

The hour has come, Emmy Noether, in which we must forever take our leave of you. Many will be deeply moved by your passing, none more so than your beloved brother Fritz, who, separated from you by half the globe, was unable to be here, and who must speak his last farewell to you through my mouth. His are the flowers I lay on your coffin. We bow our heads in acknowledgement of his pain, which it is not ours to put into words.

But I consider it a duty at this hour to articulate the feelings of your

---
56 translated from German by Ian Beaumont

German colleagues - those who are here, and those in your homeland who have held true to our goals and to you as a person. I find it apt, too, that our native tongue be heard at your graveside - the language of your innermost sentiments and in which you thought your thoughts - and which we hold dear whatever power may reign on German soil. Your final rest will be in foreign soil, in the soil of this great hospitable country that offered you a place to carry on your work after your own country closed its doors on you. We feel the urge at this time to thank America for what it has done in the last two years of hardship for German science, and to thank especially Bryn Mawr, where they were both happy and proud to include you amongst their teachers.

Justifiably proud, for you were a great woman mathematician - I have no reservations in calling you the greatest that history has known. Your work has changed the way we look at algebra, and with your many gothic letters you have left your name written indelibly across its pages. No-one, perhaps, contributed as much as you towards remoulding the axiomatic approach into a powerful research instrument, instead of a mere aid in the logical elucidation of the foundations of mathematics, as it had previously been. Amongst your predecessors in algebra and number theory it was probably Dedekind who came closest.

When, at this hour, I think of what made you what you were, two things immediately come to mind . The first is the original, productive force of your mathematical thinking. Like a too ripe fruit, it seemed to burst through the shell of your humanness. You were at once instrument of and receptacle for the intellectual force that surged forth from within you. You were not of clay, harmoniously shaped by God's artistic hand, but a piece of primordial human rock into which he breathed creative genius.

The force of your genius seemed to transcend the bounds of your sex - and in Göttingen we jokingly, but reverentially, spoke of you in the masculine, as "den Noether". But you were a woman, maternal, and with a childlike warmheartedness. Not only did you give to your students intellectually - fully and without reserve - they gathered round you like chicks under the wings of a mother hen; you loved them, cared for them and lived with them in close community.

The second thing that springs to mind is that your heart knew no malice; you did not believe in evil, indeed it never occurred to you that it could play a role in the affairs of man. This was never brought home to me more clearly than in the last summer we spent together in Göttingen, the stormy summer of 1933. In the midst of the terrible struggle, destruction and upheaval that was going on around us in all factions, in a sea of hate and violence, of fear and desperation and dejection - you went your own way, pondering the challenges of mathematics with the same industriousness as before. When you were not allowed to use the institute's lecture halls you gathered your students in your own home. Even those in their brown shirts were welcome; never for a second did you doubt their integrity. Without regard for your own fate, openhearted and without fear, always conciliatory, you went your own way. Many of us believed that an enmity had been unleashed in which there could be no pardon; but you remained untouched by it all. You were happy to go back to Göttingen last summer, where, as if nothing had happened, you lived and worked with German mathematicians striving for the same goals. You planned on doing the same this summer.

You truly deserve the wreath that the mathematicians in Göttingen have asked me to lay on your grave.

We do not know what death is. But is it not comforting to think that souls will meet again after this life on Earth, and how your father's soul will greet you? Has any father found in his daughter a worthier successor, great in her own right?

You were torn from us in your creative prime; your sudden departure, like the echo of a thunderclap, is still written on our faces. But your work and your disposition will long keep your memory alive, in science and amongst your students, friends and colleagues.

Farewell then, Emmy Noether, great mathematician and great woman. Though decay take your mortal remains, we will always cherish the legacy you left us.

Hermann Weyl

9.3 Letter to the New York Times

The following text was published on Sunday, May 5, 1935 by the New York Times, with the heading: "Professor Einstein Writes in Appreciation of a Fellow-Mathematician".

To the Editor of The New York Times:

The efforts of most human-beings are consumed in the struggle for their daily bread, but most of those who are, either through fortune or some special gift, relieved of this struggle are largely absorbed in further improving their worldly lot. Beneath the effort directed toward the accumulation of worldly goods lies all too frequently the illusion that this is the most substantial and desirable end to be achieved; but there is, fortunately, a minority composed of those who recognize early in their lives that the most beautiful and satisfying experiences open to humankind are not derived from the outside, but are bound up with the development of the individual's own feeling, thinking and acting. The genuine artists, investigators and thinkers have always been persons of this kind. However inconspicuously the life of these individuals runs its course, none the less the fruits of their endeavors are the most valuable contributions which one generation can make to its successors.

Within the past few days a distinguished mathematician, Professor Emmy Noether, formerly connected with the University of Göttingen and for the past two years at Bryn Mawr College, died in her fifty-third year. In the judgment of the most competent living mathematicians, Fräulein Noether was the most significant creative mathematical genius thus far produced since the higher education of women began. In the realm of algebra, in which the most gifted mathematicians have been busy for centuries, she discovered methods which have proved of enormous importance in the development of the present-day younger generation of mathematicians. Pure mathematics is, in its way, the poetry of logical ideas. One seeks the most general ideas of operation which will bring together in simple, logical and unified form the largest possible circle of formal relationships. In this effort toward logical beauty spiritual formulas are discovered necessary for the deeper penetration into the laws of nature.

Born in a Jewish family distinguished for the love of learning, Emmy Noether, who, in spite of the efforts of the great Göttingen mathematician, Hilbert, never reached the academic standing due her in her own country, none the less surrounded herself with a group of students and investigators at Göttingen, who have already become distinguished as teachers and investigators. Her unselfish, significant work over a period of many years was rewarded by the new rulers of Germany with a dismissal, which cost her the means of maintaining her simple life and the opportunity to carry on her mathematical studies. Farsighted friends of science in this country were fortunately able to make such arrangements at Bryn Mawr College and at Princeton that she found in America up to the day of her death not only colleagues who esteemed her friendship but grateful pupils whose enthusiasm made her last years the happiest and perhaps the most fruitful of her entire career.

Albert Einstein.
Princeton University, May 1, 1935.

9.4 Letter of Dr. Stauffer-McKee

The following letter was sent by Dr. Ruth Stauffer-McKee on October 17, 1972 to the editor of the American Mathematical Monthly, Professor H. Flanders. A carbon copy had been sent to Professor Kimberling. I am indebted to Clark Kimberling for giving me access to his private archive.

Dear Mr. Flanders,

After reading the Addendum to "Emmy Noether" in the August September issue of the American Mathematical Monthly, I was much disturbed by the apparent lack of information concerning the thirties at Princeton! Rechecking the reference to the original article which appeared in February 1972 I was even more disturbed to note that the quote was attributed to a note in the files of Bryn Mawr Alumnae Bulletin. A telephone conversation and a careful check by the Staff of the Bulletin assured me that there was nothing in the files of the Bulletin to even imply that "Mr. Einstein had never met Miss Noether."

In respect to the "thirties at Princeton", I should like to note that there was an air of continued excitement at the Institute for Advanced Study. Solomon Lefschetz, a guiding spirit who worked diligently to help the displayed mathematicians, Hermann Weyl, a leading mathematician of that time who had learned to know Miss Noether in Göttingen, and John von Neumann, then considered a brilliant young genius, were all at the Institute when Einstein arrived in December of 1933. Mrs. Wheeler, of Bryn Mawr, often told of the welcoming party which she and Miss Noether attended.

Mrs. Wheeler usually drove Miss Noether to Princeton for lectures and included Miss Noether's students in the parties. We listened to talks by these men who were the leaders in new exciting theories. It was a friendly group and after the talks everyone gathered for more talk and coffee in a long pleasant common room. There is no doubt that Einstein and Noether were acquainted. I saw them in the same group!

As regards the quote in the "addendum to 'Emmy Noether' " "inspired, if not written by Dr. Hermann Weyl" is certainly true. The writing of the obituary was a very natural occurence. Hermann Weyl was considered by the mathematicians as the mathematical leader of the time and at the peak of his productivity and he had probably the greatest knowledge and understanding of her work. Einstein had begun to slow down and Von Neumann was relatively young and still growing. It was, therefore, obvious to all the mathematicians that Weyl should write the obituary – which he did. He, furthermore, sent it to the New York Times, the

324 *Peter Roquette*

New York Times asked who is Weyl? Have Einstein write something, he is the mathematician recognized by the world. This is how Einstein's article appeared. It was most certainly "inspired" by Weyl's draft. These facts were told to me at the time by Mrs. Wheeler who was indignant that the New York Times had not recognized the mathematical stature of Hermann Weyl.

Very truly yours,

Ruth Stauffer McKee
Senior Mathematician

Bibliography

[1] Alexandroff, P., Pages from an autobiography, 1980, Russian Math. Surveys, vol. 35, 315–358

[2] Alexandroff, P., In Memory of Emmy Noether, Emmy Noether, Collected Papers, Springer, 1983, N. Jacobson, 1–11

[3] Artin, E.,Beweis des allgemeinen Reziprozitätsgesetzes, 1927, Abh. Math. Semin. Univ. Hamb., vol.5, 353–363

[4] Artin, E., The influence of J. H. Wedderburn on the development of modern algebra, 1950, Bull. Amer. Math. Soc., vol.56, 65–72

[5] Brewer, J. and Smith, M., Emmy Noether. A tribute to her life and work, Marcel Dekker, 1981, New York

[6] Brüning, J., Die Fläche des Poeten., 2007, Mittt. Deutsche Math. Vereinigung, vol.15, 33–36

[7] Curtis, C., Pioneers of representation theory: Frobenius, Burnside, Schur and Brauer, Amer. Math. Soc., 1999, History of Mathematics, Providence, R.I.

[8] Dick, A., Emmy Noether 1882–1935, Birkhäuser–Verlag, 1970, Beiheft No. 13 zur Zeitschrift "Elemente der Mathematik".

[9] Hasse, H., Über das allgemeine Reziprozitätsgesetz in algebraischen Zahlkörpern, 1925, Jahresbericht D. M. V., vol.33, 2.Abteilung, 97–101

[10] Hasse, H., Bericht über neuere Untersuchungen und Probleme aus der Theorie der algebraischen Zahlkörper. II: Reziprozitätsgesetz, B. G. Teubner, 1930, vol.6, Jahresbericht D. M. V., Ergänzungsband, Leipzig

[11] Hasse, H., Die moderne algebraische Methode, 1930, Jahresbericht D. M. V., vol.31, 22–34

[12] Hasse, H., Die Struktur der R. Brauerschen Algebrenklassengruppe über einem algebraischen Zahlkörper. Insbesondere Begründung der Theorie des Normenrestsymbols und Herleitung des Reziprozitätsgesetzes mit nichtkommutativen Hilfsmitteln, 1933, Math. Ann., vol.107, 731–760

[13] Hasse, H., Über die Klassenzahl abelscher Zahlkörper, Akademie–Verlag, 1952, Berlin

[14] Hilbert, D., Über die Theorie der algebraischen Formen, 1890, Math. Ann., vol.36, 473–534

[15] Hilbert, D.,Über die Zerlegung der Ideale eines Zahlkörpers in Primideale, 1894, Math. Annalen, 44, 1–8

[16] Jacobson, N., Structure of simple rings without finiteness assumptions, 1945, Trans. Amer. Math. Soc., vol.57, 228–245

[17] Jentsch, W., Auszüge aus einer unveröffentlichten Korrespondenz von Emmy Noether und Hermann Weyl mit Heinrich Brandt, 1986, Historia Math., vol.13, 5–12

[18] Kimberling, C., Emmy Noether., 1972, Amer. Math. Monthly, vol.79, 136–149

[19] Kimberling, C., Emmy Noether and her influence, Emmy Noether. A tribute to her life and work, Marcel Dekker, 1981, Brewer, James W. and Smith, Martha K., 3–61

[20] Klein, F., Über die Differentialgesetze für die Erhaltung von Impuls und Energie in der Einsteinschen Gravitationstheorie, 1918, vol.1918, 171–189

[21] Kosmann-Schwarzbach, Y., Les théorèmes de Noether. Invariance et lois de conservation au XXe siècle, École polytechnique, 2004, Palaiseaux

[22] Lemmermeyer, F. and Roquette, P., Helmut Hasse and Emmy Noether. Their correspondence 1925-1935. With an introduction in English, Universitäts–Verlag, 2006, Göttingen

[23] Mac Lane, S.,History of abstract algebra: origin, rise, and decline of a movement,American mathematical heritage: algebra and applied mathematics. El Pase, Tex., 1975/Arlington, Tex., 1976, 1981, vol.13, Math. Ser., 3–35, Lubbock, Tex., Texas Tech Univ.

[24] Mac Lane, S.,Mathematics at the University of Götingen 1931–1933, Emmy Noether. A tribute to her life and work, 1981, Brewer, W. and Smith, M., 65–78, New York

[25] Mertens, F., Über ganze Functionen von m Systemen von je n Unbestimmten, 1893, Monatsh. Math., vol.4, 193–228, 297–329

[26] Noether, E., Invariante Variationsprobleme, 1918, Nachr. Ges. Wiss. Göttingen 235–257

[27] Noether, E., Die arithmetische Theorie der algebraischen Funktionen einer Veränderlichen in ihrer Beziehung zu den übrigen Theorien und zu der Zahlkörpertheorie, 1919, Jb. Deutsche Math. Ver., vol.38, 192–203

[28] Noether, E. and Schmeidler, W., Moduln in nichtkommutativen Bereichen, insbesondere aus Differential- und Differenenausdrücken, 1920, Math. Zeitschr. vol.8, 1–35

[29] Noether, E., Idealtheorie in Ringbereichen, 1921, Math. Ann., vol. 83, 24–66

[30] Noether, E.,Algebraische und Differentialinvarianten, 1923, Jahresber. Dtsch. Math. Ver., vol.32, 177–184

[31] Noether, E., Abstrakter Aufbau der Idealtheorie im algebraischen Zahlkörper, 1925, Jahresber. Dtsch. Math. Ver., vol.33, Abt.2, p.102

[32] Noether, E.,Abstrakter Aufbau der Idealtheorie in algebraischen Zahl- und Funktionenkörpern, 1926, Math. Ann., vol.96, 26–61

[33] Noether, E., Hyperkomplexe Größen und Darstellungstheorie, 1929, Math. Z., vol.30, 641–692

[34] Reid, C., Courant in Göttingen and New York, Springer, 1976, New York

[35] Roquette, P., Arithmetischer Beweis der Riemannschen Vermutung in Kongruenzfunktionenkörpern beliebigen Geschlechts, 1953, J. Reine Angew. Math., vol.191, 199–252

[36] Roquette, P., History of valuation theory. Part 1, Valuation theory and its applications, vol.I, 2002, F. V. Kuhlmann et al., vol.32, Fields Institute Communications, 291–355

[37] Roquette, P., Zu Emmy Noethers Geburtstag. Einige neue Noetheriana, 2007, Mitt. Dtsch. Math.-Ver., vol.15, 15–21

[38] Roquette, P.,Zu Emmy Noethers Geburtstag. Einige neue Noetheriana, 2007, Mitt. Dtsch. Math.-Ver., vol.15, 15–21

[39] Rowe, D., The Göttingen response to relativity and Emmy Noether's theorems, The symbolic universe. Geometry and physics 1890-1930. Selected papers at a conference, 1999, Gray, J., 189–233, Oxford Univ. Press

[40] Schappacher, N., Das mathematische Institut der Universität Göttingen 1929–1950, Die Universität Göttingen unter dem Nationalsozialismus, K. G. Saur, 1987, Becker, Heinrich and andere, 345–373

[41] Schlote, K., Noether, F. – Opfer zweier Diktaturen, 1991, NTM-Schriftenreihe, vol.28, 33–41

[42] Segal, S. L.,Mathematicians under the Nazis, Princeton University Press, 2003, Princeton, NJ

[43] Siegmund-Schultze, R., Einsteins Nachruf auf Emmy Noether in der New York Times 1935, 2007, Mitt. Dtsch. Math.-Ver., vol. 15

[44] Steinitz, E., Algebraische Theorie der Körper, 1910, J. Reine Angew. Math., vol.137, 167–309

[45] Taussky–Todd, O., My personal recollections of Emmy Noether, Emmy Noether. A tribute to her life and work, M. Dekker, 1981, Brewer, J. W. and Smith, M. K., 79–92, New York

[46] Tavel, N. A., Milestones in mathematical physics, 1971, Transport Theory and Statistical Physics, vol.1, 183–207

[47] Tollmien, C., Sind wir doch der Meinung, daß ein weiblicher Kopf nur ganz ausnahmsweise in der Mathematik schöpferisch tätig sein kann. - Emmy Noether 1882-1935, Göttinger Jahrbuch, Erich Goltze, 1990, vol.38, 153–219, Göttingen

[48] van der Waerden, B. L., Nachruf auf Emmy Noether, 1935, Math. Ann., vol.111, 469–476

[49] van Dalen, Mystic, Geometer, and Intuitionist. The Life of L. E. J. Brouwer, Clarendon Press, 1999, vol.I, Oxford

[50] Weil, A., Lehr- und Wanderjahre eines Mathematikers. Aus dem Französischen übersetzt von Theresia Übelhör, Birkhäuser, 1993, Basel

[51] Weyl, H., Die Idee der Riemannschen Fläche, Teubner, 1913, Leipzig

[52] Weyl, H., Über die Gleichverteilung von Zahlen modulo Eins, 1916, Math. Annalen, vol.77, 313–352

[53] Weyl, H., Emmy Noether, 1935, Scripta math., vol.3, 201–220

[54] Weyl, H., Symmetry, Princeton Univ. Press, 1952, Princeton

[55] Weyl, H., Topologie und abstrakte Algebra als zwei Wege mathematischen Verständnisses, 1932, Unterrichtsblätter, 39, 177–188

[56] Weyl, H., The classical groups, their invariants and representations, Princeton Univ. Press, 1939, Princeton

[57] Wuensch, D., "Zwei wirkliche Kerle". Neues zur Entdeckung der Gravitationsgleichungen der Allgemeinen Relativitätstheorie durch Albert Einstein und David Hilbert, Termessos, 2005, Göttingen

Printed in the United States
by Baker & Taylor Publisher Services